MOLECULAR
BIOLOGY
INTELLIGENCE
UNIT 10

The CGRP Family: Calcitonin Gene-Related Peptide (CGRP), Amylin, and Adrenomedullin

David Poyner

Pharmaceutical Sciences Institute
Aston University

Ian Marshall

Department of Pharmacology
University College London

Susan D. Brain

Pharmacology Group and Vascular Biology
Research Centre
King's College

CRC Press
Taylor & Francis Group
Boca Raton London New York

CRC Press is an imprint of the
Taylor & Francis Group, an **informa** business

Molecular Biology Intelligence Unit

The CGRP Family: Calcitonin Gene-Related Peptide (CGRP), Amylin, and Adrenomedullin

First published 2000 by Landes Bioscience

Published 2018 by CRC Press
Taylor & Francis Group
6000 Broken Sound Parkway NW, Suite 300
Boca Raton, FL 33487-2742

© 2000 by Taylor & Francis Group, LLC
CRC Press is an imprint of Taylor & Francis Group, an Informa business

First issued in paperback 2019

No claim to original U.S. Government works

ISBN 13: 978-0-367-44752-6 (pbk)
ISBN 13: 978-1-58706-004-5 (hbk)

**Visit the Taylor & Francis Web site at
http://www.taylorandfrancis.com**

**and the CRC Press Web site at
http://www.crcpress.com**

Library of Congress Cataloging-in-Publication Data

Calcitonin Gene-Related Peptide/ [edited by] David Poyner, Ian Marshall, and Susan Brain-
    p. cm. -- (Molecular biology intelligence unit)
Includes biographical references and index
ISBN 1-58706-004-3 (alk. paper)
1.Calcitonin gene-related peptide Congresses. 2. Amylin Congresses 3. Adrenomedullin Congresses 4. Amylin--Receptors Congresses 5. Adrenomedullin--Receptors Congresses I. Poyner, David. II. Marshall Ian, 1945- III. Series
    [DNLM: 1. Calcitonin Gene-Related Peptide Congresses 2. Amyloid Congresses 3. Peptides--physiology Congresses. 4. Receptors, Calcitonin Gene-Related Peptide Congresses. WL 104 C425 1999]
Qp552.C22C48 1999
612'.015756--dc21
DNLM/DLC   99-34620
for Library of Congress   CIP

CONTENTS

EDITORS

David Poyner
Pharmaceutical Sciences Institute
Aston University
Birmingham, England, U.K.
Chapter 3

Ian Marshall
Department of Pharmacology
University College London
London, England, U.K.
Chapter 2
Appendix

Susan D. Brain
Pharmacology Group and Vascular
Biology Research Centre
King's College
London, England, U.K.
Chapters 13, 26 and 27

CONTRIBUTORS

Nambi Aiyar
Department of Biological Sciences
SmithKline Beecham Pharmaceuticals
King of Prussia, Pennsylvania, U.S.A.
Chapter 5

R.P. Allaker
Department of Oral Microbiology
St. Bartholomew's
The Royal London School of Medicine
   and Dentristy
Chapter 25

Warwick A. Arden
Vascular and Trauma/Critical
   Care Research Program
Departments of Surgery and Physiology
University of Kentucky Medical Center
Lexington, Kentucky, U.S.A.
Chapter 29

Stewart Barker
The William Harvery Research Institute
St. Bartholomew's and
The Royal London School of Medicine
   and Dentistry
Queen Mary and Westfield College
London, England, U.K.
Chapter 24

Gavin S. Bennet
Pharmacology Group and Vascular
   Biology Research Centre
King's College
London, England, U.K.,
Chapter 13

Kevin Beaumont
Amylin Pharmaceuticals Inc.
San Diego, California, U.S.A.
Chapter 9

Trinity J. Bivalacque
Departments of Pharmacology,
    Anesthesiology, and Medicine
Tulane University School of Medicine
New Orleans, Louisiana, U.S.A.
Chapters 11, 21 and 23

Stephen R. Bloom
Department of Metabolic Medicine
Imperial College School of Medicine
Hammersmith Hospital
London, England, U.K.
Chapters 10 and 22

Jason Brown
Receptor Systems and Cell Biology
    Units
GlaxoWellcome Medicines Research
    Centre
Hertfordshire, England, U.K.
Chapter 6

T. Cao
Pharmacology Group and Vascular
    Biology Research Centre
King's College London
London, England, U.K.
Chatper 26

Hunter C. Champion
Departments of Pharmacology,
    Anesthesiology, and Medicine
Tulane University School of Medicine
New Orleans, Louisiana, U.S.A.
Chapter 11, 21 and 23

George Christopoulos
Molecular Pharmacology Unit
Department of Pharmacology
The University of Melbourne
 Parkville, Victoria, Australia
Chapter 8

D.Q. Chu
Pharmacology Group and Vascular
    Biology Research Centre
King's College London
London, England, U.K.
Chapter 26

Adrian J.L. Clark
Departments of Endocrinology
St. Bartholomew's
The Royal London School of Medicine
    and Dentristy
London, England, U.S.A.
Chapter 19

Hedley A. Coppock
Department of Metabolic Medicine
Imperial College School of Medicine
Hammersmith Hospital
London, England, U.K.
Chapter 10

Roger Corder
The William Harvey Research Institute
St. Bartholomew's and
The Royal London School of Medicine
    and Dentistry
Queen Mary and Westfield College
London, England, U.K.
Chapter 24

David H. Coy
Departments of Pharmacology,
    Anesthesiology, and Medicine
Tulane University School of Medicine
New Orleans, Louisiana, U.S.A.
Chapters 11, 21 and 23

A.T. Cruchley
Molecular Signalling Group
Oral Diseases Research Centre
St Bartholomew's and the Royal
    London School of Medicine
    and Dentistry
London, England, U.K.
Chapter 35

M.J. Cumberbatch
Department of Pharmacology
Merck Sharp and Dohme Research
  Laboratories
Harlow, England, U.K.
Chapter 33

G.P. Dahl
Department of Physiology and
  Biophysics
University of Miami School of
  Medicine
Chapter 7

L. D'Este
Institute of Human Anatomy
University "La Sapienza" of Rome
Rome, Italy
Chapter 20

Ian Dickerson
Department of Physiology and
  Biophysics
University of Miami School of
  Medicine
Miami, Florida, U.S.A.
Chapter 7

Lara T. Diemel
Department of Pharmacology
Queen Mary and Westfield College
London, England, U.K.
Chapter 13

Jyoti Disa
Department of Biological Sciences
SmithKline Beecham Pharmaceuticals
King of Prussia, Pennsylvania, U.S.A.
Chapter 5

Yvan Dumont
Douglas Hospital Research Centre and
  Department of Psychiatry
McGill University
Montreal, Québec, Canada
Chapters 1 and 19

Lars Edvinsson
Department of Internal Medicine
Lund University Hospital
Lund, Sweden
Chapters 16 and 32

B.N. Evans
Department of Physiology and
  Biophysics
University of Miami School of
  Medicine
Chapter 7

William R. Ferrell
Centre for Rheumatic Diseases
Department of Medicine
Royal Infirmary
Glasgow, Scotland, U.K.
Chapter 14

Steven M. Foord
Receptor Systems and Cell Biology
  Units
GlaxoWellcome Medicines Research
  Centre
Hertfordshire, England, U.K.
Chapter 6

Neil J. Fraser
Receptor Systems and Cell Biology
  Units
GlaxoWellcome Medicines Research
  Centre
Hertfordshire, England, U.K.
Chapter 6

Neil E. Garrett
Department of Pharmacology
Queen Mary and Westfield College
London, England, U.K.
Chapter 13

Gloria Gellin
Vascular and Trauma/Critical
  Care Research Program
Departments of Surgery and Physiology
University of Kentucky Medical Center
Lexington, Kentucky, U.S.A.
Chapter 29

Mohammad A. Ghatei
Department of Metabolic Medicine
Imperial College School of Medicine
Hammersmith Hospital
London, England, U.K.
Chapters 10 and 22

Arif A. Gharzi
Department of Metabolic Medicine
Imperial College School of Medicine
Hammersmith Hospital
London, England, U.K.
Chapter 10

Peter J. Goadsby
Institute of Neurology
The National Hospital for Neurology
   and Neurosurgery
London, England, U.K.
Chapters 15 and 16

Sergio Gulbenkian
Gulbenkian Institute of Science
Oeiras, Portugal
Chapter 32

R. J. Hargreaves
Department of Pharmacology
Merck Sharp and Dohme Research
   Laboratories
Harlow, England, U.K.
Chapter 33

John Herich
Amylin Pharmaceuticals Inc.
San Diego, California, U.S.A.
Chapter 19

Wayne J.G. Hellstron
Departments of Pharmacology,
   Urology, and Medicine
Tulane University School of Medicine
New Orleans, Louisiana, U.S.A.
Chapter 23

Zs. Helyes
Department of Pharmacology and
   Pharmacotherapy
University Medical School of Pécs
Neuropharmacology Research Group of
   the Hungarian Academy of Sciences
Pécs, Hungary
Chapter 34

Peter Holzer
 Department of Experimental and
   Clinical Pharmacology
University of Graz
Graz, Austria
Chapter 12

Stephen Howitt
Pharmaceutical Sciences Institute
Aston University
Birmingham, England, U.K.
Chapter 3

J.D. Hurley
Department of Metabolic Medicine
Imperial College School of Medicine
Hammersmith Hospital
London, England, U.K.
Chapter 10

Albert L. Hyman
Departments of Pharmacology,
   Anesthesiology, and Medicine
Tulane University School of Medicine
New Orleans, Louisiana, U.S.A.
Chapter 11

Inger Jansen-Olesen
Department of Pharmacology
The Royal Danish School of Pharmacy
Copenhagen, Denmark
Chapters 16, 17 and 32

Philip J. Kadowitz
Departments of Pharmacology,
   Anesthesiology, and Medicine
Tulane University School of Medicine
New Orleans, Louisiana, U.S.A.
Chapters 11, 21 and 23

Radhika Kajekar
Johns Hopkins Asthma and Allergy
  Centre
Baltimore, Maryland, U.S.A.
Chapter 13

Supriya Kapas
Molecular Signalling Group
Oral Diseases Research Centre
St Bartholomew's and the Royal
  London School of Medicine
  and Dentistry
London, England, U.K.
Chapters 25 and 35

T. Kocsy
Department of Pharmacology and
  Pharmacotherapy
University Medical School of Pécs
Neuropharmacology Research Group
  of the Hungarian Academy
  of Sciences
Pécs, Hungary
Chapter 34

Francis Y. Lam
Department of Pharmacology
The Chinese University of Hong Kong
Shatin, Hong Kong, China
Chapter 28

Ulo Langel
Department of Neurochemistry
Stockholm University
Stockholm, Sweden
Chapter 3

Melanie G. Lee
Receptor Systems and Cell Biology
  Units
GlaxoWellcome Medicines Research
  Centre
Hertfordshire, England, U.K.
Chapter 6

Delphine Lees
The William Harvey Research Institute
St. Bartholomew's and
The Royal London School of Medicine
  and Dentistry
Queen Mary and Westfield College
London, England, U.K.
Chapter 24

Steve Legon
Department of Metabolic Medicine
Imperial College School of Medicine
London, England, U.K.
Chapter 4

A.E. Luebke
Department of Otalryngology
University of Miami School of
  Medicine
Chapter 7

T. Maeda
Department of Anatomy
Shiga University of Medical Science
Otsu, Japan
Chapter 20

J.C. Lockhart
Department of Biological Sciences
University of Paisley
Paisley, Scotland, U.K.
Chapter 14

Martin J. Main
Receptor Systems and Cell Biology
  Units
GlaxoWellcome Medicines Research
  Centre
Hertfordshire, England, U.K.
Chapter 6

J.J. McDougall
Department of Surgery
University of Calgary
Calgary, Alberta, Canada
Chapter 14

Linda M. McLatchie
Receptor Systems and Cell Biology
Units
GlaxoWellcome Medicines Research
Centre
Hertfordshire, England, U.K.
Chapter 6

L. McMurdo
Centre for Rheumatic Diseases
Department of Medicine
Royal Infirmary
Glasgow, Scotland, U.K.
Chapter 14

Dennis B. McNamara
Departments of Pharmacology,
Anesthesiology, and Medicine
Tulane University School of Medicine
New Orleans, Louisiana, U.S.A.
Chapter 11

L.O. Mnayer
Department of Molecular Biology
and Biochemistry
University of Miami School of
Medicine
Miami, Florida, U.S.A.
Chapter 7

Candace Moore
Amylin Pharmaceuticals Inc.
San Diego, California, U.S.A.
Chapter 9

Maria Morfis
Molecular Pharmacology Unit
Department of Pharmacology
The University of Melbourne
Parkville, Victoria, Australia
Chapter 8

William A. Murphy
Departments of Pharmacology,
Anesthesiology, and Medicine
Tulane University School of Medicine
New Orleans, Louisiana, U.S.A.
Chapters 11, 21 and 23

M. Nagashpour
Department of Physiology and
Biophysics
University of Miami School of
Medicine
Miami, Florida, U.S.A.
Chapter 7

J. Nemeth
Department of Pharmacology and
Pharmacotherapy
University Medical School of Pécs
Neuropharmacology Research Group
of the Hungarian Academy of
Sciences
Pécs, Hungary
Chapter 34

Ponnal Nambi
Department of Biological Sciences
SmithKline Beecham Pharmaceuticals
King of Prussia, Pennsylvania, U.S.A.
Chapter 5

Bobby D. Nossaman
Departments of Pharmacology,
Anesthesiology, and Medicine
Tulane University School of Medicine
New Orleans, Louisiana, U.S.A.
Chapter 11

N.C.B. Nyborg
Department of Pharmacology
The Royal Danish School of Pharmacy
Copenhagen, Denmark
Chapter 18

Ruth Oremus
Vascular and Trauma/Critical
Care Research Program
Departments of Surgery and Physiology
University of Kentucky Medical Center
Lexington, Kentucky, U.S.A.
Chapter 29

G. Oroszi
Pharmacology and Pharmacotherapy
University Medical School of Pécs
Neuropharmacology Research Group
   of the Hungarian Academy
   of Sciences
Pécs, Hungary
Chapter 34

Ali. A. Owji
Department of Biochemistry
Shiraz University of Medical Sciences
Shiraz, Iran
Chapter 10

K. Pahal
Molecular Signalling Group
Clinical Sciences Research Centre
St Bartholomew's and the Royal
   London School of Medicine
   and Dentistry
London, England, U.K.
Chapter 35

Katie J. Perry
Molecular Pharmacology Unit
Department of Pharmacology
The University of Melbourne
Parkville,Victoria, Australia
Chapter 8

Robert L. Pierce
Departments of Pharmacology and
   Medicine
Tulane University School of Medicine
New Orleans, Louisiana, U.S.A.
Chapter 21

Erika Pintér
Department of Pharmacology and
   Pharmacotherapy
University Medical School of Pécs
Neuropharmacology Research Group
   of the Hungarian Academy
   of Sciences
Pécs, Hungary
Chapter 34

Rémi Quirion
Douglas Hospital Research Centre
   and Department of Psychiatry
McGill University
Montreal, Québec, Canada
Chapters 1 and 19

Tindaro G. Renda
Institute of Human Anatomy
University "La Sapienza" of Rome
Rome, Italy
Chapter 20

M.I. Rosenblatt
Department of Molecular Biology and
   Biochemistry
University of Miami School of
   Medicine
Miami, Florida, U.S.A.
Chapter 7

Anette Sams
Department of Pharmacology
The Royal Danish School of Pharmacy
Copenhagen, Denmark
Chapter 17

J. Szolcsányi
Department of Pharmacology and
   Pharmacotherapy
University Medical School of Pécs
Neuropharmacology Research Group
   of the Hungarian Academy
   of Sciences
Pécs, Hungary
Chapter 34

Patrick M. Sexton
Molecular Pharmacology Unit
Department of Pharmacology
The University of Melbourne
Parkville, Victoria, Australia
Chapter 8

Shi-Hsiang Shen
Biotechnology Research Institute
National Research Council of Canada
Montreal Québec, Canada
Chapter 19

Sara L. Shepheard
Department of Pharmacology
Merck Sharp & Dohme Research
    Laboratories
Harlow, England, U.K.
Chapter 33

M. Sheykhzade
Department of Pharmacology
The Royal Danish School of Pharmacy
Copenhagen, Denmark
Chapter 18

David M. Smith
Department of Metabolic Medicine
Imperial College School of Medicine
Hammersmith Hospital
London, England, U.K.
Chapters 10 and 22

Roberto Solari
Receptor Systems and Cell Biology
    Units
GlaxoWellcome Medicines Research
    Centre
Hertfordshire, England, U.K.
Chapter 6

Ursel Soomets
Department of Biochemistry
Tartu University
Tartu, Estonia
Chapter 3

Gillian M. Taylor
Department of Metabolic Medicine
Imperial College School of Medicine
Hammersmith Hospital
London, England, U.K.
Chapters 10 and 22

Nicola Thompson
Receptor Systems and Cell Biology
    Units
GlaxoWellcome Medicines Research
    Centre
Hertfordshire, England, U.K.
Chapter 6

Nanda Tilakaratne
Molecular Pharmacology Unit
Department of Pharmacology
The University of Melbourne
Parkville, Victoria, Australia
Chapter 8

David R. Tomlinson
Department of Pharmacology
Queen Mary and Westfield College
London, England, U.K.
Chapter 13

Yiai Tong
Douglas Hospital Research Centre
Department of Psychiatry
McGill University
Montreal, Québec, Canada
Chapter 19

Pamela K. Towler
Pharmacology Group and Vascular
    Biology Research Centre
London, England, U.K.
Chapter 27

Rolf Uddman
Department of Internal Medicine
Lund University Hospital
Lund, Sweden
Chapter 14

R. Vaccaro
Institute of Human Anatomy
University "La Sapienza" of Rome
Rome, Italy
Chapter 20

Denise van Rossum
Douglas Hopsital Research Centre
Department of Psychiatry
McGill University
Verdun, Québec, Canada
Chapter 19

Run Wang
Departments of Pharmacology,
    Urology, and Medicine
Tulane University School of Medicine
New Orleans, Louisiana, U.S.A.
Chapter 23

Xian Wang
Institute of Vascular Medicine
Third Hospital
Beijing Medical University
Beijing, People's Republic of China
Chapter 30

D. J. Williamson
Department of Pharmacology
Merck Sharp and Dohme Research
    Laboratories
Harlow, England, U.K.
Chapter 33

Sunil J. Wimalawansa
Department of Internal Medicine
University of Texas Medical Branch
Galveston, Texas, U.S.A.
Chapters 20 and 31

Alan Wise
Receptor Systems and Cell Biology
    Units
GlaxoWellcome Medicines Research
    Centre
Hertfordshire, England, U.K.
Chapter 6

Franca-Maria Wisskirchen
Department of Pharmacology
University College London
London, England, U.K.
Chapter 2

Elizabeth G. Wood
The William Harvey Research Institute
St. Bartholomew's and
The Royal London School of Medicine
    and Dentistry
Queen Mary and Westfield College
London, England, U.K.
Chapter 24

Liyu Xing
Institute of Vascular Medicine
Third Hospital
Beijing Medical University
Beijing, People's Republic of China
Chapter 30

Yutong Xing
Institute of Vascular Medicine
Third Hospital
Beijing Medical University
Beijing, People's Republic of China
Chapter 30

Chandresekhar Yallampalli
Departments of Obstetrics and
    Gynecology
University of Texas Medical Branch
Galveston, Texas, U.S.A.
Chapter 31

A.L.M. Yip
Department of Pharmacology
The Chinese University of Hong Kong
Shatin, Hong Kong, China
Chapter 28

Andrew Young
Amylin Pharmaceuticals Inc.
San Diego, California, U.S.A.
Chapter 9

C. Zihni
Molecular Signalling Group
Oral Diseases Research Centre
London, England, U.K.
Chapter 25

FOREWORD

This book is based on the proceedings of the Third International Meeting on CGRP, CGRP '98, held in Shaftesbury, UK in May 1998. As such it takes its place alongside the published proceedings of the earlier two meetings; the 1992 meeting published as volume 657 of the Proceedings of the New York Academy of Sciences, and the 1995 meeting published as a special edition of the Canadian Journal of Physiology and Pharmacology. These volumes still appear to be regularly cited in the literature on CGRP and allied peptides; from personal experience I know that I find them helpful sources of data and so it is hoped that this volume might also be useful both to those working in the field and also anybody wanting an introduction to the CGRP family of peptides.

The book covers CGRP and the related peptides amylin and adrenomedullin. The first chapter by Remi Quirion provides a broad overview of the field, past and present. It is followed by two chapters dealing with the "classical" pharmacology of CGRP. In particular Ian Marshall gives an account of the pitfalls of the present generation of CGRP antagonists and the dangers of assuming that every action of CGRP is mediated by a CGRP receptor. However the bulk of this section is taken up by the recent developments concerning the cloning of receptors for CGRP and adrenomedullin. An exciting story emerged at the meeting of how a single molecule, calcitonin receptor-like receptor (CRLR) can mediate the response to both CGRP and adrenomedullin depending on the presence of different members of a family of accessory proteins, the Receptor Activity Modifying Proteins (RAMPs). Steve Legon, Nambi Aiyar and Steve Foord present their data in three chapters. In addition Ian Dickerson gives an account of another accessory protein associated with CGRP responsiveness, Receptor Component Protein (RCP).

The following sections of the book deal with the biochemistry, physiology and pharmacology of receptors for the allied peptides amylin and adrenomedullin. Patrick Sexton identifies the close connections between amylin and calcitonin receptors and Andrew Young and Kevin Beaumont deal with the role of amylin in the regulation of food intake. The molecular nature of adrenomedullin receptors is addressed in an earlier chapter by Steve Foord but in this section Dave Smith and Philip Kadowitz examine their pharmacology.

The pathophysiology of CGRP and related peptides formed a significant part of the meeting. Chapters by Sue Brain, Bill Ferrell and Peter Holzer examine the role of CGRP in vascular disorders, joint dysfunction and gastrointestinal pathology. A separate section written by

Peter Goadsby and Lars Edvinsson considers the role of CGRP in headache and migraine.

The final section of the book contains a selection of the short oral and poster communications presented at the meeting. The majority of these fit the themes developed above, and they have been ordered within this section according to whether they are broadly dealing with pharmacology (molecular or otherwise) or pathophysiology.

The book concludes with an important appendix, reflecting a workshop held to discuss the nomenclature of CGRP, amylin and adrenomedullin receptors. As the majority of the active workers in the field were present at this meeting it is hoped that the consensus view presented here might be widely acceptable.

Finally the organising committee would like to thank those who sponsored the meeting; Amylin Pharmaceuticals, Thomae Boehringer Ingelheim, the British Heart Foundation, the Wellcome Trust, and Merck, Sharp & Dohme. Valuable contributions were also made by Advanced Bioconcept, GlaxoWellcome, Novartis, Pfizer, Peninsula and Phoenix. The next CGRP meeting will be in Lund/Copenhagen in Spring 2001, organised by Lars Edvinsson and Inger Jansen-Olesen.

David Poyner
Birmingham, England, U.K.
September, 1998

CHAPTER 1

Multiple Receptors for CGRP and Related Peptides

Rémi Quirion and Yvan Dumont

1.1. Introduction

Calcitonin gene-related peptide (CGRP) was the first peptide isolated using molecular approaches instead of traditional biochemical techniques such as tissue extraction and purification. In fact, CGRP was characterized by Rosenfeld et al as a 37 amino acid peptide generated from the alternative splicing of the calcitonin gene, mostly in neurons of the central and peripheral nervous system.[1-4] Two forms (a and b) of CGRP have been characterized having almost identical biological properties (for a detailed review, see[5]). Members of the CGRP family also include amylin (AMY) and adrenomedullin (ADM) which share about 50% and 25% sequence homologies, respectively. Mammalian but more specifically fish calcitonins (salmon, for example) also share sequence homologies with CGRP-related peptides. (Fig. 1.1).

Immunohistochemical studies have revealed the broad distribution of CGRP in the brain and peripheral nervous system, CGRP being particularly abundant in sensory-motor nuclei and nerve terminals. Less is known on the discrete, detailed localization of amylin and ADM-like peptides in the CNS (see for example ref 5; D'Este et al, Sexton et al, this volume).

CGRP possesses a wide variety of biological effects in the brain as well as in peripheral tissues including most potent vasodilatory actions, pro-inflammatory and nociceptive effects, direct inotropic and chronotropic activities, modulation of locomotor behaviors, etc (Table 1.1). These various effects are mediated by the activation of specific plasma membrane receptors. This brief review summarizes key features of CGRP, amylin and ADM receptors and outlines the existence of unique, distinct classes of receptors in this peptide family. Recent extensive reviews should be consulted for additional information.[5-7]

1.2. CGRP Receptors: Putative Sub-Types

The presence of CGRP receptors has been demonstrated in many brain and peripheral tissues including the brain of various species with particularly high levels in the cerebellum[8-11] as well as the heart,[8,9,12] liver,[11] spleen,[8,11] skeletal muscle[13-15] lung,[16,17] vas deferens,[18] lymphocytes[19,20] and a variety of cell lines (SK-N-MC, HCA-7, Col 29, etc).[5-7] Apparent affinity (Kd) values (using either [^{125}I]human (h) αCGRP or [^{125}I]hCGRP$_{8-37}$) range in the low pM to low nM values.

CGRP receptors most likely belong to the family of G proteins-coupled receptors interacting with Gsα to stimulate adenylate cyclase and increase cAMP production.[5-7,21] Various reports have also described the effects of GTP and analogs on CGRP receptor agonist binding in a variety of tissues such as the cerebellum,[10,22] atrium,[22]

The CGRP Family: Calcitonin Gene-Related Peptide (CGRP), Amylin, and Adrenomedullin, edited by David Poyner. ©2000 EUREKAH.COM.

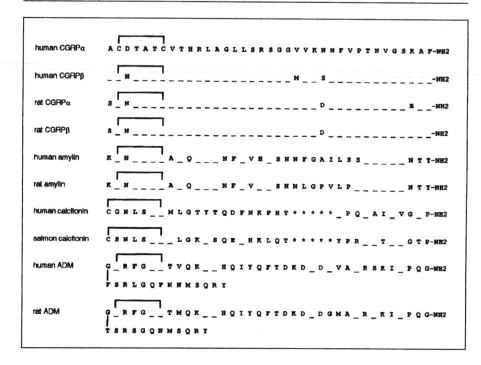

Fig. 1.1 Amino acid sequence of CGRP and related peptides. Amino acids in common to human CGRP are indicated by dash (-). Spaces inserted to allow for sequence comparison are indicated by a star (*).

lung,[17] vas deferens[22,23] and skeletal muscle.[14,15] Orphan G-protein receptor clones purportedly of the CGRP-type have been shown to be Gsα coupled.[24-26] However, their full relevance as CGRP receptors has yet to be established (see page 5).

The stimulation of CGRP receptors has also been shown to activate muscarinic K^+ channels in rat atrial cells,[27] to enhance Ca^{2+} currents in nodose ganglion neurons[28] via a pertussis toxin-sensitive mechanism again implicating a G-protein. GTPαs-sensitive increases in transmembrane Ca^{2+} currents have also been reported following CGRP application in guinea pig myocytes and bullfrog atria,[29] in accordance with the involvement of Gsα. Little is known regarding the desensitization of CGRP receptors but recent data suggest the involvement of a G-protein receptor kinase (GRK-6; Aiyar et al, this volume)

Partial biochemical purification has provided insights into the molecular structure of CGRP receptors. Early on, a putative CGRP receptor was purified from the human placenta with an apparent mass of 240kD and composed of multiple 62-68kD subunits.[24] Single binding units were then isolated from a variety of tissues including the porcine spinal cord,[30] cultured rat vascular smooth muscle cells and bovine endothelial cells,[31] guinea pig gastric smooth muscle and pancreatic acinar cells,[32] rat cerebellum[21] and lung.[33] An extensive study using a variety of rat tissues suggested an apparent molecular weight of 44kD following an enzymatic N-deglycosylation.[34] However, various other studies reported more than one cross-linked molecular weight bands in tissues such as porcine coronary arteries,[30] rat atrium,[30] porcine kidney,[35] rat liver[36] and skeletal muscle.[36] Moreover, the human cerebellum apparently

Table 1.1. Some of the purported biological effects of CGRP

- Potent vasodilatory effects
- Chronotropic and Inotropic actions
- Increase in turnover of nicotinic receptor/neuromuscular junction
- Stimulation of inflammatory processes
- Modulation of tolerance to opioids
- Hyperthermia (icv injection)
- Hypophagia (icv injection)
- Increase in locomotor behaviors (icv injection)
- Modulation of various senses (audition, vision, smell, taste; icv injection)
- Modulation of behavioral activities related to motorneurons
- Inhibition of gastric acid secretion (icv injection)

contains two CGRP binding proteins (50 and 13.7kD: 73) while Stangl et al reported three different masses of 60, 54 and 17 kD, respectively with evidence that the higher molecular weight components were glycosylated.[37] The presence of a low molecular weight component is most interesting in the context of the recent identification of the RAMPs (Foord et al, Ch. 6, this volume).

1.2.1. The CGRP$_1$ Receptor Sub-Type

The existence of CGRP receptor subtypes was proposed on the basis of the differential antagonistic potencies of C-terminal fragments to block the action of CGRP, and the comparative agonistic properties of linear agonists (such as [Cys(acetamidomethyl)$_{2,7}$]haCGRP; [CysACM]CGRP) in a variety of tissues (Table 1.2). Early on, we reported that C-terminal fragments such as CGRP$_{8-37}$, CGRP$_{9-37}$ and to a lesser extent CGRP$_{12-37}$ behaved as relatively potent, competitive antagonists of CGRP-induced inotropic and chronotropic effects in the guinea pig atria while being much less effective (at least ten times) in blocking the effects of CGRP in

the rat vas deferens.[18,38] In contrast, the linear agonist [CysACM]CGRP was a weak agonist (EC$_{50}$ = 100nM) in the vas deferens while being mostly inactive in the atrial preparations.[38] Similar results were obtained using either rat or guinea pig atrial and vasa deferentia preparations indicating that the observed differential profile of activity are not merely species-related.[39] On that basis, we proposed to classify CGRP$_{8-37}$-sensitive (pA$_2$ 7.0 and higher) linear agonist-insensitive CGRP receptors as the CGRP$_1$ sub-type, the reverse profile being observed for the CGRP$_2$ sub-type (Table 1.2). Shorter C-terminal fragments such as CGRP$_{19-37}$,[40] CGRP$_{23-37}$ and Tyr$_0$-CGRP$_{28-37}$[41,42] as well as modified fragments [Asp^{31}, Pro^{34}, Phe^{35}] CGRP$_{27-37}$ and [Pro^{34}, Phe^{35}] CGRP$_{27-37}$[43] also behave as CGRP$_1$ receptor antagonists with potencies similar to that of CGRP$_{8-37}$ (Poyner et al, this volume). The development of non-peptide antagonists of the CGRP$_1$ receptors should be reported in the very near future.[44]

Various groups have now reported on the antagonists properties of C-terminal CGRP fragments, especially CGRP$_{8-37}$, in a variety of in vitro and in vivo bioassays.[18,39,45-53] For example, upon intracerebro ventricular (icv) administration in the rat, CGRP$_{8-37}$ was shown to antagonize the analgesic and anorexic effects of haCGRP without modifying CGRP-induced hyperthermia.[53] However, the apparent pA$_2$ values of CGRP$_{8-37}$ in various assays are highly variable with values ranging between 6.0 to greater than 9.0 (see Marshall et al, Ch. 2, this volume for a detailed discussion). Hence, additional tools such as non-peptide antagonists are anxiously awaited to exclude the possible involvement of various peptidases or poor peptide tissue penetration as factors explaining these discrepant pA$_2$ values. In any case, it would appear that pA$_2$ values in the range of 7.0-8.0 are likely to reflect the true antagonistic potential of CGRP$_{8-37}$ for the CGRP$_1$ receptor, lower and higher values being mostly artifactual or linked to the unique characteristics of the preparation under study. The use of linear

Table 1.2. Characteristics of putative CGRP and related peptides receptor sub-types

	CGRP$_1$	CGRP$_2$	Amylin	ADM
Order of potency of homologs	αCGRP≥βCGRP>ADM> Amylin	βCGRP≥αCGRP>ADM> Amylin	Amylin≥salmon calcitonin>αCGRP βCGRP≥ADM	ADM>αCGRP≥βCGRP> Amylin
Preferential agonist	--	[CysACM^{27}]αCGRP	--	--
Antagonists	CGRP$_{8-37}$ (pA$_2$ 7.0-8.0) [Tyr^{27} Asp^{31} Pro^{34} Phe^{35}]-CGRP$_{27-37}$ (pA$_2$ 7.0-8.0)	CGRP$_{8-37}$ (pA$_2$<6.0)	AC 187	ADM$_{22-52}$[1]
Coupling	Gs	Gs	Gs	Gs
Enriched tissue	Atria/Spleen/Lung Pulmonary artery SK-N-MC cells	Vas deferens/Liver Col 29 cells	Nucleus Accumbens Skeletal muscle	Lung/Vascular smooth muscle
Molecular information	CRLR plus RAMP1[2]	--	?[3]	CRLR plus RAMP2[2]

[1]; Some evidence have suggested the possible existence of ADM receptor sub-types. Also, the genuine antagonistic properties of ADM$_{22-52}$ have yet to be fully established

[2]; See Foord et al, Ch.2, this volume, for detailed information on CRLR and RAMPs. In addition the cloning of another ADM receptor has been reported.[75]

[3]; See Sexton et al, Ch. 8, this volume, on molecular features of amylin receptors.

agonists such as [CysACM]CGRP (mostly inactive on the $CGRP_1$ sub-type) should also be considered to help in establishing further the CGRP receptor sub-type(s) present in a given tissue or cell line.[39]

The cloning of CGRP receptors was expected to resolve issues as to the existence of sub-types and the molecular nature of the $CGRP_1$ receptor. In fact, it originally confused the situation. Indeed, the cloning of two purported $CGRP_1$ receptors has been reported over the past two years.[25,26] However, their sequence homology is minimal (around 30%) raising some doubts as to their respective $CGRP_1$-like nature. First, RDC-1 (originally isolated as an orphan G-protein coupled receptor from the dog thyroid)[54,55] was proposed to be a $CGRP_1$ receptor on the basis of its sensitivity to CGRP and $CGRP_{8-37}$.[25] At the same time, another orphan receptor isolated from the rat lung and know as the "calcitonin receptor like receptor," or CRLR[56] with an unknown pharmacology[57,58] was subsequently characterized as a $CGRP_1$ receptor.[26,59] However, both clones have proved to be extremely difficult to successfully transfect and express in transiently or permanently expressing cell lines (HEK 293, COS-7, CHO, etc). Moreover, the anatomical distribution of these two genes, especially in the CNS, failed to closely resemble that of CGRP receptor binding sites reported earlier using an agonist or an antagonist[60] as radioligands (see Tong et al, this volume for in situ hydridisation (ISH) of RDC-1 and Oliver et al,[61] for both RDC-1 and CRLR ISH in the rat CNS). Taken together, these data raised doubts as to the genuine nature of these two clones as $CGRP_1$ receptors.

Most interestingly, however, recent data on the cloning of accessory proteins[62] especially the RAMPs[63] may help to resolve, at least partly, this issue. Indeed, McLatchie et al have demonstrated that RAMP1 must be cotransfected along with CRLR to generate a functional $CGRP_1$-like receptor in transfected HEK 293 cells;[63] the transfection of either one of those being unable to generate an active $CGRP_1$ receptor (see

Foord et al, Chapter 6, this volume for details). Moreover, the cotransfection of CRLR with another RAMP, RAMP2, generated a func- tional adrenomedullin receptor. These new exciting data demonstrate that the pharmacology of a given G-protein receptor clone can be switched from one type to another depending on the accompanying RAMP. These findings open a totally new chapter on the pharmacology of G-protein coupled receptors and accessory proteins that are likely to extend beyond CGRP receptors.

The cotransfection of RDC-1 with either RAMP1 or 2 failed to generate a functional receptor. It thus appears that RDC-1 is not a functionally relevant CGRP receptor or that a yet to be characterized RAMP-like protein is required for this G-protein coupled orphan receptors to behave as a CGRP receptor. Finally, it is likely that CRLR and RAMP1 do not represent all CGRP binding sites detected in the CNS as the cerebellum[5,60] for example, is highly enriched with specific sites but not in either CRLR or RAMP1 mRNAs.[63] Hence, it would appear that additional $CGRP_1$-like receptor clones have yet to be identified.

1.2.2. The $CGRP_2$ Receptor Sub-Type

In addition to the lower antagonistic potency of $CGRP_{8-37}$ in certain preparations, the use of linear agonists such as [CysACM]CGRP[38,39] and [Cys (Et^{2,7})]-haCGRP[64] suggested the existence of another receptor sub-type, in addition to the better characterized $CGRP_1$ receptor. Indeed, these analogues behave as relatively weak but active agonists in the rat and guinea pig vasa deferentia while being mostly inactive in atrial preparations.[38,39,64] The reverse situation is seen for $CGRP_{8-37}$, this fragment being most potent at antagonizing the effects of CGRP in the atria while being almost inactive in the vas deferens. These two distinct profiles suggested to us the existence of at least two classes of CGRP receptors termed $CGRP_1$ and $CGRP_2$, respectively (Table 1.2). Stangl et al also reported that CGRP and [CysACM]CGRP produced CGRP$_{8-37}$-insensitive increases in cAMP formation in the liver suggesting an enrichment with the putative CGRP$_2$ sub-type in this tissue.[34] The guinea pig urinary bladder[65] and a human adrenocarcionma cell line (Col-29)[66] may also be enriched with CGRP$_2$ receptors and could be used as starting material toward the cloning of this receptor sub-type. Data in that regard are anxiously awaited.

1.2.3. An Atypical Receptor Sub-Type

In addition to the existence of the proposed CGRP$_1$ and CGRP$_2$ receptor sub-types, another receptor binding site with a unique ligand selectively profile may exist in some brain areas, especially in the nucleus accumbens.[8,67] These sites are recognized with high nM affinities by CGRP, amylin (rat or human) and salmon (but not mammalian) calcitonin. Adrenomedullin, in contrast, failed to demonstrate significant affinity for these sites[68] (Table 1.2). Moreover, the ontogenic maturation of this receptor protein is very slow compared to that of specific [^{125}I]haCGRP binding sites in other regions of the rat brain.[8] The nature of the preferential endogeneous ligand of these sites remains to be fully established since CGRP, fish calcitonins and amylin possess high affinity. The cloning of this receptor site is also awaited to confirm its unique pharmacological profile and to establish its functional significance.

1.3. Amylin Receptors

Evidence for the existence of unique amylin receptors are reviewed in detail by Sexton et al in this volume. We briefly summarize here key data regarding their CNS distribution and characteristics.

Various groups[69-71] including ours[72] have reported on the presence of amylin receptors in the CNS. Specific binding sites with high pM affinity were detected in various areas of the mammalian brain, especially in the nucleus accumbens. In vitro receptor autoradiography revealed that specific [^{125}I]amylin binding sites are particularly abundant in the olfactory tubercle, nucleus accumbens, tail of the caudate putamen, amygdaloid body, and various hypothalamic and brainstem nuclei.[72] In contrast to CGRP receptors, the cerebellum is devoid of specific amylin binding sites.[70,72] These amylin binding sites are highly sensitive to both amylin and salmon calcitonin while αCGRP, CGRP$_{8-37}$ and [CysACM]CGRP are much less potent (see ref 72 and unpublished results). Moreover, both [^{125}I]Bolton Hunter (BH)-amylin and [^{125}I]salmon calcitonin binding sites are similarly distributed in the CNS[70,72] suggesting that previously characterized salmon calcitonin receptors in mammalian brain may in fact represent amylin receptors. However, the cloning of unique amylin receptors has yet to be reported, cloned genes thus far belonging to the calcitonin receptor family. It may be that a calcitonin receptor gene cotransfected with a RAMP-like protein behaves as a unique amylin receptor (see Sexton et al, Ch. 8, this volume).

1.4. Adrenomedullin Receptors

Comparatively less information is currently available on detailed characteristics of specific adrenomedullin receptors (see other chapters in this volume on this topic). This situation is likely to be related to the fact that early studies indicated that adrenomedullin was acting via CGRP$_{8-37}$ sensitive-CGRP receptors to induce its effects negating the needs for unique adrenomedullin receptors. However, more recent data using a variety of fragments and analogues of adrenomedullin and CGRP have clearly shown that distinct adreno-medullin receptors are expressed in a variety of brain areas and peripheral tissues[68] (see also Smith et al, Ch. 10 and Kadowitz et al, Ch. 11, this volume). In fact, in the rat CNS, adrenomedullin is a rather weak competitor against both [^{125}I]hCGRPα and [^{125}I]BH amylin receptor binding sites suggesting that it has to act via its own receptors if having any significant physiological role in the CNS.[68] The availability of specific adreno-medullin radioligands confirms and extends these findings.[73-75] Moreover, using both agonist and antagonist radioligands, we recently showed that specific adrenomedullin binding sites, distinct from CGRP receptors, are expressed in a variety of tissues including the brain and the lung.[73] The ligand selectivity profile of these specific adrenomedullin sites was clearly distinct from that of CGRP receptors with adrenomedullin derivatives being most potent while CGRP related analogues were much weaker. Moreover, distinct binding kinetics and ligand binding profiles are observed when using agonist vs antagonist radiolabelled probes suggesting the possible existence of heterogeneous populations of adrenomedullin receptors.[73]

The existence of unique adrenomedullin receptors was demonstrated further with the cloning of an adrenomedullin-preferring receptor.[76] This clone belongs to the family of seven transmembrane domains-rhodopsin receptor super family. Northern blot analysis of this transcript showed positive signals in the lung, adrenal, ovary, heart, spleen, brain cortex and cerebellum.[76] However, in subsequent studies, it was difficult to successfully transfect and express this clone in a variety of cell lines and in situ hybridization studies demonstrated its very restricted distribution in the CNS.[61] This is rather similar to observations made earlier with the CGRP₁ receptor clones. Again, the recent cloning of the RAMPs[63] may help to resolve, at least partly, this puzzle. Indeed, it was recently shown that a functional adrenomedullin receptor can be derived from the cotransfection of CRLR and RAMP2 in HEK293 cells, a similar observation failing to be seen with the adrenomedullin clone[76] and any of the currently known RAMPs.[63] It thus suggests that an adrenomedullin receptor can be generated by the coexpression of RAMP2 and CRLR[63] while the other clone[76] either belongs to the orphan sub-group or requires a yet to be discovered RAMP-like molecule to be functionally relevant.

1.5. Conclusion

It is now evident that multiple subtypes of receptors exist for CGRP and related peptides including distinct receptor or receptor complexes for CGRP, amylin and adrenomedullin. A unique feature of this family resides in the requirement for a jointly expressed protein or RAMP to generate a functional receptor. While a single G-protein coupled receptor protein (CRLR) can display different pharmacological profiles depending on the coexpressed RAMP (CGRP-like with RAMP1; adrenomedullin-like with RAMP2), it is likely that other clones have yet to be characterized, the sum of CRLR plus RAMP1 and 2 not corresponding fully to the known distribution of either CGRP or adrenomedullin binding sites in the mammalian CNS. Moreover, the cloning of the putative CGRP₂ subtype and the atypical receptor of the nucleus accumbens is anxiously awaited as final proof of their uniqueness and relevance. The recent cloning of the RAMPs and the expected availability of potent non-peptide antagonists should prove most useful to elucidate further the relevance of CGRP and related peptides in the maintenance of normal body homeostasis.

References

1. Amara SG, Jonas V, Rosenfeld MG et al. Alternative RNA processing in calcitonin gene expression generates mRNAs encoding different polypeptide products. Nature 1982; 296:240-244.
2. Amara SG, Arriza JL, Leff SE et al. Expression in brain of a messenger RNA encoding a novel neuropeptide homologous to calcitonin generelated peptide. Science 1985; 229:1094-1097.
3. Rosenfeld MG, Amara SG, Evans RM. Alternative RNA processing: Determining neuronal phenotype. Science 1984; 225:1315-1320.
4. Rosenfeld MG, Emerson RB, Yeakley JM et al. Calcitonin gene-related peptide: A neuropeptide generated as a consequence of tissue-specific, developmentally regulated alternative RNA processing events. Ann NY Acad Sci 1992; 657:1-17.

5. vanRossum D, Hanisch UK, Quirion R. Neuroanatomical localization, pharmacological characterization and functions of CGRP and related peptide receptors. Neurosci Biobehav Rev 1997; 21:649-678.

6. Wimalawansa SJ. Calcitonin gene-related peptide and its receptors: Molecular genetics, physiology, pathophysiology, and therapeutic potentials. Endo Rev 1996; 17:533-585.

7. Poyner DR. Pharmacology of receptors for calcitonin gene-related peptide and amylin. TIPS 1995; 16:424-428.

8. Dennis T, Fournier A, Guard S et al. Calcitonin gene-related peptide (hCGRPα) binding sites in the nucleus accumbens. Atypical structural requirements and marked phylogenic differences. Brain Res 1991; 539:59-66.

9. Yoshizaki H, Takamiya M, Okada T. Characterization of picomolar affinity binding sites for [^{125}I]-human calcitonin gene-related peptide in rat brain and heart. Biochem Biophys Res Commun 1987; 146:443-451.

10. Chatterjee TK, Fisher RA. Multiple affinity forms of the calcitonin gene-related peptide receptor in rat cerebellum. Mol Pharmacol 1991; 39:798-804.

11. Nakamuta H, Fukuda Y, Koida M et al. Binding sites of calcitonin gene-related peptide (CGRP); abundant occurrence in visceral organs. Jpn J Pharmacol 1986; 42:175-180.

12. Chatterjee TK, Moy JA, Fisher RA. Characterization and regulation of high affinity calcitonin gene-related peptide receptors in cultured neonatal rat cardiac myocytes. Endocrinology 1991; 128:2731-8.

13. Jennings CGB, Mudge AW. Chick myotubes in culture express high-affinity receptors for calcitonin gene-related peptide. Brain Res 1989; 504:199-205.

14. Roa M, Changeux JP. Characterization and developmental evolution of a high-affinity binding site for calcitonin gene-related peptide on chick skeletal muscle membrane. Neuroscience 1991; 41:563-570.

15. Takamori M, Yoshikawa H. Effect of calcitonin gene-related peptide on skeletal muscle via specific binding site and G protein. J Neurol Sci 1989; 90:99-109.

16. Mak JCW, Barnes PJ. Autoradiographic localization of calcitonin gene-related peptide (CGRP) binding sites in human and guinea pig lung. Peptides 1988; 9:957-963.

17. Umeda Y, Arisawa M. Characterization of calcitonin gene-related peptide (CGRP) receptors in guinea pig lung. J Pharmacol 1989; 51:377-384.

18. Dennis T, Fournier A, Cadieux A et al. hCGRP8-37, a calcitonin gene-related peptide antagonist revealing calcitonin gene-related peptide receptor heterogeneity in brain and periphery. J Pharmacol Exp Ther 1990; 254:123-128.

19. McGillis JP, Humphreys S, Rangnekar V et al. Modulation of B lymphocyte differentiation by calcitonin gene-related peptide (CGRP). I. Characterization of high-affinity CGRP receptors on murine 70Z/3 cells. Cell Immunol 1993; 150:391-404.

20. Umeda Y, Arisawa M. Characterization of the calcitonin gene-related peptide receptor in mouse T lymphocytes. Neuropeptides 1989; 14:237-242.

21. Chatterjee TK, Moy JA, Cai JJ et al. Solubilization and characterization of a guanine nucleotide-sensitive form of the calcitonin gene-related peptide receptor. Mol Pharmacol 1993; 43:167-175.

22. vanRossum D, Ménard DP, Quirion R. Effect of guanine nucleotides and temperature on calcitonin gene-related peptide receptor binding sites in brain and peripheral tissues. Brain Res 1993; 617:249-257.

23. Tache Y, Raybould H, Wei JY. Central and peripheral actions of calcitonin gene-related peptide on gastric secretory and motor function. Adv Exp Med Biol 1991; 298:183-198.

24. Foord SM, Craig RK. Isolation and characterisation of a human calcitonin-gene-related peptide receptor. Eur J Biochem 1987; 170:373-379.

25. Kapas S, Clark AJL. Identification of an orphan receptor gene as a type 1 calcitonin gene-related peptide receptor. Biochem Biophys Res Commun 1995; 217:832-838.

26. Aiyar NJ, Rand IC, Elshourbagy NA et al. A cDNA encoding the calcitonin gene-related peptide type 1 receptor. J Biol Chem 1996; 271:11325-11329.

27. Kim D. Calcitonin gene-related peptide activates the muscarinic-gated K$^+$ cur-

rent in atrial cells. Pflugers Arch 1991; 418:338-345.

28. Wiley JW, Gross RA, Macdonald RL. The peptide CGRP increases a high-threshold Ca^{2+} current in rat nodose neurones via a pertussis toxin-sensitive pathway. J Physiol (Lond.) 1992; 455:367-381.

29. Raybould HE. Inhibitory effects of calcitonin gene-related peptide on gastrointestinal motility. Ann NY Acad Sci 1992; 657:248-257.

30. Sano Y, Hiroshima O, Yuzuriha T et al. Calcitonin gene-related peptide binding sites of porcine cardiac muscles and coronary arteries; Solubilization and characterization. J Neurochem 1989; 52:1919-1924.

31. Hirata Y, Takagi Y, Takata S et al. Calcitonin gene-related peptide receptor in cultured vascular smooth muscle and endothelial cells. Biochem Biophys Res Commun 1988; 151:1113-1121.

32. Honda T, Zhou Z-C, Gu Z-F et al. Structural analysis of CGRP receptors on gastric smooth muscle and pancreatic acinar cells. Am J Physiol (Gastrointest Liver Physiol) 1993; 27:1142-1152.

33. Bhogal R, Smith DM, Purkiss P et al. Molecular identification of binding sites for calcitonin gene-related peptide (CGRP) and islet amyloid polypeptide (IAPP) in mammalian lung: species variation and binding of truncated CGRP and IAPP. Endocrinology 1993; 133:2351-2361.

34. Stangl D, Muff R, Schmolck C et al. Photoaffinity labelling of rat calcitonin gene-related peptide receptors and adenylate cyclase activation: Identification of receptor subtypes. Endocrinology 1993; 132:744-750.

35. Aiyar N, Nambi P, Griffin E et al. Identification and characterization of calcitonin gene-related peptide receptors in porcine renal medullary membranes. Endocrinology 1991; 129:965-969.

36. Chantry A, Leighton B, Day AJ. Cross-reactivity of amylin with calcitonin-gene-related peptide binding sites in rat liver and skeletal muscle membranes. Biochem J 1991; 277:139-143.

37. Stangl D, Born W, Fischer JA. Characterization and photoaffinity labelling of a calcitonin gene-related peptide receptor solubilized from human cerebellum. Biochemistry 1991; 30:8605-8611.

38. Dennis T, Fournier A, St-Pierre S et al. Structure-activity profile of calcitonin gene-related peptide in peripheral and brain tissues. Evidence for receptor multiplicity. J Pharmacol Exp Ther 1989; 251:718-725.

39. Quirion R, vanRossum D, Dumont Y et al. Characterization of $CGRP_1$ and $CGRP_2$ receptor subtypes. Ann NY Acad Sci 1992; 657:88-105.

40. Rovero P, Giuliani S, Maggi CA. CGRP antagonist activity of short C-terminal fragments of human αCGRP, CGRP (19-37), CGRP (23-37). Peptides 1992; 13:1025-1027.

41. Chakder S, Rattan S. [Tyr^0]-calcitonin gene-related peptide 28-37 (rat) as a putative antagonist of calcitonin gene-related peptide responses on opossum internal anal sphincter smooth muscle. J Pharmacol Exp Ther 1990; 253:200-206.

42. Maton PN, Pradhan T, Shou Z-C et al. Activities of calcitonin gene-related peptide (CGRP) and related peptides at the CGRP receptor. Peptides 1990; 11:485-489.

43. Rist B, Entzeroth M, Beck-Sickinger AG. From micromolar to nanomolar affinity-A systematic approach to identify the binding site of CGRP at the human calcitonin gene-related peptide 1 receptor. J Med Chem 1998; 41:117-123.

44. Aiyar N. Personal Communication. 3rd International CGRP Meeting Shaftesbury, England, May 1998.

45. Gardiner SM, Compton AM, Kemp PA et al. Antagonistic effect of human alpha-CGRP (8-37) on the in vivo regional haemodynamic actions of haCGRP. Biochem Biophys Res Commun 1990; 171:938-943.

46. Han S-P, Naes L, Westfall TC. Inhibition of periarterial nerve stimulation-induced vasodilation of the mesenteric arterial bed by CGRP (8-37) and CGRP receptor desensitization. Biochem Biophys Res Commun 1990; 168:786-791.

47. Hughes SR, Brain SD. A calcitonin gene-related peptide (CGRP) antagonist (CGRP8-37) inhibits microvascular responses induced by CGRP and

capsaicin in skin. Br J Pharmacol 1991; 104:738-742.

48. Wang F, Millet I, Bottomly K et al. Calcitonin gene-related peptide inhibits interleukin 2 production by murine T lymphocytes. J Biol Chem 1992; 267:21052-21057.

49. Evangelista S, Tramontana M, Maggi CA. Pharmacological evidence for the involvement of multiple calcitonin gene-related peptide (CGRP) receptors in the antisecretory and antiulcer effect of CGRP in rat stomach. Life Sci 1992; 50:13-18.

50. Maggi CA, Chiba T, Giuliani S. Human alpha-calcitonin gene-related peptide (8-37) as an antagonist of exogenous and endogenous calcitonin gene-related peptide. Eur J Pharmacol 1991; 192:85-88.

51. Mimeault M, Fournier A, Dumont Y et al. Comparative affinities and antagonistic potencies of various human calcitonin gene-related peptide fragments on calcitonin gene-related peptide receptors in brain and periphery. J Pharmacol Exp Ther 1991; 258:1084-1090.

52. Longmore J, Hogg JE, Hutson PH et al. Effects of two truncated forms of human calcitonin-gene related peptide: implications for receptor classification. Eur J Pharmacol 1994; 265:53-59.

53. Jolicoeur FB, Ménard D, Fournier A et al. Structure-activity analysis of CGRP's neurobehavioral effects. Ann NY Acad Sci 1992; 657:155-163.

54. Libert F, Parmentier M, Lefort A et al. Selective amplification and cloning of four new members of the G protein-coupled receptor family. Science 1989; 244:569-572.

55. Libert F, Parmentier M, Lefort A et al. Complete nucleotide sequence of a putative G protein coupled receptor: RDC1. Nucl Acid Res 1990; 18:1917-1925.

56. Njuki F, Nicholl CG, Howard A et al. A new calcitonin-receptor-like sequence in rat pulmonary blood vessels. Clin Sci 1993; 85:385-388.

57. Chang CP, Pearse RV, O'Connel S et al. Identification of a seven transmembrane helix receptor for corticotropin-releasing factor and sauvagine in mammalian brain. Neuron 1993; 11:1187-1195.

58. Fluhmann B, Muff R, Hunziker W et al. A human orphan calcitonin receptor-like structure. Biochem Biophys Res Commun 1995; 206:341-347.

59. Han Z-Q, Copock HA, Smith DM et al. The interaction of CGRP and adrenomedullin with a receptor expressed in the rat pulmonary vascular endothelium. J Mol Endocrinol 1997; 18:267-272.

60. vanRossum D, Ménard DP, Fournier A et al. Binding profile of a selective calcitonin gene-related peptide (CGRP) receptor antagonist ligand, [^{125}I-Tyr]hCGRP$_{8-37}$, in rat brain and peripheral tissues. J Pharmacol Exp Ther 1994; 269:846-853.

61. Oliver KR, Wainwright A, Heavens RP et al. Distribution of novel CGRP$_1$ receptor and adrenomedullin receptor mRNAs in the rat central nervous system. Mol Brain Res 1998; 57:149-154.

62. Luebke AE, Dahl GP, Roos BA et al. Identification of a protein that confers calcitonin gene-related peptide responsiveness to oocytes by using a cystic fibrosis transmembrane conductance regulator assay. Proc Natl Acad Sci USA 1996; 93:3455-3460.

63. McLatchie LM, Fraser NJ, Main MJ et al. RAMPs regulate the transport and ligand specificity of the calcitonin-receptor-like receptor. Nature 1998; 393:333-339.

64. Dumont Y, Fournier A, St-Pierre S et al. A potent and selective CGRP$_2$ agonist, [Cys(Et)^{2,7}]hCGRPα: comparison in prototypical CGRP$_1$ and CGRP$_2$ in vitro bioassays. Can J Physiol Pharmacol 1997; 75:671-676.

65. Giuliani S, Wimalawansa SJ, Maggi CA. Involvement of multiple receptors in the biological effects of calcitonin gene-related peptide and amylin in rat and guinea-pig preparations. Br J Pharmacol 1992; 107:510-514.

66. Cox HM, Tough IR. Calcitonin gene-related peptide receptors in human gastrointestinal epithelia. Br J Pharmacol 1994; 113:1243-1248.

67. Sexton PM, McKenzie JS, Mendelsohn FAO. Evidence for a new subclass of calcitonin/calcitonin gene-related peptide binding site in rat brain. Neurochem Int 1988; 12:323-335.

68. vanRossum D, Ménard DP, Chang JK et al. Comparative affinities of human adrenomedullin for [^{125}I]hCGRPα and [^{125}I]BH-amylin specific binding sites in

the rat brain. Can J Physiol Pharmacol 1995; 73:1084-1088.

69. Beaumont K, Kenney MA, Young AA et al. High affinity amylin binding sites in rat brain. Mol Pharmacol 1993; 44:493-497.

70. Sexton PM, Paxinos G, Kenney MA et al. *In vitro* autoradiographic localization of amylin binding sites in rat brain. Neuroscience 1994; 62:553-567.

71. Veale PR, Bhogal R, Morgan DG et al. The presence of islet amyloid polypeptide/calcitonin gene-related peptide/salmon calcitonin binding sites in the rat nucleus accumbens. Eur J Pharmacol 1994; 262:133-141.

72. vanRossum D, Ménard D, Fournier A et al. Autoradiographic distribution and receptor binding profile of [^{125}I]Bolton Hunter rat amylin binding sites in the

rat brain. J Pharmacol Exp Ther 1994; 270:779-787.

73. Dumont Y, Jacques D, Rémillard J et al. Evidence for the existence of specific [^{125}I]adrenomedullin receptors in the rat brain: comparison with [^{125}I]CGRP and [^{125}I]adrenomedullin (22-50) binding sites. Soc Neurosci Abst 1997; 23:395.

74. Eguchi S, Hirata Y, Kano H et al. Specific receptors for adrenomedullin in cultured rat vascular smooth muscle cells. FEBS Lett 1994; 340:226-230.

75. Owji AA, Smith DM, Coppock HA et al. An abundant and specific binding site for the novel vasodilator adrenomedullin in the rat. Endocrinology 1995; 136:2127-2134.

76. Kapas S, Catt KJ, Clark AJL. Cloning and expression of cDNA encoding a rat adrenomedullin receptor. J Biol Chem 1995; 270:25344-25347.

CHAPTER 2

CGRP Receptor Heterogeneity: Use of CGRP$_{8-37}$

Ian Marshall and Franca-Maria Wisskirchen

2.1. Introduction

CGRP produces a wide range of central and peripheral effects which involve short-term and long-term actions. Soon after the identification of CGRP it was established that the peptide did not appear to act indirectly via the release of other biologically active mediators.[1] Therefore it is widely accepted that many effects of CGRP are mediated through specific membrane- based receptors and up to now receptor classification has been forced to rely on radioligand binding and isolated tissue studies. The recent identification of specific binding sites for the related peptides amylin and adrenomedullin has made it more difficult to determine which receptors are involved in a given situation.

2.2. Classification of CGRP Receptors

A major problem in this area has been the lack of selective agonists and antagonists. The first CGRP receptor antagonist was human (h) αCGRP$_{8-37}$.[2] In the absence of more potent and selective antagonists, CGRP$_{8-37}$ remains in widespread use although shorter and less potent C-terminal fragments have been described.[3-5] Agonists may also be helpful in identifying receptors. There are a number of closely related structural forms of CGRP (e.g., human and rat α and βCGRP) but these are usually similar in biological activity.[6,7] However acetamidomethylcysteine$_{2,7}$ hαCGRP ([CysACM]CGRP) has been suggested to discriminate between subtypes of CGRP receptors (see also Quirion and Dumont, Ch. 1, this volume)

2.2.1. CGRP Receptor Heterogeneity

Pharmacological evidence is not consistent with the presence of just one CGRP receptor. For instance the antagonist hαCGRP$_{8-37}$ displays a wide range of affinities (Table 2.1). While being inactive in guinea pig trachea and mouse aorta, hαCGRP$_{8-37}$ is a weak antagonist in the rat vas deferens. It has around 10-fold higher potency in the guinea pig right atrium and even higher affinity in the opossum internal anal sphincter and human SK-N-MC cells. Even restricting values to vascular tissues there is at least a 1,000 range of antagonist affinities. For example these extend from mouse aorta $<6.0^9$, through guinea pig basilar artery 7.0^{18} to canine lingual artery $9.1.^{19}$ This diversity is seen not only in isolated tissues but is equally apparent from radio-ligand binding experiments. The pK$_i$ values for hαCGRP$_{8-37}$ are 6.1 in rat neonatal cardiac membranes,[20] 6.6 in opossum internal anal sphincter,[15] 7.5 in pig renal medullary membranes,[21] 8.7 in human SK-N-MC cells[17] and 9.3 in guinea pig vas deferens.[10] It is very likely that at least part of this range may be due to receptor heterogeneity.

The CGRP Family: Calcitonin Gene-Related Peptide (CGRP), Amylin, and Adrenomedullin, edited by David Poyner. ©2000 EUREKAH.COM.

Table 2.1. Antagonist affinities for haCGRP$_{8-37}$ against CGRP responses in a selection of different species and preparations to demonstrate the range of pA$_2$ values

Species	Preparation	pA$_2$ value	Reference
Guinea pig	Trachea	<5	Bhogal et al[8]
Mouse	Aorta	<6	Quirion et al[9]
Rat	vas deferens	6.2	Mimeault et al[10]
Pig	coronary artery	6.7	Gray et al[11]
Rat	adipocytes	6.9	Casini et al[12]
Guinea pig	ileum	7.0	Dennis et al[13]
Guinea pig	right atrium	7.2	Mimeault et al[10]
Rat	mesenteric vasculature	7.4	Nuki et al[14]
Opossum	internal anal sphincter	7.8	Chakder & Rattan[15]
Rat	cardiomyocytes	8.0	Bell & McDermott[16]
Human	SK-N-MC cells	8.7	Longmore et al[17]

Recently receptors for CGRP have been sub-classified into two types[3,9,10,13] (see also Quirion and Dumont, Ch. 1, this volume). This is based mainly on the differing antagonist affinities for hαCGRP$_{8-37}$ but also on the properties of the weaker antagonist hαCGRP$_{12-37}$ and the weak agonist [CysACM]CGRP. Thus while hαCGRP$_{12-37}$ antagonizes CGRP responses in the guinea pig atrium it is even less potent in the rat vas deferens. This "tissue selectivity" is also exhibited by [CysACM]CGRP although in this case it is an agonist in the vas but not in the atrium.[7] On the basis of these results CGRP receptors have been sub-classified into CGRP$_1$ and CGRP$_2$ receptors, typified by receptors for the peptide in the guinea pig atrium and rat vas deferens respectively. The CGRP$_1$ receptor is characterized by a higher affinity for the antagonist hαCGRP$_{8-37}$, an

affinity for hαCGRP$_{12-37}$ and the lack of activity of [CysACM]CGRP. In contrast at the CGRP$_2$ receptor hαCGRP$_{8-37}$ has a lower affinity, hαCGRP$_{12-37}$ is inactive and [CysACM] hαCGRP is a weak agonist.

Although this scheme can account for some of the findings there are a number of issues which remain unresolved. For example the all-or-none difference in both hαCGRP$_{12-37}$ and [CysACM]CGRP between the two receptor subtypes has not been investigated using higher concentrations of these compounds. Further the significance of affinity values for hαCGRP$_{8-37}$ higher than that proposed for the CGRP$_1$ receptor is not addressed (Table 2.1). A third point is that the difference in affinity for hαCGRP$_{8-37}$ between the guinea pig atrium and rat vas deferens is less clear with values of 6.6 in the vas[13] and 6.9 in the atrium.[23] Additionally some of the difference in potency of the CGRP antagonist may reflect unequal distribution of tissue peptidases rather than receptor subtypes.[17] Finally the prototypical CGRP$_1$ and CGRP$_2$ receptors are in different species, the guinea pig and the rat respectively, and the extent of species differences in CGRP receptors is unclear.

2.2.2. Pharmacological Characterization of CGRP Receptors in Rat Smooth Muscles

Because of the issues raised above, a more systematic investigation of the receptors mediating the effects of CGRP in a number of isolated smooth muscle preparations from one species has been carried out. The four tissues from the rat were the vas deferens, the pulmonary artery, the thoracic aorta and the internal anal sphincter.

The Rat Vas Deferens

In the vas deferens hαCGRP, hβCGRP and rβCGRP all inhibited twitch responses of the field stimulated preparation with about equal potency while rat amylin had about one tenth relative potency and [CysACM]CGRP was at least 3,000-fold weaker than the parent peptide.[24] This potency was unaltered by incubation with a mix of peptidase inhibitors (amastatin, bestatin, captopril phosphoramidon and thiorphan, all at 10^{-6}M).

The antagonist affinity of hαCGRP$_{8-37}$ in the vas was initially studied to establish equilibrium conditions. A similar affinity was found after either 2 or 20 minutes incubation but surprisingly a much lower one was obtained after 60 minutes.[24] This could reflect peptide breakdown or some other mechanism reducing its concentration in the biophase over a long period of time. Subsequent 20 minute incubations showed that the affinity of hαCGRP$_{8-37}$ was independent of the agonist with values between 5.6 and 6.0 against hαCGRP, hβCGRP, rβCGRP and rat amylin (Table 2.2). In addition the closely related antagonist hβCGRP$_{8-37}$ also competitively inhibited hαCGRP inhibition of

twitch responses of the vas with an affinity close to that of hαCGRP$_{8-37}$.[24] It is unlikely that peptidases might be reducing peptide activity as the mix of 5 peptidase inhibitors (see above) did not alter either the potency of hαCGRP or of ha CGRP$_{8-37}$.

The Rat Pulmonary Artery

The phenylephrine-evoked tone of endothelium-intact rat pulmonary artery was reduced by hα CGRP and this was dependent on the presence of an intact endothelium. Extreme care was taken to ensure the minimum of damage to the tissue otherwise the maximum relaxation to hαCGRP (100% of evoked tone) was reduced. HαCGRP, hβCGRP and human adrenomedullin were of similar potency while rat amylin and [CysACM] CGRP were about 50 and 3,000 times less potent respectively.[24] Therefore [CysACM]CGRP had similar potency in the pulmonary artery to that in the vas deferens.

HαCGRP$_{8-37}$ was around 10 times more potent in the pulmonary artery than in the vas (affinity around 7.0; Table 2.2) against both hαCGRP and hβCGRP and a similar value was obtained after either 2 or 20 minute equilibration.[24] A similar affinity was found for hβCGRP$_{8-37}$ against hαCGRP. The relaxation of phenylephrine-evoked tone by adrenomedullin was not antagonized by hαCGRP$_{8-37}$ suggesting it was not mediated through CGRP receptors. Pre-incubation with a mix of peptidase inhibitors (see above) did not alter the potency of either the agonist hαCGRP or of the antagonist hαCGRP$_{8-37}$.

These data have established the presence of two apparently homogenous populations of CGRP receptors, one in the vas and the other in the pulmonary artery. The affinity of hαCGRP$_{8-37}$ in the artery is in good agreement with the value for a CGRP$_1$ receptor while that in the vas is consistent with a CGRP$_2$ receptor. Therefore this data supports the proposed CGRP classification and has done so within a single species. The effect of the antagonist was independent of the agonist in both tissues. In addition these receptor subtypes were also distinguished

Table 2.2. Antagonist affinity estimates for human CGRP$_{8-37}$ against CGRP and related peptides in rat vas deferens, internal anal sphincter (IAS), pulmonary artery and thoracic aorta

Agonist	Antagonist	pA$_2$/pK$_B$±sem*value			
		Vas deferens	IAS	Pulmonary artery	Aorta
hαCGRP	haCGRP$_{8-37}$	6.0	5.7±0.3*	6.9	<5
hβCGRP	haCGRP$_{8-37}$	5.7	ND	7.1±0.1*	<5
rβCGRP	haCGRP$_{8-37}$	5.8±0.1*	5.8±0.2*	ND	<5
rat amylin	haCGRP$_{8-37}$	5.8±0.1*	ND	7.2±0.1*	ND
hαCGRP	hβCGRP$_{8-37}$	5.6	6.1±0.1*	7.1±0.1*	ND
human adrenomedullin	haCGRP$_{8-37}$	ND	ND	<6	ND

ND: not determined

by a second, structurally related antagonist, hβCGRP$_{8-37}$ which showed the same difference in affinity as the human α-form. Finally the different forms of CGRP were of similar potency in a given tissue. Therefore the sequence differences among the tested group of closely related peptides did not produce any difference in receptor subtype selectivity.

There are two further points worth noting from these experiments. Firstly, inhibition of peptidases did not alter the potency of either agonist or antagonist in the artery or the vas. Therefore the difference in the affinity of hαCGRP$_{8-37}$ between the two tissues is unlikely to be explained by differences in peptide degradation and supports the idea of receptor subtypes. Secondly, [CysACM]CGRP has been suggested as a selective agonist at the CGRP$_2$ receptor with around 1% of the potency of hαCGRP.[3] However, it was 3000 times less potent than the parent compound at both the CGRP$_1$ receptor in the pulmonary artery and at the CGRP$_2$ receptor in the vas deferens. Further, the low agonist potency in the vas was not increased by the presence of peptidase inhibitors. Therefore the reason for this much lower potency than that originally reported in this tissue[3] and subsequently used as one of the criteria for establishing the presence of a CGRP$_2$ receptor remains unclear. However, it suggests that the use of this compound as a tool to subtype CGRP receptors is at present still open to doubt.

The Rat Thoracic Aorta

Relaxation of noradrenaline-evoked tone in the aorta was dependent on the presence of an intact endothelium as shown previously.[25,26] In this respect the tissue resembled the pulmonary artery and was even more sensitive to loss of endothelium. Invariably the maximum relaxation in the second control curve was lower than the first and the apparent potency of CGRP was reduced, so it proved necessary to use only a single curve in each preparation. Under these conditions similar potency and maximum response (72-80% relaxation) was

found with hαCGRP, hβCGRP, rβCGRP and adrenomedullin. Rat amylin was about 25-times less potent and [CysACM]CGRP was at least 1000-fold less potent than hαCGRP. Incubation of hαCGRP$_{8-37}$ at a high concentration (10^{-5}M) for either 2, 20 or 60 minutes failed to antagonise relaxation to hαCGRP, hβCGRP or rβCGRP (Table 2.2). Neither effects of the agonists nor the lack of effect of the antagonist was altered by the presence of peptidase inhibitors.

This result does not fit with the CGRP$_1$ and CGRP$_2$ receptor classification as the affinity of hαCGRP$_{8-37}$ would be lower than 5. Recently, a higher affinity was reported in the aorta[27] but it is not clear from the methods whether control and antagonist data was obtained from single or separate tissues. Thus the higher affinity could result in part from second curves to CGRP in this tissue, the shift in the concentration-effect curve in the presence of the antagonist being due to the loss of endothelium reducing the effect of the agonist.

This is not the first time that a very low affinity for hαCGRP$_{8-37}$ has been reported although most studies do not use concentrations above 10^{-6}M or 3×10^{-6}M. For example in the mouse aorta,[9] guinea pig urinary bladder[28] and guinea pig vas deferens[29] these concentrations of hαCGRP$_{8-37}$ did not antagonize CGRP responses. It is generally concluded that these are CGRP$_2$ receptors although no affinity for hαCGRP$_{8-37}$ was demonstrated. If the value is below 5 in these tissues as found in the rat aorta, this could suggests another group of CGRP receptors. An alternative possibility is that these reflect the action of CGRP at a non-CGRP receptor. This is probably not an amylin or calcitonin receptor as amylin is much weaker than CGRP and calcitonin is inactive. In addition the equal potency of CGRP with adrenomedullin make an adrenomedullin receptor unlikely although not impossible.

The Rat Internal Anal Sphincter

A high affinity for hαCGRP$_{8-37}$ (7.8) was reported in the internal anal sphincter of the opossum.[15] To see if the value reflected some tissue factor or might be due to species differences in CGRP receptors, this smooth muscle preparation was investigated in the rat. The potency of hαCGRP in relaxing spontaneously developed tone was similar to that found in the pulmonary artery or vas deferens. The inhibition curves were reproducible in the sphincter. While hβCGRP and rβCGRP were equally potency, [CysACM] CGRP was inactive up to 10^{-5}M.

The antagonist hαCGRP$_{8-37}$ had an affinity just below 6 against hαCGRP and rβCGRP respectively (Table 2.2). A similar value was obtained with hβCGRP. The presence of peptidase inhibitors did not alter either agonist or antagonist potency.

These findings support the presence of a homogenous group of CGRP receptors in the rat internal anal sphincter which is the same as that found in the rat vas deferens, CGRP$_2$ receptors. The affinity of the CGRP antagonist was 100-times higher in the opossum internal anal sphincter suggesting closer similarity to a CGRP$_1$ receptor. However, it is likely that at least part of the discrepancy is due to species differences in CGRP receptors between the rat and opossum. The surprising conclusion is that it may be difficult to decide whether a given antagonist affinity value represents a CGRP$_1$ or CGRP$_2$ receptor unless both have been found in a given species. For example, is the value of 7.8 in the opossum anal sphincter a high value at a CGRP$_1$ receptor or an even higher value at a CGRP$_2$ receptor? While the former is more likely, the latter possibility cannot be excluded. Unfortunately antagonist affinity is the sole criterion for differentiating subtypes once it is decided that [CysACM]CGRP is not a useful tool.

2.3. Limitations to the Usefulness of CGRP$_{8-37}$ in Subclassifying CGRP Receptors

There are a number of factors which can affect the apparent affinity of an antagonist. It is important to rule out as many of these as possible so that any remaining differences may be ascribed to receptor subtypes.

2.3.1. Lack of Equilibrium Conditions

The lack of equilibrium conditions for either the agonist or the antagonist may lead to false conclusions. It is well known that mechanisms removing agonist from the biophase necessarily prevent the establishment of equilibrium between an agonist and antagonist at a receptor. While there is no evidence of an uptake mechanism for CGRP there is evidence for other processes. For example in the aorta the continual loss of endothelium during an experiment leads to an ever decreasing potency and/or maximum response. This makes it dangerous to draw conclusions from more than one concentration-relaxation curve to CGRP in a given isolated aorta. It is also known that in some tissues there may be tachyphylaxis to CGRP through mechanisms which have not been elucidated but are probably partly beyond the receptor level. This is the case in the opossum anal sphincter[15] but was not found in this tissue from the rat.

The affinity value for an antagonist may be low because the antagonist has had insufficient time to equilibrate with the agonist at the receptors. In the present experiments this has been checked using variable periods of incubation with the antagonist. A spurious low value may be found also with a long incubation time if there is some mechanism lowering the concentration of available antagonist eg., if it is degrading with time. Something along these lines appears to be happening in the vas deferens where between 20 and 60 minutes of incubation the antagonist affinity fell from 6.0 to an unmeasurable level, certainly less than 5.0. Degradation of peptides in tissues is well known although no evidence was found for such a mechanism in the present tissues after incubation of agonist and antagonist with a cocktail of peptidase inhibitors. Therefore the reason for the loss of activity of hαCGRP$_{8-37}$ after 1h incubation is unknown. This did not occur to the same extent in the pulmonary artery where values of 7.0 and

6.6 were found for the antagonist after 20 and 60 min incubation respectively.[24]

2.3.2. Species Differences in Receptors

At present there is relatively little information on whether there are species differences in CGRP receptors. The affinity values for hαCGRP$_{8-37}$ in rat and guinea pig tissues are similar and both support the CGRP$_1$ and CGRP$_2$ sub-classification although additional values remain to be interpreted. However, the higher value in the opossum, 100-fold above that in the same tissue in the rat, suggests that species differences probably exist. The cloning of CRLR and RAMPs from different species over the next few years may help to resolve this issue.

2.3.3. Cross-Reactivity with Other Receptors

The agonist CGRP and the antagonist hαCGRP$_{8-37}$ may interact with other receptors. Use of the antagonist should help to decide if an effect is mediated via a CGRP receptor. However, given that hαCGRP$_{8-37}$ is a relatively weak antagonist in some tissues, it is unclear whether there may be other CGRP receptor subtypes for which this antagonist has an even lower affinity. An example of this is the effect of CGRP in the rat aorta. Conversely the antagonist may have affinity for other related receptors although until transfected cell lines expressing a homogenous population of e.g., CGRP$_2$ or adrenomedullin receptors becomes available it will remain difficult to draw firm conclusions.

2.3.4. Additional Effects of hαCGRP$_{8-37}$

The antagonist may have confounding effects in some situations. For example hαCGRP$_{8-37}$ at high concentrations (above 10^{-6}M) caused relaxation of phenylephrine-induced tone in the pulmonary artery i.e., an effect like that of CGRP. This was not seen in the rat aorta, vas deferens or internal anal sphincter. One possible reason for this is that the antagonist had its highest affinity in the pulmonary artery and therefore this additional effect might not have been seen in other tissues given that it began to occur around 10-fold higher concentrations than the antagonist affinity at CGRP receptors. Where the effects of high concentrations of hαCGRP$_{8-37}$ were included in the analysis a Schild slope regression less than unity was obtained.[24] When experiments in the pulmonary artery in which this relaxant effect of the antagonist occurred were excluded, a Schild slope close to 1.0, consistent with competitive antagonism was found.

The vasodilator effect of hαCGRP$_{8-37}$ might be due to the compound being a partial agonist in the pulmonary artery. Alternatively, the effect might be mediated via a non-CGRP receptor. Cross-reaction with receptors for CGRP homologues such as amylin, adrenomedullin or calcitonin seem unlikely since either the peptides were much weaker (amylin and calcitonin) or hαCGRP$_{8-37}$ was ineffective (against adrenomedullin). However, an interaction with other receptors remains a possibility. For example, hαCGRP$_{8-37}$ has been suggested to be an agonist at tachykinin NK$_1$ receptors in rabbit iris sphincter.[30] Activation of NK$_1$ receptors by substance P causes vasodilation in the pulmonary circulation of many species including the guinea pig[31] as well as in the vascular system of the rat.[32,33] An NK$_1$ receptor mediated dilator response in the rat pulmonary artery would be consistent with the lack of effect of hαCGRP$_{8-37}$ on twitch responses in the vas deferens since this tissue contains only NK$_2$ receptors.[34] Therefore it is clear that hαCGRP$_{8-37}$ does not behave as a "pure" competitive antagonist and this may complicate the identification of CGRP receptors and their subtypes.

2.4. Conclusion

The sub-classification of CGRP receptors into CGRP$_1$ and CGRP$_2$ has had some success in rationalising observed data. However, there has been a tendency to abuse the classification. For example if the highest concentration of $h\alpha CGRP_{8-37}$ used in a study is $10^{-6}M$ and this has not antagonised an effect of CGRP it has been concluded that the effect is not via a $CGRP_1$ receptor. As discussed above even this conclusion may be flawed depending on the species used, experimental conditions etc. However, much worse is the following conclusion that given the receptor is not one subtype it must be the other, therefore a $CGRP_2$ receptor. The literature has a number of these conclusions where there is no evidence to support the claim, only a lack of effect of $h\alpha CGRP_{8-37}$ at the concentration tested. Hopefully, the present experiments and results will have illustrated the need in these situations to try even higher concentrations of antagonist so that either, the possibility of a $CGRP_2$ receptor is confirmed by seeing some antagonism of CGRP, or the antagonist remains ineffective making the conclusion of a $CGRP_2$ receptor unlikely.

It is clear that there are affinity values for $h\alpha CGRP_{8-37}$ both above and below the $CGRP_1$-$CGRP_2$ range. The high values (above 7.5) are found in tissues from the dog and opossum, species where the comparative pharmacology of CGRP receptors is non-existent! Why these high values occur in some cell lines is unclear. Of particular interest bearing in mind potential therapeutic applications of CGRP antagonists in e.g., migraine, are the affinity values for $h\alpha CGRP_{8-37}$ at human CGRP receptors. Here there is a mixture of both higher and lower values (Table 2.3). However both values for isolated cerebral blood vessels are in the $CGRP_1$-$CGRP_2$ range found in the rat and guinea pig.

Affinity values below the $CGRP_2$ range might reflect actions of CGRP at a non-CGRP receptor but it is possible that these may reflect a subtype for which $h\alpha CGRP_{8-37}$ has a very low affinity. If there are CGRP receptors that are not recognised by $h\alpha CGRP_{8-37}$ it is impossible to characterise them further with the tools available at the moment.

Finally it is obvious that many of the problems presented in trying to understand CGRP receptor subtypes arise from both our lack of knowledge of receptor structure and from the limitations of the available drugs. It is not possible to be completely sure of the selectivity of either CGRP or its antagonist when making conclusions. If a CGRP receptor is not necessarily defined by antagonism of an effect of the peptide by $h\alpha CGRP_{8-37}$, what is left? Certainly it cannot be assumed that CGRP receptors are those membrane sites acted on by the peptide itself since CGRP can act at other receptors. For example a $CGRP_3$ binding site has been suggested which has a very atypical pharmacology (this is also known as a C3 binding site and may represent an amylin receptor; see Chapter 8 by Sexton et al., this volume). While $h\alpha CGRP_{8-37}$ has given us an idea of CGRP receptor subtypes it cannot provide all the answers.

The wide range of affinities for $h\alpha CGRP_{8-37}$ is apparent with results from different techniques (radioligand binding and isolated tissue experiments), across different species and tissues or within a given tissue type e.g., vascular tissues. In some cases it is unclear as to whether equilibrium conditions have been established and whether the breakdown of peptides may be occurring. A more rigorous approach to the pharmacology of CGRP receptors is required which takes these factors into account including the possibility of species differences. Finally, while $h\alpha CGRP_{8-37}$ remains our prime antagonist, more information is needed on its selectivity of action and additional biological effects, particularly in the vasculature.

References

1. Marshall I, Craig RK. The cardiovascular effects and mechanism of action of the calcitonin gene-related peptides (CGRP). In: Vanhoutte PM, ed. Vasodilatation: Vascular smooth muscle, autonomic nerves and endothelium. New York: Raven Press, 1988:81-88.

2. Chiba T, Yamaguchi A, Yamatani T et al. Calcitonin gene-related peptide receptor antagonist human CGRP-(8-37). Am J Physiol 1989; 256:E331-E335.

Table 2.3. Affinity values for hαCGRP$_{8-37}$ derived either (a) using hαCGRP responses in functional experiments or (b) in competition with [^{125}I]hαCGRP or^{125}I[Tyr^0]rαCGRP specific binding using human preparations

A. Functional studies

Preparation	pA$_2$/pK$_B$*	Reference
Pial artery	5.7**	Marshall et al[35]
Cerebral artery	6.7*	Jansen et al[36]
COL 29 colonic epithelium cells#	<6*	Cox & Tough[37]
HCA-7 colonic epithelium cells#	6.4**	Cox & Tough[37]
Umbilical vein endothelial cells#	6.5	Kato et al[38]
Glioma membranes	7.5**	Robberecht et al[39]
SK-N-MC cells#	7.8-8.7	Muff et al[40] Entzeroth et al[41] Longmore et al[17] Zimmermann et al[42]

B. Radio-ligand binding experiments

Preparation	pK$_i$	Reference
Glioma membranes	7.5	Robberect et al[39]
SK-N-MC whole cells#	8.6	Muff et al[40] Zimmerman et al[42]
SK-N-MC cell membranes#	9.1	Longmore et al[17]

** pK$_B$ values estimated from literature data
# data obtained from cultured cells

3. Dennis T, Fournier A, St-Pierre S et al. Structure-activity profile of calcitonin gene-related peptide in peripheral and brain tissues. Evidence for receptor multiplicity. J Pharmacol Exp Ther 1989; 251: 718-725.

4. Chakder S, Rattan S. [Tyr°]- calcitonin gene-related peptide 28-37 (rat) as a putative antagonist of calcitonin gene-related peptide responses on opossum internal anal sphincter smooth muscle. J Pharmacol Exp Ther 1990; 253: 200-206.

5. Rovero P, Giuliani S, Maggi CA. CGRP antagonist activity of short C-terminal fragments of human αCGRP, CGRP (23-37) and CGRP(19-37). Peptides 1992; 13:1025-1027.

6. Marshall I, Al-Kazwini SJ, Roberts PM et al. Cardiovascular effects of human and rat calcitonin gene-related peptides compared in the rat and other species. Eur J Pharmacol 1986; 123:207-216.

7. Marshall I, Al-Kazwini SJ, Holman JJ et al. Human and rat alpha-calcitonin gene-related peptide but not calcitonin cause mesenteric vasodilatation in rats. Eur J Pharmacol 1986; 123:217-222.

8. Bhogal R, Sheldrick RLG, Coleman RA et al. The effect of IAPP and CGRP on guinea pig tracheal smooth muscle in vitro. Peptides 1994; 15:1243-1247.

9. Quirion R, Van Rossum D, Dumont Y et al. Characterization of $CGRP_1$ and $CGRP_2$ receptor subtypes. Ann NY Acad Sci 1992; 657:88-105.

10. Mimeault M, Fournier A, Dumont Y et al. Comparative affinities and antagonistic potencies of various human calcitonin gene-related peptide fragments on calcitonin gene-related peptide receptors in brain and periphery. J Pharmacol Exp Ther 1991; 258:1084-1090.

11. Gray DW, Marshall I, Bose C et al. Subtypes of the calcitonin gene-related peptide (CGRP) receptor in vascular tissues. Br J Pharmacol 1991; 102:189P.

12. Casini A, Galli G, Salzano R et al. Calcitonin gene-related peptide increases the production of glycosaminoglycans but not of collagen type I and III in cultures of rat fat-storing cells. Life Sci 1991; 49:PL163- PL168

13. Dennis T, Fournier A, Cadieux A et al. hCGRP8-37, a calcitonin gene-related peptide antagonist revealing calcitonin gene-related peptide receptor heterogeneity in brain and periphery. J Pharmacol Exp Ther 1990; 254:123-128

14. Nuki C, Kawasaki H, Takasaki K et al. Structure-activity study of chicken calcitonin gene-related peptide (CGRP) on vasorelaxation in rat mesenteric resistance vessels. Jpn J Pharmacol 1994; 65:99-105.

15. Chakder S, Rattan S. Antagonism of calcitonin gene-related peptide (CGRP) by human CGRP-(8-37): Role of CGRP in internal anal sphincter relaxation. J Pharmacol Exp Ther 1991;256:1019-1024.

16. Bell D, McDermott BJ. Calcitonin gene-related peptide stimulates a positive contractile response in rat ventricular cardiomyocytes. J Cardiovasc Pharmacol 1994; 23:1011-1021.

17. Longmore J, Hogg JE, Huston PH et al. Effects of two truncated forms of human calcitonin-gene related peptide: Implications for receptor classification. Eur J Pharmacol 1994; 265:53-59.

18. O'Shaughnessy CT, Waldron GJ, Connor HE. Lack of effect of sumatriptan and UK-14,304 on capsaicin-induced relaxation of guinea-pig isolated basilar artery. Br J Pharmacol 1993; 108:191-195.

19. Kobayashi D, Todoki K, Ozono Set al. Calcitonin gene-related peptide mediating neurogenic vasorelaxation in the isolated canine lingual artery. Jpn J Pharmacol 1995; 67:329-339.

20. Chatterjee TK, Moy JA, Fisher RA. Characterization and regulation of high affinity calcitonin gene-related peptide receptors in cultured neonatal rat cardiomyocytes. Endocrinology 1991; 128:2731-2738.

21. Aiyar N, Nambi P, Griffin E et al. Identification and characterisation of calcitonin gene-related peptide receptors in porcine renal medullary membranes. Endocrinology 1991; 129:965-969.

22. Poyner DR, Andrew DP, Brown D et al. Pharmacological characterization of a receptor for calcitonin gene-related peptide on rat, L6 myocytes. Br J Pharmacol 1992; 105:441-447.

23. Maggi CA, Chiba T, Giuliani S. Human α-calcitonin gene-related peptide (8-37) as an antagonist of exogenous and endogenous calcitonin gene-related peptide. Eur J Pharmacol 1991; 192:85-88.

24. Wisskirchen FM, Burt RP, Marshall I. Pharmacological characterization of CGRP receptors mediating relaxation of the rat pulmonary artery and inhibition of twitch responses of the rat vas deferens. Br J Pharmacol 1998; 123:1673-1683.

25. Gray DW, Marshall I. Human α-calcitonin gene-related peptide stimulates adenylate cyclase and relaxes rat thoracic aorta by releasing nitric oxide. Br J Pharmacol 1992; 107:691-696.

26. Gray DW, Marshall I. Nitric oxide synthesis inhibitors attenuate calcitonin gene-related peptide endothelium-dependent vasorelaxation in rat aorta. Eur J Pharmacol 1992; 212:37-42.

27. Yoshimoto R, Mitsui-Saito M, Ozaki H et al. Effect of adrenomedullin and calcitonin gene-related peptide on contractions of the rat aorta and porcine coronary artery. Br J Pharmacol 1998; 123:1645-1654.

28. Giuliani S, Wimalawansa SJ, Maggi CA. Involvement of multiple receptors in the biological effects of calcitonin gene-related peptide and amylin in rat and guinea-pig preparations. Br J Pharmacol 1992; 107:510-514.

29. Tomlinson AE, Poyner DR. Multiple receptors for calcitonin gene-related peptide and amylin on guinea-pig ileum and vas deferens. Br J Pharmacol 1996; 117:1362-1368.

30. Andersson SE, Almegard B. CGRP (8-37) and CGRP (32-37) contract the iris sphincter in the rabbit eye; antagonism by spantide and GR82334. Regul Pept 1993; 49:73-80.

31. Saria A, Martling CR, Yan Z et al. Release of multiple tachykinins from capsaicin-sensitive sensory nerves in the lung by bradykinin, histamine, dimethylphenyl piperazinium, and vagal nerve stimulation. Am Rev Respir Dis 1988; 137:1330-1335.

32. Maggi CA, Guiliani S, Santicioli P et al. Comparison of the effects of substance P and substance K on blood pressure, salivation and urinary bladder motility in urethane-anesthetized rats. Eur J Pharmacol 1985; 113:291-294.

33. Couture R, Laneuville O, Guimond C et al. Characterization of the peripheral action of neurokinins and neurokinin receptor selective agonists on the rat cardiovascular system. Naunyn Schmiedebergs Arch Pharmacol 1989; 340:547-557.

34. Tousignant C, Dion S, Drapeau G et al. Characterization of pre- and postjunctional receptors for neurokinins and kinins in the rat vas deferens. Neuropeptides 1987; 9:333-343.

35. Marshall I, Tilford NS, Bose C et al.. The effect of human α-calcitonin gene-related peptide (CGRP) in human cerebral arteries. Br J Pharmacol 1991; 102:188P.

36. Jansen I, Mortensen A, Edvinsson L. Characterization of calcitonin gene-related peptide (CGRP) receptors in human cerebral vessels. Ann NY Acad Sci 1992; 657:435-440.

37. Cox HM, Tough IR. Calcitonin gene-related peptide receptors in human gastrointestinal epithelia. Br J Pharmacol 1994; 113:1243-1248.

38. Kato J, Kitamura K, Kangawa K et al. Receptors for adrenomedullin in human vascular endothelial cells. Eur J Pharmacol 1995; 289:383-385.

39. Robberecht P, De Neef P, Woussen-Colle MC et al. Presence of calcitonin gene-related peptide receptors coupled to adenylate cyclase in human gliomas. Regul Peptides 1994; 52:53-60.

40. Muff R, Stangl D, Born W et al.. Comparison of a calcitonin gene-related peptide receptor in a human neuroblastoma cell line (SK-N-MC) and a calcitonin receptor in a human breast carcinoma cell line (T47D). Ann NY Acad Sci 1992; 657:106-116.

41. Entzeroth M, Doods HN, Wieland HA et al. Adrenomedullin mediates vasodilation via CGRP1 receptors. Life Sci 1995; 56:PL19-PL25.

42. Zimmermann U, Fischer JA, Muff R. Adrenomedullin and calcitonin gene-related peptide receptors interact with the same receptor in cultured human neuroblastoma SK-N-MC cells. Peptides 1995; 16:421-424.

CHAPTER 3

Structure Activity Relationship for CGRP

David Poyner, Stephen Howitt, Ursel Soomets and Ulo Langel

3.1. Introduction

The study of structure-activity relationships (SARs) has traditionally played an important role in understanding how a ligand interacts with its receptor. Information derived from such studies has aided drug development and given important insights into the architecture of ligand binding sites on receptors. In this chapter data on the SAR for calcitonin gene-related peptide (CGRP) will be reviewed.

3.2. Structure and Sequence of CGRP

3.2.1. Structure

Inspection of the structure of human αCGRP, a 37 amino acid peptide, reveals a number of obvious structural features (Fig. 3.1). There is a disulfide bridge between residues 2 and 7 at the N-terminus. Residues 8-18 have the potential to form an amphipathic α-helix. The C-terminus is amidated.[1]

The solution structure of CGRP and fragments has been examined by circular dichroism (cd) and nuclear magnetic resonance (NMR).[1-8] In summary, for full length CGRP there is evidence for the predicted α-helix; however the structure is not particularly stable in aqueous solution, and is also difficult to demonstrate in $CGRP_{8-37}$. The α-helix is probably stabilised by interactions with other parts of the CGRP molecule, particularly the N-terminus.[4,9] Of course, the favoured conformation of a peptide in aqueous solution is not necessarily a guide to its conformation at the receptor. As will be seen below, there is good evidence that the presence of a α-helix is indeed important for CGRP-receptor interactions.

Several studies have suggested that there are turns in other parts of the molecule, although the details vary from group to group. Based on NMR and modelling studies of a $CGRP_{19-37}$ fragment, a possible β-turn has been postulated between residues 19-22.[5] However, an NMR study of a CGRP suggested a γ-turn limited to $Ser_{19}-Gly_{21}$.[7] A variety of structures have been suggested for the C-terminus of CGRP. Boulanger et al have produced NMR evidence for a hydrogen bond between the amide carbonyl oxygen of Val_{28} and Thr_{31}, using dimethyl sulfoxide (DMSO) as solvent.[9] By contrast, Sagoo et al found NMR evidence for a βI turn which included residues 31 and 33, and a loop between Thr_{30} and Phe_{37}.[5] None of these features were seen by Breeze et al or O'Connell et al, working with a water/trifluoroethanol solvent mix.[3,6] Modelling studies suggest that the C-terminus does have a propensity for forming turns, and a βII'-turn between Val_{32} and Lys_{35} or a γ-turn between Val_{32} and Ser_{34} have been proposed.[10]

The CGRP Family: Calcitonin Gene-Related Peptide (CGRP), Amylin, and Adrenomedullin, edited by David Poyner. ©2000 EUREKAH.COM.

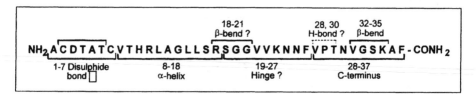

Fig. 3.1. Possible regions of secondary structure in CGRP. It should be noted that some of these features are very speculative!

Perhaps the most important conclusion to be drawn from this rather confusing set of data is that the C-terminus has the potential to form a number of different but distinctive conformations, but these are likely to be very sensitive to environmental considerations. The implications of this will be considered below.

3.2.2. Naturally Occuring Variants

CGRP occurs in two forms (α and β) in rat and man, and single forms of the peptide have been identified from at least six other species, including chick, salmon, trout and frog, *Rana ridibunda*. These show a minimum of 76% identity, and between α and β variants the identity is over 90%.[11,12] Although there are a few reports of differences in potency between these different forms,[13,14] generally they are rather similar (see ref. 15 for a review of early data on agonist potency ratios). The similarity between endogenous forms of CGRP means that they are of limited use when defining SARs.

3.2.3. Related Peptides

CGRP is frequently considered to be part of a family of bioactive peptides including calcitonin, amylin and adrenomedullin (see Legon, this volume for discussion of this issue).[11,16] It has some 47% identity with amylin. Although a different length from both adrenomedullin and calcitonin, it has 24% and 28% homology with human adrenomedullin and salmon calcitonin. There is limited but significant cross-reactivity between these peptides. In addition to homology at the level of primary structure, they all have a conserved, disulphide-bonded ring at their

N-termini, potential amphipathic α-helices and amidated C-termini. Comparison of these structures can give useful information about CGRP SARs. Thus the regions of greatest homology are concentrated in the areas corresponding to the amphipathic α-helix and the extreme C-terminus of CGRP.

3.3. Detailed Structure Activity Relationships

3.3.1. The Basic Framework; Studies on CGRP Fragments

There is good evidence that CGRP might fit at least partly into the "message-address" concept of peptide signalling molecules, where there are discrete domains for binding and receptor activation, as has been proposed for other peptides.[17] Removal of the first seven residues, including the disulfide-bonded loop leaves $CGRP_{8-37}$, an antagonist.[18] Accordingly, residues 8-37 would seem to be predominantly involved with binding and not receptor interactions. By contrast N-terminal fragments such as $CGRP_{1-12}$ act as agonists, albeit at high concentrations.[19] This would seem to confirm that the region of the peptide needed for receptor activation is found in the N-terminus. It would be wrong to press these results too far. The affinity of $CGRP_{8-37}$ varies according to the system under study (it is of course used to define the two classes of CGRP receptor), but is generally at least an order of magnitude less potent than CGRP itself (see ref. 11 for review). This suggests that the N-terminus does contribute something towards binding. Although in most systems $CGRP_{8-37}$ appears to behave as a simple antagonist, at high concentrations it can show some agonist-like behaviour (see ref. 11 for review). The agonist-like activity of N-terminal fragments has not been widely demonstrated and CGRP$_{1-7}$ itself is either inactive or even a weak agonist.[19] However, despite these reservations, the division of CGRP into an N-terminal domain predominantly involved in receptor activation, and a C-terminal domain mainly involved in receptor binding is a useful concept that will be followed in this article. Furthermore, the C-terminus can itself be subdivided into three regions based on structural data (see pages 25 and 26): 8-18, an amphipathic α-helix; 19-27, a mixture of bends and random coil and 28-37, the more structured extreme C-terminus.

3.3.2. The N-terminus: Residues 1-7

The disulfide bond between cysteines at positions two and seven gives the N-terminus a relatively rigid structure.[3] The presence of the intact disulfide bond is very important; reduction and derivatisation destroys agonist activity at the classical CGRP$_1$ receptor and considerably reduces potency at the CGRP$_2$ receptor.[19] Although there is some disagreement about whether [acetamidomethylcysteine$_{2,7}$]-CGRP ([CysACM]CGRP) is always a weak but selective CGRP$_2$ agonist, there is no question that it is much less potent than CGRP itself. In radioligand binding studies [CysACM]CGRP binds with high affinity, but these do not always give a true reflection of a compound's affinity at CGRP receptors as measured in functional assays.[11,19] Recently it has been suggested that [CysEt] CGRP can activate CGRP$_2$ receptors with a potency approaching that of CGRP itself, but it remains largely inactive on CGRP$_1$ receptors (see Quirion et al, Ch. 1, this volume).[20] The conclusion remains that an intact disulfide bond is very important for receptor activation at CGRP$_1$ receptors, even if it is less important at CGRP$_2$ receptors. The geometry imposed by the disulfide is essential for activity; if it is replaced by an amide linkage, all agonist activity is abolished at both CGRP receptor subtypes.[19]

Although most studies do point to the crucial importance of the disulphide ring for receptor activation, there are a few suggestions that other factors may also be important. One study has reported that cyclo-CGRP$_{1-7}$ is actually a weak antagonist (apparent pK$_B$ of 6.6) against the positive inotropic effects of CGRP in the guinea pig right atrium, a classical CGRP$_1$-type tissue. Although this effect was not observed against the inotropic effect measured on the left atrium, another CGRP$_1$-type tissue,[19] it does raise the possibility that the disulphide ring by itself is unable to activate some types of CGRP receptor. Hakala et al[21] have produced evidence that analogues that either have shortened disulfide rings or which lack them entirely can activate CGRP receptors. Although these are interesting observations, as yet no confirmation has been reported, and no antagonists were used in the original study to confirm that the peptides were indeed acting on CGRP receptors.

N-terminal to the disulphide ring is a single amino acid, which in naturally occurring CGRP forms is either alanine or serine; small amino acids. Extension of the N-terminus by tyrosine produces rather variable results; there is frequently (but not invariably) a decrease in potency, usually less than an order of magnitude (see ref. 11 for review). Extension of the N-terminus by introduction of a biotin group causes a drastic loss of potency, at least in rat L6 skeletal myocytes, which express a CGRP$_1$ receptor.[22] It seems reasonable to conclude that whilst limited N-terminal extension is tolerated, this cannot be taken too far and that the N-terminus is likely to reside in some kind of conformationally restricted space on the receptor.

3.3.3. The α-Helix: Residues 8-18

The amphipathic α-helix that is formed by residues 8-18 is a key feature of CGRP, as noted in the pioneering study of Lynch and Kaiser.[1] Similar regions of potential helix can be identified in amylin, calcitonin and adrenomedullin and this helps explain why these peptides or their derivatives are active at CGRP receptors. If residues 8-18 are removed, then the resulting fragment is generally 50-100 fold less potent than $CGRP_{8-37}$,[22-24] although in some systems $CGRP_{19-37}$ can show anomalous behaviour.[25] However residues 8-18 by themselves do not bind to the receptor with appreciable affinity;[26] they must be just one of several regions responsible for high affinity binding. There is some dispute as to where exactly the helix ends. The introduction of a helix destabilising proline residue at position 16 causes at least a ten fold decrease in affinity of $CGRP_{8-37}$ suggesting that it probably does extend towards residue 18.[27,28]

Detailed work has been done to analyse the effects of truncations and substitutions in the region of residues 8-18.[29,30] For $CGRP_{8-37}$ acting at the $CGRP_1$ receptors found in the guinea-pig heart, Val_8 is relatively unimportant. Deletion of Thr_9 causes about a five-fold loss of potency relative to $CGRP_{8-37}$. His_{10} produces a fragment with about a tenth of the potency of $CGRP_{8-37}$. Continuing deletions of Arg_{11} and Leu_{12} cause further but smaller reductions in affinity. This approximately correlates with the mean helical content of the fragments, and probably largely represents an effect on helix stability. L-Alanine substitution of these individual residues reveals small (two to three fold) reductions in affinity except at position 11 (arginine), where there is about a five fold reduction in potency. There is also a 40% reduction in helical content with this analogue, an interesting result since the substitution would normally be expected to increase helical content.

In mammalian CGRP forms, position 14 is a glycine; in other species it is an aspartic acid. In some studies, chick CGRP has been reported to be more potent than its mammalian counterparts.[31] Since glycine is usually considered to be a helix breaker, it has been suggested that the aspartic acid manifests its effects by increasing helical content. Replacement of Gly_{14} by aspartate,

asparagine, alanine or aminoisobutyrate in CGRP produced less than a two-fold decrease in potency on the $CGRP_1$ receptor of porcine coronary artery. There was no predictable change in helical content as determined by cd, which varied from a 30% increase to a 15% decrease (the aspartate substitution failed to increase helical content), nor did potencies correlate with this parameter. Only when proline was introduced was there a marked decrease in affinity, and this was associated with a 25-30% decrease in helical content.[32] Interestingly, substitution of Leu_{16} by proline also caused at least a ten fold reduction in potency.[27,28]

Replacement of Ser_{17} by alanine increases the potency of $CGRP_{8-37}$ on the guinea-pig atrium two fold. The authors correlated this not with any changes in the structure of residues 8-18 but with changes in the structure of the C-terminus between residues 28 and 37.[8]

The information reviewed above is complex, but has several implications. The data with [Ala_{17}] $CGRP_{8-37}$ demonstrates that the helix may influence affinity by altering other sections of the peptide. The stability of the helix is determined not just by the residues within 8-18, but also by interactions with other portions of the molecule. Helical content is certainly a factor in determining affinity, but there is not necessarily an exact correlation: modest decreases of 10 or 20% probably make little difference by themselves. Other factors are also significant. The physical size of the N-terminus of fragments such as $CGRP_{8-37}$ or $CGRP_{12-37}$ may be important; if it is too short, affinity will suffer. Amphipathicity is an important consideration. A CGRP derivative (mastoparan-$CGRP_{28-37}$) where the entire region is replaced by the amphipathic peptide mastoparan, binds with identical affinity as its parent compound ($CGRP_{8-18, 28-37}$),[24] and a similar finding was made by Lynch and Kaiser.[1] If the arginines at positions 11 and 18 are both replaced simultaneously with alanines (thereby greatly reducing amphipathicity), the analogue binds with 100 fold lower affinity than $CGRP_{8-37}$.[22,24] The available data from alanine scanning mutations suggests that the contribution of any single residue to affinity is limited; rather it is the global properties of size, helical content and charge distribution which work together to determine the affinity of any analogue.

3.3.4. Bends and Coils: Residues 19-27

The middle section of CGRP is the least explored portion of the molecule. Deletion of the entire segment causes a seven fold decrease in affinity.[24] Whilst this implies the region has a role in promoting high affinity binding, the reduction is less than found for deletion of other parts of the molecule. As noted above, several structural studies have suggested that the region begins with a turn, but the geometry of this is unclear.[3,5,7] Substitution of bend favoring residues such as proline or a beta turn dipeptide at position 19 (serine) causes no loss of potency of $CGRP_{8-37}$ at $CGRP_1$ receptors, indicating that a turn can be tolerated at this point.[28] $CGRP_{8-21,28-37}$, an analogue where the bend is retained, binds with a similar affinity to $CGRP_{8-37}$. However $CGRP_{8-18, 21-37}$ where the bend is selectively removed also binds with the same affinity as $CGRP_{8-37}$. One interpretation of this evidence is that residues 19-27 predominantly function as a hinge in the binding of CGRP. The turn that seems to occur at the start of this section may normally fulfil the role of the hinge; however, if it is deleted then the rest of the residues have sufficient flexibility to take over this function. Boulanger et al have also produced evidence that the main role of this part of CGRP is to influence the conformation of other parts of the molecule. Alanine substitutions at residues 20 and 21 alter the affinity of $CGRP_{8-37}$; however their NMR structures suggest this effect is via changes in the structure of the C-terminus.[8]

3.3.5. The C-terminus: Residues 8-37

There is now considerable evidence that the C-terminus of CGRP is crucial for high affinity binding. The data of Boulanger et al.[8] has been reviewed above. Although most (but not all)[14] studies have shown that isolated fragments such as $CGRP_{28-37}$ have very low affinity,[22,25,26] $[Asp_{31}, Pro_{34}, Phe_{35}]$ $CGRP_{27-37}$ binds to the $CGRP_1$ receptor on human SK-N-MC cells with a Kd of about 30nM.[33] As noted above, there is no agreement on the structure of this part of CGRP, and the results of detailed structure-activity studies are not always easy to interpret.

Matters are fairly clear at the C-terminus. There is agreement that removal of the terminal phenylalanine residue results in complete loss of high affinity binding, and of course further truncations also produce largely inactive species.[6,26,34,35] O'Connell et al could find no obvious effects of the deletion on peptide structure and so concluded the residue might interact directly with the receptor.[6] What little affinity fragments such as $[Tyr_0]$ $CGRP_{27-37}$ have is rapidly lost by N-terminal deletions.[33]

A major study by Rist et al of $CGRP_{27-37}$ concluded that thr_{30}, val_{32}, gly_{33} and phe_{37} were crucial to receptor binding; replacement by alanine, phenylalanine, proline or the corresponding D-amino acids all caused major decreases in affinity. Remarkably thr_{30} could not even be replaced by serine. Ala_{36} also showed considerable sensitivity; replacement by phenylalanine or glycine caused a considerable fall in potency. Substitution of val_{28} and ser_{34} was possible with little change in affinity; removal of the charge at position 35 (provided by either lysine or glutamic acid, depending on species) by substitution of leucine or phenylalanine made little difference to affinity. Asn_{31} could be replaced by aspartic acid and leucine but not by glutamine, alanine or phenylalanine, suggesting size rather than charge was an important feature of this position. A proline scan produced the very interesting analogue $[Pro_{34}]CGRP_{27-37}$ with a 10 fold increase in affinity (see below). Multiple replacement studies demonstrated that if asn_{31} was substituted by aspartic acid (itself a largely neutral substitution in terms of affinity), then affinity was improved by the presence of a positively charged or neutral amino acid at position 35. Combined with the effects of proline, this lead to the design of the high affinity ligand [Asp_{31},Pro_{34},Phe_{35}]$CGRP_{27-37}$.

The effects of pro_{34} point to the possibility of a bend playing an important role in promoting high affinity binding of this analogue. Hakala et al synthesised peptides where disulfide bridges were introduced into the C-terminus to mimic different turns; the analogue with a $\beta II'$ turn between val_{32} and glu_{35} had a 40 fold higher affinity than an analogue with a βI turn between asn_{31} and ser_{34}.[21] This is consistent with the observation that introducing a beta turn dipeptide between residues 33 and 34 maintains high affinity binding. Against this, Rist et al have argued that a proline at position 34 will either favour a βI turn from gly_{33} to ala_{36}, or a βIV turn from val_{32} to glu_{35}. It is not clear how any of this relates to the observation of Boulanger et al that a hydrogen bond between the carbonyl of val_{28} and the amide of thr_{30} is needed for high affinity binding of $CGRP_{8-37}$. Since [Asp_{31},Pro_{34},Phe_{35}] $CGRP_{29-37}$ (where the above mentioned hydrogen bond obviously cannot be formed) binds with a Kd of about 200 nM, the interaction must at best be of secondary importance for this series of fragments. It may be significant that the studies of Rist and coworkers were carried out on $CGRP_{27-37}$ and its derivatives. In the absence of the rest of the molecule, this might adopt conformations that are not seen or important for longer fragments such as $CGRP_{8-37}$. This may be relevant to the debate about the secondary structure of this part of the molecule.

In summary, studies have demonstrated that the C-terminus is essential for high affinity binding. Residues 30 to 37 seem particularly important, and there seems to be good evidence that these are folded into some type of bend. It remains to be determined whether individual residues such as val_{32} make direct contacts with the receptor or are important in maintaining secondary structure; it would be surprising if some at least were not in direct contact with the receptor.

3.4. Conclusion

Although much detail remains to be resolved, there does seem to be a reasonable working model to explain the interactions of CGRP at its receptors, at least as far as $CGRP_1$ receptors are concerned. Activation of the receptor is done by determinants located in the N-terminus, predominantly associated with the disulfide bonded loop formed by residues 1-7. The remainder of the molecule is concerned with binding. One possibility is that the sequence of $CGRP_{8-37}$ can be divided into three domains. The C-terminus, residues 28-37 contain most of the amino acids which make important individual contacts with the receptor. The α-helix, residues 8-18, normally works with the C-terminus in promoting high affinity binding. However this structure relies predominantly on its overall physicochemical properties, particularly its helicity and amphipathicity, and may make fewer specific contacts with the receptor. Residues 19-27 predominantly act as a spacer, allowing the other two domains to adopt the correct conformation for high affinity binding. This model is speculative, but testable.

Numerous questions remain. Little work has been done on $CGRP_2$ receptors, largely because of their low affinity for $CGRP_{8-37}$ and derivatives. In broad terms the SAR between the two receptors appears similar, but much more work is needed on this issue. There is little information on the extent to which individual residues or structures make direct contact with the receptor, or whether their role is to maintain a particular conformation of the peptide. More information on this should come now that a CGRP receptor has been cloned (Aiyar et al, Legon and Foord et al, this volume). The whole issue of the preferred conformation of CGRP is unclear; the peptide seems to be able to adopt numerous conformations in solution. It is interesting to speculate whether this means that it might also show several different modes of binding to the receptor. This is particularly relevant to studies with small CGRP fragments, which could adopt conformations not shown by the parent molecule. Finally, with the reports of non-peptide antagonists, a whole new field is about to open up.[36]

Acknowledgments

This work was assisted by support from the Wellcome Trust.

References

1. Lynch B, Kaiser ET. Biological properties of two models of calcitonin gene-related peptide with idealized amphiphilic alpha helices of different lengths. Biochemistry 1988; 27:7600-7607.

2. Manning MC. Conformation of the α-form of human calcitonin gene-related peptide (CGRP) in aqueous solution as determined by circular dichroism spectroscopy. Biochem Biophys Res Commun 1989; 160:388-392.

3. Breeze AL, Harvey TS, Bazzo R et al. Solution structure of human calcitonin gene-related peptide by 1H NMR and distance geometry with restrained molecular dynamics. Biochemistry 1991; 30:575-582.

4. Hubbard JAM, Martin SR, Chaplin LC et al. Solution structures of calcitonin gene-related peptide analogues of calcitonin gene-related peptide and amylin. Biochem J 1991; 275:785-788.

5. Sagoo JK, Bose C, Beeley NRA et al. Structural studies on the [bu^t-Cys^{18}](19-37)-fragment of human β-calcitonin gene-related peptide. Biochem J 1991; 280:147-50.

6. O'Connell JP, Kelly SM, Raleigh DP et al. On the role of the C-terminus of α-calcitonin gene-related peptide (α-CGRP). The role of desphenylaninamide[37]-αCGRP and its interaction with the CGRP receptor. Biochem J 1993; 30:575-582.

7. Boulanger Y, Khiat A, Chen Y et al. Structure of human calcitonin gene-related peptide (hCGRP) and of its antagonist hCGRP8-37 as determined by NMR and molecular modelling. Peptide Res 1995; 8:206-213.

8. Boulanger Y, Khiat A, Larocque A et al. Structural comparison of alanine substituted analogues of the calcitonin gene-related peptide 8-37. Int J Peptide Protein Res 1996; 47:477-483.

9. Mimeault M, Quirion R, Dumont Y et al. Structure-activity study of hCGRP8-37, a calcitonin gene-related peptide receptor antagonist. J Med Chem 1992; 35:2163-2168.

10. Hakala JML, Vihavainen S. Modelling the structure of the calcitonin gene-related peptide receptor. Protein Engineering 1997; 7:1069-1075.

11. Poyner DR. Molecular pharmacology of receptors for calcitonin gene-related peptide, amylin and adrenomedullin. Biochem Soc Trans 1997; 25:1032-1036.

12. Conlon JM, Tonon MC, Vaudry H. Isolation and structural characterization of calcitonin gene-related peptide from the brain and intestine of the frog, Rana ridibunda. Peptides 1993; 14:581-586.

13. Jansen I. Characterization of calcitonin gene-related peptide receptors in guinea-pig basilar artery. Neuropeptides 1992; 21:73-79.

14. Chakder S, Rattan S. [Tyr^0]-calcitonin gene-related peptide 28-37 (rat) as a putative antagonist of calcitonin gene-related peptide response on opossum internal anal sphincter smooth muscle. J Pharmacol Exp Ther 1990; 253:200-206.

15. Poyner DR. Calcitonin gene-related peptide; multiple actions, multiple receptors. Pharmacol Ther 1992; 56:23-51.

16. Wimalawansa SJ. Calcitonin gene-related peptide and its receptors; molecular genetics, physiology, pathophysiology and therapeutic potentials. Endocrine Rev 1996; 17:533-85.

17. Schwyzer R. Membrane assisted molecular mechanisms of neurokinin receptor subtype selection. EMBO J 1987; 6:2255-2259.

18. Chiba T, Yamaguchi A, Yamatani T et al. Calcitonin gene-related peptide antagonist CGRP(8-37). Am J Physiol 1989; 256:E331-E335.

19. Dennis T, Fournier A, St Pierre S et al. Structure-activity profile of calcitonin gene-related peptide in peripheral and brain tissues. Evidence for receptor multiplicity. J Pharmacol Exp Ther 1989; 251:718-725.

20. Dumont Y, Fournier A, St Pierre S et al. A potent and selective CGRP(2) agonist, [Cys(Et)(2,7)]hCGRP alpha: comparison in prototypical CGRP(1) and CGRP(2) in vitro bioassays. Can J Physiol Pharmacol 1997; 75:671-676.

21. Hakala JML, Valo T, Vihavainen S et al. Constrained analogues of the calcitonin gene-related peptide. Biochem Biophys Res Commun 1994; 202:497-503.

22. Howitt SG, Poyner DR. The selectivity and structural determinants of peptide antagonists at the CGRP receptor of rat, L6 myocytes. Br J Pharmacol 1997; 121:1000-1004.

23. Rovero P, Giuliani S, Maggi CA. CGRP antagonist behaviour of short C-terminal fragments of human αCGRP, CGRP (19-37) and CGRP(23-37). Peptides 1992; 13:1025-1027.

24. Poyner DR, Soomets U, Howitt SG et al. Structural determinants for binding to CGRP receptors expressed by human SK-N-MC and Col 29 cells: Studies with chimeric and other peptides. Br J Pharmacol 1998; 124:1659-1666.

25. Tomlinson AE, Poyner DR. Multiple receptors for calcitonin gene-related peptide on guinea-pig ileum and vas deferens. Br J Pharmacol 1996; 117:1362-1368.

26. Poyner DR, Andrew D, Brown D et al. Pharmacological characterization of a receptor for calcitonin gene-related peptide on rat, L6 skeletal myocytes. Br J Pharmacol 1992; 105:441-447.

27. Wisskirchen FM, Doyle PM, Gough SL et al. Structure activity relationship of analogues of the antagonist hα CGRP$_{8-37}$ in rat prostatic vas deferens. Br J Pharmacol 1994; 112:239.

28. Wisskirchen FM, Doyle PM, Gough SL et al. Conformational restraints to find biologically relevant structures of CGRP$_{8-37}$ in rat prostatic vas deferens, pulmonary artery and internal anal sphincter. Br J Pharmacol 1996; 120:209.

29. Mimeault M, Fournier A, Dumont Y et al. Comparative affinities and antagonist potencies of various human calcitonin gene-related peptide fragments on calcitonin gene-related peptide receptors in brain and periphery. J Pharmacol Exp Ther 1991; 258:1084-90.

30. Mimeault M, Quirion R, Dumont Y et al. Structure activity study of hCGRP$_{8-37}$, a calcitonin gene related peptide antagonist. J Med Chem 1993; 35:2163-2168.

31. Nuki C, Kawasaki H, Takasaki K et al. Structure activity study of chicken calcitonin gene-related peptide (CGRP) on rat mesenteric vessels. Jpn J Pharmacol 1994; 65:99-105.

32. Li Z, Matsuura JE, Waugh DJJ et al. Structure-activity studies on position 14 of human α-calcitonin gene-related peptide. J Med Chem 1997; 40:3071-3076.

33. Rist B, Entzeroth M, Beck-Sickinger AG. From micromolar to nanomolar: A systematic approach to identify the binding site of CGRP at the human calcitonin gene-related peptide 1 receptor. J Med Chem 1998; 41:117-123.

34. Smith DD, Li J, Wang Q et al. Synthesis and biological activity of C-terminally truncated fragments of human A-calcitonin gene-related peptide. J Med Chem 1993; 36:2536-2541.

35. Zaidi M, Brain SD, Tippins JR et al. Structure-activity relationship of human calcitonin gene-related peptide. Biochem J 1990; 269:775-780.

36. Daines RA, Sham KKC, Taggart JJ et al. Quinine analogues as non-peptide calcitonin gene-related (CGRP) receptor anatagonists. Bioorg Med Chem Lett 1997; 2673-2676.

CHAPTER 4

The Binding of CGRP and Adrenomedullin to a Cloned Receptor

Steve Legon

4.1. Introduction

Calcitonin gene-related peptide (CGRP) and adrenomedullin (ADM) are members of a small family of regulatory peptides which is often referred to as the calcitonin family. This group consists of calcitonin (CT), α and β CGRPs, amylin and ADM though there are indications in the literature suggesting that there might be further members of this family yet to be discovered.[1-4] This peptide family is remarkable in that it has been linked with three of the most significant conditions associated with ageing-osteoporosis (CT), hypertension (CGRP, ADM) and type 2 diabetes (amylin). Although the links may be somewhat tenuous they cannot be ignored and there has been considerable interest in isolating the receptors for these peptides. Despite this, until 1991 there had been no report of the cloning of a receptor cDNA for any of these peptides. At that time, a paper was published describing the characterisation of the pig CT receptor cDNA.[5] This paper accompanied one describing a similar receptor, the opossum parathyroid hormone receptor[6] and slightly followed publication of the characterisation of the rat secretin receptor.[7] Although their ligands are dissimilar, these cDNAs are clearly related and define a new class of seven transmembrane domain receptors which show no similarity with the previously characterised 'rhodopsin-like' receptors beyond possession of seven transmembrane domains.

Secretin and its receptor provide a good example of how regulatory peptides and their receptors evolve in parallel. Secretin is a member of an extensive family of peptides including glucagon, parathyroid hormone (PTH), corticotrophin releasing factor (CRF), Glucagon like peptide (GLP)-1, GLP-2, vasoactive imtestinal polypeptide (VIP), gastrointestinal polypeptide (GIP), pituitary adenylate cyclase activating peptide (PACAP) and gonadotrophin hormone releasing-hormone (GHRH). These appear to have evolved from a single progenitor peptide by a process of exon and gene duplication and divergence and in parallel with this, a family of related receptors has evolved (see Fig. 4.1). The logic behind our approach to isolating receptor cDNAs from the CT family has therefore been that if CT is similar to the other members of the CT family then the CT receptor will be similar to the receptors for CGRP, ADM and amylin. This sequence similarity can be used as the basis for isolating receptor cDNAs.

The assumption that CT is related to the other members of 'its' family by common descent from a progenitor peptide is

The CGRP Family: Calcitonin Gene-Related Peptide (CGRP), Amylin, and Adrenomedullin, edited by David Poyner. ©2000 EUREKAH.COM.

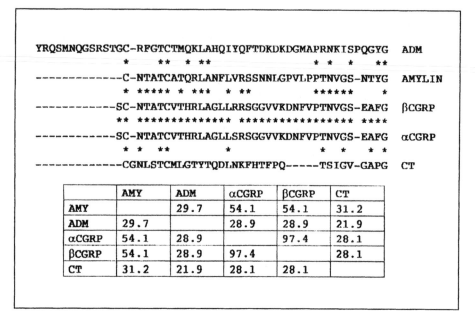

Fig. 4.1. The CT family of peptides. Amino acid sequences of rat ADM, amylin, aCGRP, bCGRP and CT are aligned for maximal similarity. The terminal glycine which gives rise to the amidated C terminus is included. The table shows sequence identity between each pair of peptides expressed as % of the shorter peptide identical with the longer peptide, with no penalty for gaps.

sometimes questioned. The rat peptides are shown in Figure 4.1. CGRP and amylin have over half their residues in common whereas they have only about 30% sequence identity with ADM and CT. However, the fact that CT is an alternate product of the α CGRP gene, makes it easy to see how the two peptides could have evolved from a common ancestor. The alternative view is that CT came to resemble CGRP by a process of convergent evolution. If this were the case then one would suppose that the force driving this convergence was a common receptor and the argument indicating that the present day CT receptor should resemble the receptors for CGRP, ADM and amylin would still hold. We therefore decided to make use of the pig CT receptor sequence to clone related cDNA sequences from the rat. Our minimum aim was to isolate the rat homologue of the porcine sequence but our hope was that this procedure might lead to the isolation of receptor cDNAs for the other members of the CT family of peptides.

4.2. Isolation of a Rat cDNA Related to the Calcitonin Receptor

The cDNA sequences of the opossum PTH receptor and pig CT receptor were published in adjacent papers in Science in 1991.[5,6] These two receptors have 32% sequence identity and it is assumed that conserved motifs must represent important structural/functional features. Two such regions were selected (Fig. 4.2) and used as the basis for the synthesis of a pair of mixed sequence PCR primers. Rat hypothalamic and kidney cDNA were amplified and the PCR products were gel fractionated. A broad band of approximately 450 bp was seen from both tissues. The hypothalamus derived DNA was cloned into an M13 vector for sequence analysis. Two major species were found and both encoded polypeptides similar to the corresponding part of the pig CT receptor. One had 74% identity with the pig sequence and proved to be part of the rat CT receptor cDNA. The other predicted

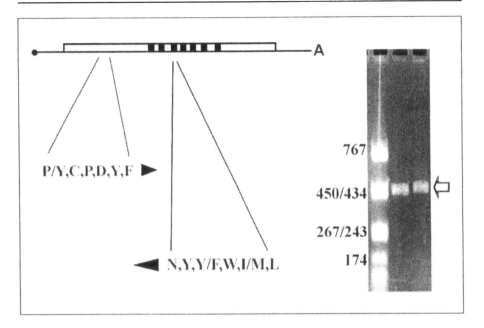

Fig. 4.2. Amplification of hypothalamic cDNA. 100 ng of rat hypothalamic cDNA was amplified using primers based on the indicated amino acid sequences which are common to the pig CT receptor and opossum PTH receptor. Amplified DNA was analysed on a 1% agarose gel either before (left) or after (right) digestion with EcoR1. Marker lane is a Hae 111 digest of the plasmid Bluescript.

polypeptide had only 53% identity with pig CT receptor and was referred to as the calcitonin receptor like receptor (CRLR).[8] In further experiments, clones were isolated from a rat lung cDNA library and the complete coding sequence of CRLR was determined and deposited in the Genbank database (X70658). This sequence has also been described by Chang et al[9] and the human equivalent has also been isolated.[10,11]

The predicted amino acid sequence of rat CRLR is 463 amino acids in length and is a typical member of the 'secretin family' of seven transmembrane domain receptors (Fig. 4.3). It has an amino terminal extracellular domain of 137 amino acids and an intracellular tail of 76 amino acids. The first 27 amino acids are not well conserved between the rat and human sequences and may represent a signal peptide. There are eight cysteine residues in the extracellular domain (two in the signal peptide) and six

potential sites for N linked glycosylation, only four of which are conserved in man. In addition there are 11 sites which have the potential to become phosphorylated in the intracellular domain. Analysis of sequences in the Genbank database revealed that CRLR is most closely related to the CT receptor with the PTH and CRF receptors being more distantly related (Fig. 4.4). There are no orphan receptors similar to CRLR and no unidentified entries in the est databases with substantial similarity to CRLR.

4.3. CRLR

4.3.1. Localization of the Calcitonin Receptor Related Receptor

The first step in the analysis of CRLR was to determine where it was expressed. A northern blot was prepared with 18 rat tissues and probed with one of the rat CT receptor PCR clones and with a

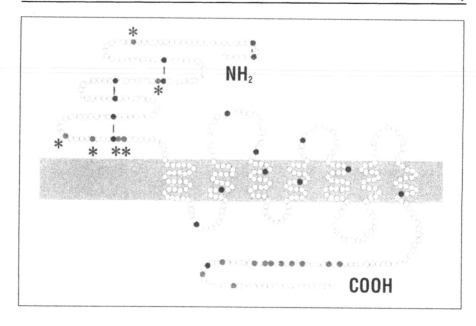

Fig. 4.3. Features of the CRLR protein. The amino acid sequence of rat CRLR is shown with the extracellular domain at the top. Cysteine residues are in black, potential phosphorylation sites are shown in half tone and potential glycosylation sites are shown in half tone with stars.

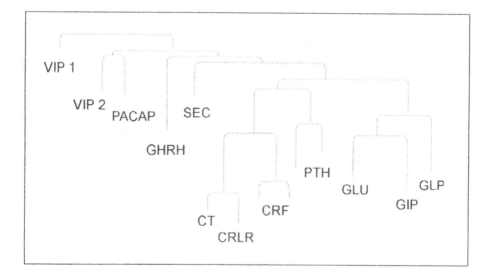

Fig. 4.4. Evolution of CRLR and related receptors. Figure was assembled using the 'Growtree' programme from the University of Wisconsin Genetics Computing Group (gcg) suite of programmes.

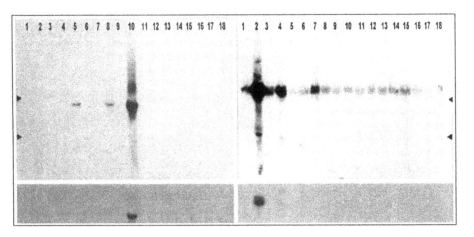

Fig. 4.5. Distribution of CT and CRLR mRNAs. 25 μg of total rat RNA was fractionated on duplicate formaldehyde agarose gels, blotted onto Hybond nylon membranes and probed with α^{32}P dCTP labelled rat CT receptor or CRLR sequences. Track order : thyroid, lung, heart, spleen, kidney, liver, spinal cord, medulla + pons, cerebellum, hypothalamus, cerebral cortex, olfactory bulb, colon, cecum, ileum, jejunum, duodenum & stomach. Left panel, CT receptor probe; right panel, CRLR probe. Upper panels, 18 hour exposure; lower panels, 2 hour exposure. Data reproduced from Njuki et al[8] by permission of The Biochemical Society & Portland Press.

corresponding CRLR probe (Fig. 4.5). CRLR mRNA is 4-5,000 bases in length and gives a significantly broader band than does the CT receptor probe. CRLR is very abundant in the lung but can be detected on longer exposure in all the tissues tested. The distribution is not consistent with the distribution of CT or amylin binding sites in the body and is not entirely concordant with the expected distribution of CGRP and/or ADM receptors. For instance, both liver and cerebellum are major binding sites for CGRP and yet have only modest amounts of CRLR mRNA.

In the lung, ^{32}P labelled RNA probes were prepared and used for in situ hybridisation. Signals were seen over the blood vessels and not the airways or alveoli (Fig. 4.6 A&B). A polyclonal antiserum was raised against the C-terminal hexapeptide and was used for a more detailed localisation. Immunoreactivity was detected in the endothelial cells lining the small blood vessels and in the inter-alveolar capillaries (Fig. 4.6C). Such a distribution would be

consistent with a role for CRLR as an ADM receptor.

4.3.2. Biological Activity of the Calcitonin Receptor Related Receptor

The full length CRLR sequence was cloned into pcDNA1 and transfected into COS-7 cells for ligand binding studies. None of the peptides of the CT family were shown to bind specifically to the CRLR transfected cells (Fig. 4.7). ADM gave similar levels of 'specific' binding to both CRLR- and vector transfected cells making it impossible to assay at that time. We later discovered that the binding was to the plastic and could be reduced to acceptable levels by pre-coating the plates with poly-D lysine (Fig. 4.7, ADM+). Similar results with ADM binding to COS-7 cells on non-coated plates are described in a recent paper commenting on the identification of another sequence as an ADM receptor.[12] It should be noted that in the work of Kapas et al[13] bound ADM was

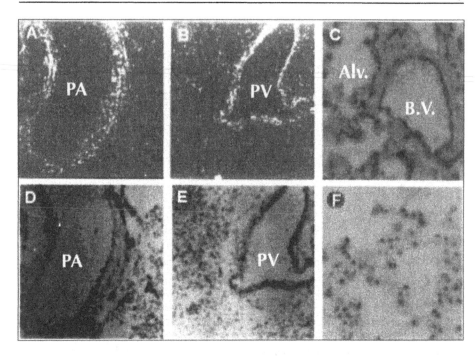

Fig. 4.6. Localisation of CRLR in the lung. In situ hybridisation, panels A,B, D & E. Sections of rat lung were probed with ^{32}P labelled antisense rat CRLR RNA. A & B are darkfield photomicrographs; D & E are haemotoxylin stained brightfield photomicrographs. Pulmonary arterioles and venules are indicated (PA, PV). Immunostaining, panels C & F. A rabbit polyclonal antibody directed against the C-terminal heptapeptide of rat CRLR was used to localise CRLR immunoreactivity in paraffin sections of rat lung using a peroxidase labelled streptavidin detection method. Upper panel, immune serum; lower panel, pre-immune serum. Data first presented in modified forms in Njuki et al[8] (A,B,D&E) and Han et al[15] (C&F) and are reproduced by permission of The Biochemical Society/Portland Press and The Society for Endocrinology Ltd.

measured after detaching the cells and centrifuging them free of unbound ADM.

Our conclusions at this point had to be either that our experiments were technically flawed or that the ligand for CRLR was not a known member of the CT family. The latter conclusion seemed the more plausible when receptors for the most likely candidates (ADM and CGRP) were described.[13,14] However a subsequent publication from Aiyar and colleagues (see ref. 11 and Chapter 5, this volume) demonstrating CGRP receptor activity for human CRLR caused us to reconsider our conclusions. The crucial difference between our experiments and theirs appeared to lie in their use of a human

embryonic kidney cell line known as 293. These cells are described as having an epithelial-type morphology and have been immortalised by transformation with fragmented adenovirus (Ad5) DNA. Two possibilities occurred to us. Perhaps these cells were able to supply CRLR with an accessory factor required for its functional expression or, on the other hand, perhaps the particular CRLR transformed line isolated by Aiyar et al was expressing an endogenous receptor. We felt it was important to demonstrate receptor activity with an unselected population of transfected cells. To raise the sensitivity of detection we measured the elevation of intracellular cAMP seen in response to ligand.

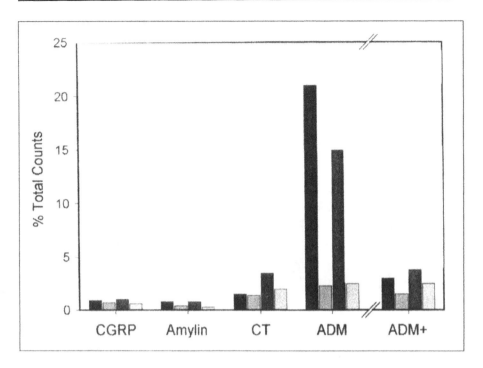

Fig. 4.7. Binding of ligands to CRLR in COS-7 cells. COS-7 cells were transiently transfected with rat CRLR in pcDNA1 using lipofectin and assayed for their ability to bind the indicated ligands after 3 days. In each group, the first column represents binding of ^{125}I labelled peptide in the absence of competitor and the second column, binding in the presence of 1 μM unlabelled peptide. The third and fourth columns represent binding of vector transfected cells, with and without cold competitor. The data at the right (ADM+) is from a separate experiment in which the plate was first pre-treated with poly D-lysine to reduce background ADM binding.

The critical experiment is shown in Figure 4.8 where 293 cells have been transiently transfected by rat CRLR or vector and challenged with the CT family of peptides. Both CGRP and ADM give an elevation of cAMP levels in CRLR transfected cells that is not seen with vector transfected cells and which is abolished by the antagonist, $CGRP_{8-37}$. The EC_{50} values we measured (4 and 20 nM respectively) are typical of a $CGRP_1$ receptor.

One feature, which remains puzzling, is that we have been unable to demonstrate convincing levels of CGRP binding with transiently transfected cells. This is in part because the cells have a low level of endogenous binding and partly because the

absolute levels of binding have been lower than we might have expected. As it was important to demonstrate binding and yet crucial to avoid selecting a permanently transformed cell line, we transfected 293 cells with CRLR and selected transformants with G418 until about 200 discrete colonies were visible. These were pooled and grown up for use in the binding studies shown in Figure 4.9. These CRLR transformed 293 cells bound both CGRP (IC_{50} 1.2 nM) and ADM (IC_{50} 11 nM).

Our conclusions from this work were that CRLR was behaving as a $CGRP_1$ receptor in the 293 cell line and that these cells were supplying some component necessary for the correct processing, membrane

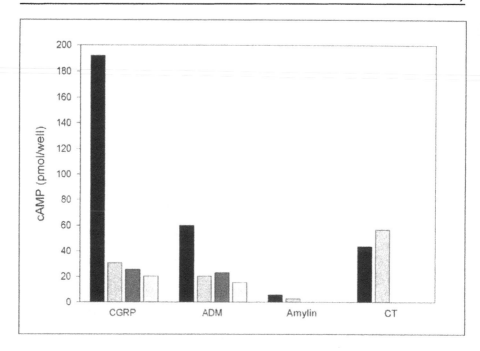

Fig. 4.8. Elevation of cAMP in transiently transfected 293 cells. Intracellular cAMP was measured in 293 cells transiently transfected with rat CRLR in pcDNA3.1 or with vector only and challenged with the indicated peptides. Within each group, the first column is CRLR transfected cells; the second column, vector transfected and, where appropriate, the third and fourth columns are assays from correspondingly transfected cells exposed to peptide plus 1 μM CGRP$_{8-37}$. Data first presented in modified form in Han et al[15] and reproduced by permission of The Society for Endocrinology Ltd.

insertion or functioning of the receptor.[15] Although this appears to be an inevitable conclusion from the data we obtained there are nonetheless some features of the distribution which are hard to reconcile with the binding data. There are tissues (liver and cerebellum) and cells (SK-N-MC) where high levels of CGRP binding are seen with only low levels of CRLR so there remains a strong possibility that there are other CGRP receptors in the body. Furthermore, one would not have anticipated that the major CGRP receptor in the lung would be located on the vascular endothelial cells unless its function were simply as a clearance receptor. The work reported at the recent CGRP '98 conference did much to throw light on these questions and we can begin to resolve some of these problems.

4.4. CGRP/Adrenomedullin Receptors and Their Associated Proteins

The major development in this area has been the discovery of the RAMP proteins (see ref. 16 and Chapter 6 in this volume). The hypothetical accessory factor that we supposed allowed our 293 cells to support CRLR receptor activity is presumably the RAMP 1 protein. Our assumption would be that the 293 cells used by ourselves and by Aiyar et al must be expressing significant levels of the RAMP 1 protein with little RAMP 2 or RAMP 3. This would support terminal glycosylation, export to the cell membrane and CGRP receptor activity. COS-7 cells (and the 293T cells used by Foord et al.) must be supposed to lack sufficient RAMP to allow any activity. Interestingly, the

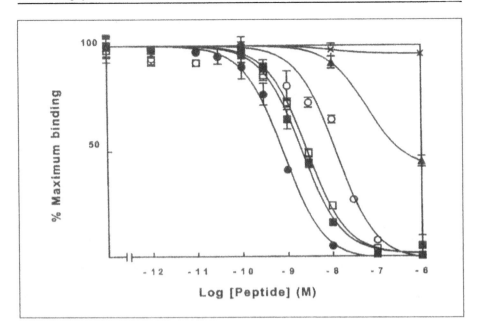

Fig. 4.9. Binding of ligands to stably transfected 293 cells. A mixed population of 293 cells stably transfected with rat CRLR in pcDNA3.1 was used for ligand binding studies with ^{125}I-labelled CGRP and the following unlabelled peptides: CGRP, filled squares; CGRP$_{8-37}$, open squares; ADM, open circles; amylin, filled triangles and CT, crosses. The filled circles represent a control with SK N MC cells expressing their endogenous CGRP receptor. Data first presented in Han et al[15] and reproduced by permission of The Society for Endocrinology Ltd.

lung, where most CRLR is expressed, lacks RAMP 1 but expresses high levels of both RAMP 2 and RAMP 3.[16] In the lung then, one would assume that CRLR is incompletely glycosylated and functions as an ADM receptor. This is consistent with its presence in the vascular endothelial cells and this form of the receptor may account for the majority of the CRLR expressed in the body.

Despite this, the evidence strongly indicates that CRLR is a dual receptor, functioning as a CGRP$_1$ receptor or as an ADM receptor according to whether RAMP 1 or RAMP 2/3 are coexpressed. It would however be rash to assume that CRLR/RAMP 1 is the only receptor for CGRP. Although we note that others (Foord and co-workers) can detect a trace of CRLR mRNA in the human SK-N-MC cell line, the amount seems to us too low to account for the substantial CGRP binding seen with these cells. There is a similar disparity between mRNA levels and binding in liver and cerebellum and it remains our belief that there may be further CGRP receptors to be discovered. We have been unable to detect further CRLR-like sequences in these tissues and therefore suspect that any further CGRP receptors will prove to be only distantly related to CRLR.

Acknowledgments

I would like to acknowledge the work of all my collaborators in this project, in particular, Fred Njuki who initiated the work and Dr Z-Q Han who was responsible for cloning and characterising the complete CRLR cDNA.

References

1. Lasmoles F, Jullienne A, Day F et al. Elucidation of the nucleotide sequence of chicken calcitonin mRNA: Direct evidence for the expression of a lower vertebrate calcitonin-like gene in man and rat. EMBO J 1985; 4(10):2603-2607.

2. Fischer JA, Tobler PH, Henke H et al. Salmon and human calcitonin-like peptides coexist in the human thyroid and brain. J Clin Endocrinol Metab 1983; 57(6):1314-1316.

3. Gropp C, Luster W, Havemann K. Salmon and human calcitonin like material in lung cancer. Br J Cancer 1985; 51(6):897-901.

4. Hilton JM, Mitchelhill KI, Pozvek G et al. Purification of calcitonin-like peptides from rat brain and pituitary. Endocrinology 1998; 139(3):982-992.

5. Lin HY, Harris TL, Flannery MS et al. Expression cloning of an adenylate cyclase-coupled calcitonin receptor. Science 1991; 254(5034):1022-1024.

6. Juppner H, Abou Samra AB, Freeman M et al. A G protein-linked receptor for parathyroid hormone and parathyroid hormone-related peptide. Science 1991; 254(5034):1024-1026.

7. Ishihara T, Nakamura S, Kaziro Y et al. Molecular cloning and expression of a cDNA encoding the secretin receptor. EMBO J 1991; 10(7):1635-1641.

8. Njuki F, Nicholl CG, Howard A et al. A new calcitonin-receptor-like sequence in rat pulmonary blood vessels. Clin Sci Colch 1993; 85(4):385-388.

9. Chang CP, Pearse RV2, O'Connell S et al. Identification of a seven transmembrane helix receptor for corticotropin-releasing factor and sauvagine in mammalian brain. Neuron 1993; 11(6):1187-1195.

10. Fluhmann B, Muff R, Hunziker W et al. A human orphan calcitonin receptor-like structure. Biochem Biophys Res Commun 1995; 206(1):341-347.

11. Aiyar N, Rand K, Elshourbagy NA et al. A cDNA encoding the Calcitonin Gene-related Peptide Type 1 Receptor. J Biol Chem. 1996; 271(19):11325-11329.

12. Kennedy SP, Sun D, Oleynek JJ et al. Expression of the rat adrenomedullin receptor or a putative human adrenomedullin receptor does not correlate with adrenomedullin binding or functional response. Biochem Biophys Res Commun. 1998; 244(3):832-837.

13. Kapas S, Catt KJ, Clark AJL. Cloning and expression of cDNA encoding a rat adrenomedullin receptor. J Biol Chem 1995; 270(43):25344-25347.

14. Kapas S, Clark AJL. Identification of an orphan receptor gene as a type 1 calcitonin gene-related peptide receptor. Biochem Biophys Res Commun. 1995; 217(3):832-838.

15. Han ZQ, Coppock HA, Smith DM et al. The interaction of CGRP and adrenomedullin with a receptor expressed in the rat pulmonary vascular endothelium. J Mol Endocrinol 1997; 18(3):267-272.

16. McLatchie LM, Fraser NJ, Main et al. RAMPs regulate the transport and ligand specificity of the calcitonin-receptor-like receptor. Nature 1998; 393(6683):333-339.

CHAPTER 5

CGRP Receptors, Structure and Function

Nambi Aiyar, Jyoti Disa and Ponnal Nambi

5.1. Introduction

Calcitonin gene related peptide (CGRP) is a sensory neuropeptide with potent cardiovascular effects that include positive inotropic and chronotropic actions, systemic vasodilation and hypotension in animals and in man.[1-3] CGRP is distributed throughout the central and peripheral nervous systems and possesses diverse biological actions including regulatory effects on central sympathetic outflow, insulin secretion, neuronal differentiation and endothelial cell proliferation. The strategic location of CGRP within the sensory networks of virtually all organs suggests a potential role for this peptide in diverse physiological and pathophysiological processes. These diverse actions of CGRP are mediated by cell surface receptors that are specific for CGRP and their coupling to the activation of second messenger systems. Evidence suggests that CGRP receptors are coupled to the activation of adenylyl cyclase through GTP binding proteins (G-proteins) and belong to the ever increasing family of seven transmembrane G-protein-coupled receptors. CGRP receptors have been identified and characterized in several tissues, including brain, heart, endothelial and vascular smooth muscle cells. These receptors have been classified into $CGRP_1$ and $CGRP_2$ subtypes according to the pharmacological definition proposed by Dennis and colleagues.[4] The fragment $CGRP_{8-37}$ is a potent antagonist at $CGRP_1$ receptors, whereas it is a weak antagonist at $CGRP_2$ receptors. In addition, a linear analog, acetoamidomethyl cysteine$_{2,7}$ CGRP ([CysACM]CGRP), has been reported to be a selective agonist of $CGRP_2$ receptors in functional assays, but has no effect at $CGRP_1$ receptors. Although functionally $CGRP_{8-37}$ and [CysACM]CGRP show differential effects between $CGRP_1$ and $CGRP_2$ receptors, both ligands display equal affinity for inhibiting [^{125}I] CGRP binding to the two receptors.[5] $CGRP_1$ receptors have been shown to be present in lung, kidney, spleen and atria. Evidence suggests that vas deferens and liver display $CGRP_2$ receptors. A number of excellent review articles have been published on CGRP and its receptor.[6-8] This review summarizes the characterization of the cloned $CGRP_1$ receptor. Recent findings from our laboratory have been highlighted in this review.

5.2. Molecular Biology

5.2.1. Sequence

The cloning of CGRP receptors has been the object of intense research by a number of investigators. Recently, these receptors have been cloned from human,[9] pig[10] and rat.[11] These receptors show 91-95% identity at the amino acid level among species. Related (calcitonin-like) receptors[12,13] were reported previously but investigators were unable to identify the ligand. It is important to note that the expression of functional CGRP receptors could be accomplished only in human embryonic kidney 293 (HEK-293) cells. We have cloned a human CGRP type 1 receptor using an expressed sequence tag (EST) derived from a human synovial tissue cDNA library. This incomplete 800-bp cDNA insert was used as a probe to screen a human lung cDNA library to isolate a clone with a complete open reading frame. Three potential in-frame ATG codons preceded the open reading frame of the protein, however, translation from the second or third ATG codon encoded a protein with a size consistent with that predicted for the cloned rat calcitonin (CT)-like "orphan" receptor, with which it shares 91% amino acid sequence homology and thus likely represents a receptor ortholog. Also, since the third ATG most closely approximates a Kozak consensus translation initiation site,[14] it is probably the translation initiation codon. Consequently, the cDNA encodes a protein of 461 amino acids, sharing several features that are common with other G-protein-coupled receptors.[15] Most prominent is the existence of seven hydrophobic regions of 16 to 28 amino acids each, which are likely membrane spanning domains that form the seven transmembrane motif found among G-protein-coupled receptors. In addition, a series of 52 amino acid residues highly conserved among the recently described subfamily of G-protein-coupled receptors including CT, secretin, parathyroid, glucagon, and other receptors[16] are also interspersed within the sequence. Among this subfamily, the cloned receptor shares its greatest sequence identity, 55.5%, with the human CT receptor. Furthermore, in the amino terminal domain, there are several sites for post-translational modifications including three asparagine residues

within the consensus sites for glycosylation, and a potential cleavage site for an N-terminal hydrophobic sequence that may be a signal peptide. In addition, the receptors in this group, including the CGRP receptors, possess a large N-terminal extracellular domain, and six conserved cysteine residues in the N-terminal extracellular domain that are found in other receptors that interact with relatively large peptide hormones. The deduced amino acid sequence of the human lung CGRP receptor is 93% and 91% identical to porcine and rat CGRP receptors, respectively.[10,11] Fluorescent in situ hybridization studies have localized the human CGRP receptor gene to chromosome 2q.32.2. A hypothetical model of the secondary structure of the human CGRP receptor is shown in Figure 5.1. This model is based on the arrangement of the seven transmembrane domains within the plasma membrane, forming a central ligand binding pocket as was first proposed for adrenoreceptors. The deduced amino acid sequence of human CGRP receptor and its comparison with porcine and rat CGRP receptors are shown in Figure 5.2. All these CGRP receptors have similar length ranging from 461 to 474 amino acids with large N-terminal hydrophobic domains (120 amino acids) that are likely to represent leader sequences.

Kapas and Clark[17] reported the characterization of an orphan receptor, RDC1 (originally cloned from canine thyroid), as a $CGRP_1$ receptor. Addition of CGRP or adrenomedullin (ADM) to COS cells expressing RDC1 resulted in an increase in cAMP formation. $CGRP_{8-37}$ inhibited CGRP-mediated cAMP accumulation in a noncompetitive manner. On the other hand, ADM-mediated cAMP formation was not affected by $CGRP_{8-37}$. [^{125}I]CGRP bound specifically to this receptor with a Kd of 9.3 nM. The authors mentioned that the northern blot analysis of RDC1 revealed that the messenger RNA for this receptor was present in heart and kidney, but no signal was found in lung. This observation was not in agreement with the high density

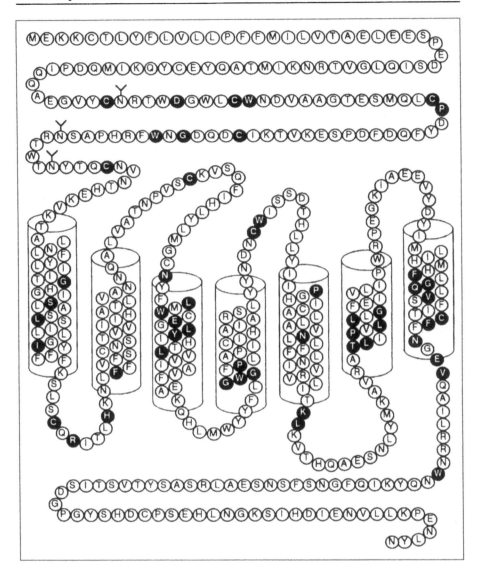

Fig. 5.1. Hypothetical model on the secondary structure of recombinant human CGRP receptor. Y represents sites for glycosylation. ● represents amino acids that are conserved within this family of receptors.

of CGRP receptors present in lung from all species tested, including human, rat, rabbit, cow and pig. The amino acid sequence of this putative CGRP receptor clone (RDC1) shares 30% homology with the rat ADM receptor (also reported by the same group),[18]

but displays very little homology with CT receptors.

5.2.2. Tissue Localization

Tissue localization of CGRP receptor was performed using dot blot analysis of

Amino Acid Alignment of CGRP Receptors

	1				50
Human	MEKKCTLYFL	VLLPFFMILV	TAEL.EESPE	DSIQLGVTRN	KIMTAQYECY
Porcine	MEKKYILYFL	FLLPFFMILV	IA ETEEENPD	DLIQLTVTRN	KIMTAQYECY
Rat	MDKKCTLFL	FLLLLNMALI	AAESEEGA..	NQTDLGVTRN	KIMTAQYECY

	51				100
Human	QKIMQDPIQQ	AEGVYCNRTW	DGWLCWNDVA	AGTESMQLCP	DYFQDFDPSE
Porcine	QKIMQDPIQQ	TEGIYCNRTW	DGWLDWNDVA	AGTESMQHCP	DYFQDFDPSE
Rat	QKIMQDPIQQ	GEGLYCNRTW	DGWLCWNDVA	AGTESMQYCP	DYFQDFDPSE

	101				150
Human	KVTKICDQDG	NWFRHPASNR	TWTNYTQCNV	NTHEKVKTAL	NLFYLTIIGH
Porcine	KVTKICDQDG	NWFRHPESNR	TWTNYTQCNI	NTHEKVQTAL	NLFYLTIIGH
Rat	KVTKICDQDG	NWFRHPDSNR	TWTNYTLCNN	STHEKVKTAL	NLFYLTIIGH

	151				200
Human	GLSIASLLIS	LGIFFYFKSL	SCQRITLHKN	LFFSFVCNSV	VTIIHLTAVA
Porcine	GLSIASLLIS	LGIFFYFKSL	SCQRITLHKN	LFFSFVCNSI	VTIIHLTAVA
Rat	GLSIASLLIS	LGIFFYFKSL	SCQRITLHKN	LFFSFVCNSI	VTIIHLTAVA

	201				250
Human	NNQALVATNP	VSCKVSQFIH	LYLMGCNYFW	MLCEGIYLHT	LIVVAVFAEK
Porcine	NNQALVATNP	VSCKVFQFIH	LYLMGCNYFW	MLCEGIYLHT	LIVVAVFAEK
Rat	NNQALVATNP	VSCKVSQFIH	LYLMGCNYFW	MLCEGIYLHT	LIVVAVFAEK

	251				300
Human	QHLMWYYFLG	WGFPLIPACI	HAIARSLYYN	DNCWISSDTH	LLYIIHGPIC
Porcine	QHLMWYYFLG	WGFPLIPACI	HAVARRLYYN	DNCWISSDTH	LLYIIHGPIC
Rat	QHLMWYYFLG	WGFPLLPACI	HAIARSLYYN	DNCWISSDTH	LLYIIHGPIC

	301				350
Human	AALLVNLFFL	LNIVRVLITK	LKVTHQAESN	LYMKAVRATL	ILVPLLGIEF
Porcine	AALLVNLFFL	LNIVRVLITK	LKVTHQAESN	LYMKAVRATL	ILVPLLGIEF
Rat	AALLVNLFFL	LNIVRVLITK	LKVTHQAESN	LYMKAVRATL	ILVPLLGIEF

	351				400
Human	VLIPWRPEGK	IAEEVYDYIM	HILMHPQGLL	VSTIFCFFNG	EVQAILRRNW
Porcine	VLIPWRPEGK	IAEEVYDYIM	HILVHYQGLL	VSTIYCFFNG	EVQAILRRNW
Rat	VLFPWRPEGK	VAEEVYDYVM	HILMHYQGLL	VSTIFCFFNG	EVQAILRRNW

	401				450
Human	NQYKIQFGNS	FSNSEALRSA	SYTVSTISDG	PGYSHDCPSE	HLNGKSIHDI
Porcine	NQYKIQFGNS	FSHSDALRSA	SYTVSTISDG	AGYSHDYPSE	HLNGKSIHDM
Rat	NQYKIQFGNG	FSHSDALRSA	SYTVSTISDV	QGYSHDCPTE	HLNGKSIQDI

	451	466
Human	ENVLLKPENL	YN - - - -
Porcine	ENIVIKPEKL	YD - - - -
Rat	ENVALKPEKM	YDLVM

Fig. 5.2. Amino acid alignment of CGRP receptors. Comparison of the amino acid sequences of human, porcine and rat CGRP receptors. Differences in amino acids are represented by shaded areas.

polyA$^+$ RNA isolated from 20 different normal human tissues, and abundant message for CGRP receptor was detected in the lung followed by heart. In addition, CGRP receptor expression was also detected in varying amounts in kidney, ovary, placenta, prostate, spleen, stomach and uterus. Northern blot analysis using the full length cDNA as the hybridization probe revealed mRNA species of approximately 7.5, 5.5 and 3.5 kb present predominantly in the lung and, to a lesser degree, in the heart.[9] In situ analysis of mouse lung and heart tissues demonstrated specific expression of the CGRP$_1$ receptor mRNA in alveolar cells of the lung and myocytes of the heart. Antiserum raised against the C-terminal heptapeptide of the recombinant rat CGRP receptor was used to locate the CGRP receptor in lung, and the results showed that it was primarily located in the pulmonary vasculature.[11] Using reverse transcriptase polymerase chain reaction (RT-PCR), Edvinsson et al.[19] have demonstrated the presence of CGRP receptor mRNA in trigeminal ganglia which is suggestive of a role for CGRP in migraine.

5.2.3. Binding Studies

Membranes prepared from untransfected HEK-293 cells displayed very little specific binding for [^{125}I]CGRP (data not shown), whereas membranes prepared from HEK-293 cells transfected with human, porcine and rat CGRP receptors cDNA displayed high affinity, low density binding sites for [^{125}I]CGRP. The apparent dissociation constants (K_d) for the human, porcine and rat CGRP receptors were 19, 38 and 1200 pM, respectively.[9-11] The densities of binding sites were 86 fmol/mg protein, 160 fmol/mg protein and 1200 fmol/10^6 cells for human, porcine and rat CGRP receptors, respectively. HEK-293 cells expressing recombinant porcine CGRP receptors (HEK-293PR) were used as a model system to study the pharmacological and functional properties of recombinant CGRP receptors. The specificity of CGRP binding was assessed by examining the ability of CGRP and related peptides to compete for [^{125}I] CGRP binding to these receptors (Fig. 5.3). A number of CGRP analogs and fragments were potent in competing for [^{125}I] CGRP binding to the recombinant CGRP receptor. Both α and β human CGRP were equipotent in the competition binding assay. The relative order of potencies for the C-terminal fragments of CGRP was CGRP$_{8-37}$ > CGRP$_{9-37}$ >CGRP$_{10-37}$ > CGRP$_{11-37}$. This rank order correlates well with that observed for these fragments in rat brain.[5] The linear peptides, [CysACM]CGRP and ADM, were 30 and 200 fold weaker than the native peptide. Salmon CT displayed negligible affinity for this receptor. Unrelated peptides such as angiotensin II and endothelin 1 did not compete for the binding (data not shown). The binding affinity of [^{125}I]CGRP and the pharmacological profiles of the competing ligands for the recombinant receptor were very similar to that reported for endogenous CGRP receptors present in membranes prepared from human SK-N-MC cells. Moore et al.[20] observed that the PCR product of SK-N-MC cells was identical to the human CGRP receptor clone. In all cases, the peptides displayed monophasic competition curves with Hill coefficients which were not significantly different from unity, indicative of interaction with a single class of binding sites.

The CGRP receptor possesses several features that are common to other G-protein-coupled receptors. G-protein-coupled receptors typically exist in interconvertible high and low affinity states, depending on receptor G-protein association. The ability of guanine nucleotides to decrease the affinity of an agonist has been demonstrated for a number of seven transmembrane G protein-coupled receptors. This has been shown as a decrease in affinity when the radioligand used is an agonist or a shift in the agonist competition curve from high to low affinity when the radioligand used is an antagonist. [^{125}I]human αCGRP binding to membranes prepared from rat liver, skeletal muscle, guinea

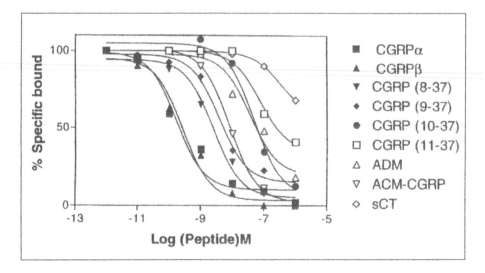

Fig. 5.3. Displacement of [125I]CGRP binding to HEK-293PR cell membranes by CGRP and its related peptide. Competition of [125I]CGRP binding to recombinant porcine CGRP receptor (HEK-293PR) by various peptides.

pig lung, rat cerebellum and SK-N-MC cells has been shown to be sensitive to guanine nucleotides.[21,22]

To assess CGRP receptor G-protein interactions in HEK 293 cells expressing recombinant porcine CGRP receptors, competition binding experiments were performed with [125I]CGRP$_{8-37}$ and unlabeled CGRP in the presence and absence of a nonhydrolyzable GTP analog, GTPγS. GTPγS shifted the CGRP competition curve to the right, resulting in an increase in IC$_{50}$ and Ki values from 0.8 to 3.5 nM and 0.21 to 1.5 nM, respectively. These results show that the binding of agonists to recombinant porcine CGRP receptors was regulated by guanine nucleotides, presumably by interaction with guanine nucleotide binding protein.

5.3. Functional Studies

5.3.1. Activation of Adenylyl Cyclase

To determine the functional coupling of the recombinant CGRP receptor, agonist-stimulated cAMP accumulation was determined in HEK-293 cells transfected with CGRP receptor as well as vector alone.

Exposure of vector-transfected 293 cells to 1μM CGRP resulted in less than a 2 fold increase in the accumulation of cAMP (data not shown), whereas addition of CGRP to 293 cells expressing the recombinant receptor resulted in a 30-60 fold increase in cAMP accumulation, depending on the concentration of CGRP used. The threshold, half maximal, and maximal concentrations of CGRP required to stimulate cAMP accumulation in these cells were 0.1, 0.9, and 10 nM, respectively. These results agreed well with similar studies performed using SK-N-MC cells. Similarly, CGRP elicited a concentration-dependent increase in adenylyl cyclase activity in membranes prepared from the HEK-293 cells expressing recombinant CGRP receptor (Fig. 5.4). Maximal and half-maximal activation were observed at 100 nM and 2.5 nM CGRP, respectively. Similar to human αCGRP, human βCGRP also stimulated the adenylyl cyclase activity, with an EC$_{50}$ value of 0.4±0.12 nM (Fig. 5.4). A number of related peptides were tested for their effects on recombinant CGRP receptor expressed in HEK-293 cells. At high concentrations, ADM increased cAMP levels (Fig. 5.5)

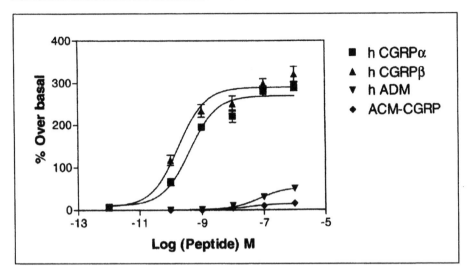

Fig. 5.4. Activation of adenylyl cyclase by CGRP, ADM and ACM-CGRP in HEK-293PR cell membranes.

which was not surprising since this agonist has been reported to interact with CGRP receptor very weakly. [CysACM]CGRP (selective agonist for $CGRP_2$ receptors) was inactive in stimulating adenylyl cyclase activity in HEK-293PR cell membranes (Fig. 5.4). In contrast, human CT failed to interact with the recombinant CGRP receptor since this ligand-stimulated cAMP accumulation was identical in CGRP receptor-transfected as well as untransfected 293 cells, indicating that HEK-293 cells display endogenous CT receptors.

The CGRP receptor antagonist, $CGRP_{8-37}$, has been shown to selectively antagonize the responses mediated by $CGRP_1$ receptor subtypes. Other truncated C-terminal fragments such as $CGRP_{9-37}$, $_{10-37}$ and $_{11-37}$ have been found to be much weaker than $CGRP_{8-37}$ (Fig. 5.5). $CGRP_{11-37}$ was ineffective in antagonizing the CGRP effect. These fragments by themselves had very little effect on the accumulation of cAMP in HEK-293 cells or activation of adenylyl cyclase in HEK-293 cell membranes expressing the recombinant porcine CGRP receptor. Addition of increasing concentrations of $CGRP_{8-37}$ to HEK-293 cells expressing recombinant CGRP receptors

caused a parallel rightward shift in the CGRP concentration-response curve for the accumulation of cAMP, indicating competitive inhibition of CGRP-mediated response with a calculated pA_2 of 7.57.[12] These data further suggest that the receptor being studied belongs to the $CGRP_1$ subtype.

5.3.2. Mobilization of Intracellular Calcium

The newly described family of G-protein-coupled receptors, which includes secretin/VIP, has been shown to increase both cAMP and cytosolic calcium levels. We also examined CGRP-mediated $[Ca^{2+}]_i$ release using HEK-293 cells expressing recombinant human and porcine CGRP receptors. In both cells, CGRP mediated a significant increase $[Ca^{2+}]_i$ (Fig. 5.6).[23] The increase in intracellular calcium release was dependent upon the concentration of CGRP used, with an EC_{50} value of 1.63 nM. The maximal response was observed at 100 nM of human αCGRP with porcine as well as human CGRP receptors. As controls, pCDN vector-transfected HEK-293 cells were used to study the effect of CGRP on $[Ca^{2+}]_i$ mobilization. CGRP did not cause an increase in $[Ca^{2+}]_i$ release in cells transfected

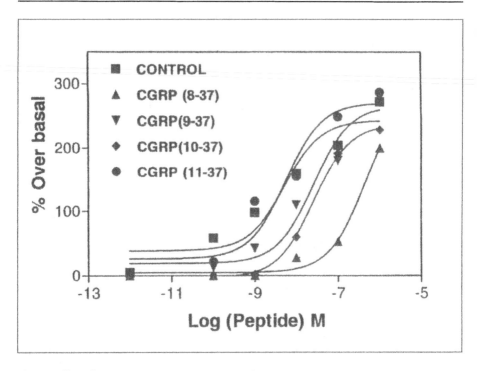

Fig. 5.5. Effect of CGRP (8-37), (9-37), (10-37) and (11-37) on CGRP-activated adenylyl cyclase in HEK-293PR cell membranes.

with vector alone. The CGRP receptor antagonist, $CGRP_{8-37}$, by itself had no effect on $[Ca^{2+}]_i$ release but inhibited the CGRP-stimulated $[Ca^{2+}]_i$ release. Increase in $[Ca^{2+}]_i$ was specific for CGRP since CGRP-related peptides such as [CysACM] CGRP, ADM and CT did not have any effect on $[Ca^{2+}]_i$ release in HEK-293 cells expressing recombinant CGRP receptors (human or porcine). In addition, exposure of these cells to forskolin did not cause an increase in $[Ca^{2+}]_i$ release, suggesting that the stimulatory effect of CGRP on $[Ca^{2+}]_i$ release may not be secondary to the activation of the adenylyl cyclase pathway. Furthermore, treatment of $[^3H]$ myoinositol-prelabeled HEK-293 cells expressing recombinant CGRP receptors with 100 nM human αCGRP also resulted in a 2.5 fold increase (2800 dpm to 7200 dpm) of inositol phosphate accumulation. Taken together, these results indicate that the recombinant human or porcine CGRP receptor expressed

in HEK-293 cells is coupled to the activation of adenylyl cyclase and phospholipase C. Our data demonstrate that the concentration-response curves of CGRP for the stimulation of cAMP accumulation and intracellular calcium release are very similar, indicating that the same receptor efficiently couples to both effector systems. We also observed that pretreatment of HEK-293 cells expressing recombinant CGRP receptors with cholera toxin (CTX) (0.2 mg/ml, 2 hours) did not change the basal $[Ca^{2+}]_i$ level but inhibited CGRP-mediated $[Ca^{2+}]_i$ release by 70%. In addition, CTX pretreatment of HEK-293 cells expressing recombinant CGRP receptors resulted in a 7.5 fold increase in basal adenylyl cyclase activity, with no change in intracellular calcium release. CGRP-induced $[Ca^{2+}]_i$ release as well as activation of adenylyl cyclase in recombinant CGRP receptors in HEK-293 cells were not affected by pertussis toxin (PTX) (100 ng/ml, 18 hours) pretreatment

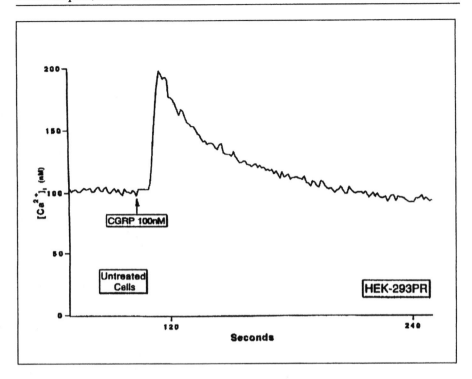

Fig. 5.6. Stimulation of intracellular calcium release by CGRP in HEK-293PR cells.

(data not shown). These results suggest that CTX-sensitive G-protein(s) is (are) involved in CGRP-mediated $[Ca^{2+}]_i$ release as well as CGRP-mediated activation of adenylyl cyclase. Also, these results show that the recombinant human as well as porcine CGRP receptor can independently increase cAMP production as well as intracellular calcium release.

5.3.3. Extracellular Acidification

Exposure of HEK-293 cells expressing recombinant CGRP receptors to human αCGRP or human βCGRP resulted in a significant increase in extracellular acidification rate as measured by a cytosensor microphysiometer. Repeated additions of CGRP resulted in progressive loss of responsiveness, presumably due to the desensitization of the receptor. However, maintaining the cells in buffer in the absence of CGRP for 20-30 min resulted in a total recovery of original response when challenged with CGRP.

5.3.4. Receptor Regulation

Using recombinant CGRP receptors, we also studied CGRP-mediated desensitization of CGRP receptors and the role of second messenger-activated kinases as well as G-protein receptor-coupled kinases in this process. Pretreatment of HEK-293 cells expressing recombinant porcine CGRP receptors to CGRP resulted in ~60% decrease in CGRP-stimulated adenylyl cyclase activity. In addition, there was a five fold rightward shift in the concentration-response curve of CGRP for stimulating adenylyl cyclase activity, suggesting a decrease in affinity of CGRP for the stimulation of adenylyl cyclase following desensitization.[24] Pretreatment of these cells with CGRP had no effect on adenylyl cyclase responsiveness to isoproterenol or

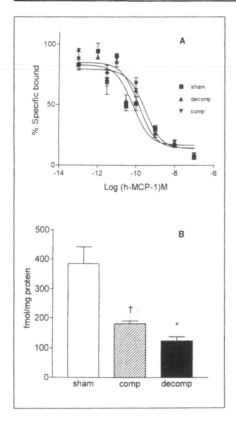

Figure 5.7. A. Competition binding curves of human MCP-1 and human 125I-MCP-1 in rat heart membranes. B. 125I-MCP-1 binding to rat heart membranes as calculated from the homologous competitve binding data. *p<0.05 decompensated vs sham. †p<0.05 for compensated vs sham.

forskolin, suggesting the desensitization was homologous in nature. This process appears to involve G protein-coupled receptor kinases, specifically GRK-6.

5.4. Conclusion

The availability of these clones will facilitate the investigations on the potential roles of CGRP receptor and its possible participation in various physiological and pathophysiological processes. The structural knowledge and function of CGRP receptors could also facilitate the rational drug design aimed at producing selective and potent compounds.Ack

Anowledgments

We thank Sue Tirri for expert secretarial assistance.

References

1. Brain SD, Williams TJ, Tippins JR et al. Calcitonin gene-related peptide is a potent vasodilator. Nature 1985; 313:54-56.
2. Gennari C, Fischer JA. Cardiovascular action of calcitonin gene-related peptide in humans. Calcif Tissue Int 1985; 37:581-584.
3. Bell D, McDermott BJ. Calcitonin gene-related peptide in the cardiovascular system: characterization of receptor populations and their (patho)physiological significance. Pharmacological Reviews 1996; 48:254-288.
4. Dennis TB, Fournier A, St. Pierre S et al. Structure-activity profile of calcitonin gene-related peptide in peripheral and brain tissues. Evidence for receptor multiplicity. J Pharmacol Exp Ther 1989; 251:718-725.
5. Quirion R, Van Rossum D, Dumont Y et al. Characterization of CGRP$_1$ and CGRP$_2$ receptor subtypes. Ann NY Acad Sci 1992; 657:88-105.
6. Poyner DR. Calcitonin gene-related peptide: multiple actions, multiple receptors. Pharmacol Ther 1992; 56:23-51.
7. Wimalawansa SJ. Calcitonin gene-related peptide and its receptors: Molecular genetics, physiology, pathophysiology, and therapeutic potentials. Endocrine Reviews 1997; 17:533-585.
8. Van Rossum D, Hanisch U, Qurion R. Neuroanatomical localization, pharmacological characterization and functions of CGRP-related peptides and their receptors. Neuroscience and Behavioral Reviews 1997; 21:649-678.
9. Aiyar N, Rand K, Elshourbagy NA et al. cDNA encoding the calcitonin gene-related peptide type 1 receptor. J Biol Chem 1996; 271:11325-11329.
10. Elshourbagy NA, Adamou JE, Swift AM et al. Molecular cloning and characterization of the porcine calcitonin gene-related peptide receptor. Endocrinology 1998; 139:1678-1683.

11. Han Z-Q, Coppock HA, Smith DM et al. The interaction of CGRP and adrenomedullin with a receptor expressed in the pulmonary vascular endothelium. J Mol Endocrinology. 1997; 18:267-272.

12. Fluhmann B, Muff R, Hunziker W et al. A human orphan calcitonin receptor-like structure. Biochem Biophys Res Commun 1995; 206:341-347.

13. Njuki F, Nicholl CG, Howard A et al. A new calcitonin-receptor-like sequence in rat pulmonary blood vessels. Clinical Sci 1993; 85:385-388.

14. Kozak M. Adherence to the first-AUG rule when a second AUG codon follows closely upon the first. Proc Natl Acad Sci (USA) 1995; 92:2662-2666.

15. Dohlman HG, Thomer J, Caron MG et al. Model systems for the study of seven-transmembrane segment receptors. Annu Rev Biochem 1991; 60:653-688.

16. Segre GV, Goldring SR. Receptors for secretin, calcitonin, parathyroid hormone (PTH)/PTH-related peptide, vasoactive intestinal peptide, glucagon-like peptide 1, growth hormone-releasing hormone, and glucagon belong to a newly discovered G-protein-linked receptor family. Trends Endocrinol Metab 1993; 4:309-314.

17. Kapas S, Clark AJL. Identification of an orphan receptor gene as a type 1 calcitonin gene-related peptide receptor. Biochem Biophys Res Commun 1995; 217:832-838.

18. Kapas S, Catt KJ, Clark AJL. Cloning and expression of cDNA encoding a rat adrenomedullin receptor. J Biol Chem 1995; 270:25344-25347.

19. Edvinsson L, Cantera L, Jansen-Olesen I et al. Expression of calcitonin gene-related peptide1 receptor mRNA in human trigeminal ganglia and cerebral arteries. Neuroscience Letters 1997; 229:209-211.

20. Moore JB, Xu J, Look R et al. Comparison of cloned calcitonin gene-related peptide (CGRP1) receptor expressed in HEK 293 cells with binding sites on human neuroblastoma SK-N-MC cells. FASEB J 1997; 11:337.

21. Semark JE, Middlemiss DN, Hutson PH. Comparison of calcitonin gene-related peptide receptors in rat brain and a human neuroblastoma cell line, SK-N-MC. Mol Neuropharmacology 1992; 2:311-317.

22. Aiyar N, Disa J, Siemens IR et al. Differential effects of guanine nucleotides on [125I]-hCGRP(8-37) binding to porcine lung and human neuroblastoma cell membranes. Neuropeptides 1997; 31:99-103.

23. Disa J, Lysko PG, Stadel J et al. Recombinant human CGRP receptor couple to phospholipase C as well as adenylyl cyclase. The Endocrine Society 79th annual meeting, Minneapolis, Minnesota Abstract: P1-148, 1997.

24. Aiyar N, Disa J, Dang K et al. Desensitization of recombinant porcine calcitonin gene-related peptide (CGRP) receptor. A possible role for G protein-coupled receptor kinase. FASEB J 1997; 11:1130.

CHAPTER 6

Receptor Activity Modifying Proteins(RAMPS) and CRLR Define the Receptors for CGRP and Adrenomedullin

Steven M. Foord, Neil J. Fraser, Martin J. Main, Alan Wise, Linda M. McLatchie, Jason Brown, Nicola Thompson, Roberto Solari and Melanie G. Lee

6.1. Introduction

The calcitonin family of peptides comprises five known members: calcitonin, amylin, two calcitonin-gene-related peptides (αCGRP and βCGRP) and adrenomedullin (ADM). The calcitonin receptor was cloned in 1991[1] and it was thought that the receptors for the other members of the peptide family would be similar (as the receptors for opioid peptides or chemokines are similar). In 1993 the Calcitonin-Receptor-Like Receptor, or CRLR, was identified.[2,3] It has an overall identity to the calcitonin receptor of 55% (Fig. 6.1A). Other members of the calcitonin peptide family were clearly candidate ligands for CRLR but it did not appear to confer CGRP receptors on a number of cell lines.[4] Professor Jan Fischer and Dr Walter Born sent us their cDNA encoding CRLR to test in oocytes from *Xenopus laevis*. An advantage of expressing receptors in oocytes is that their activation can be detected by electrophysiological techniques. Receptors that activate Gq, mobilise intracellular Ca^{2+} leading to the activation of an endogenous chloride channel and depolarization but receptors that activate Gs or Gi produce no electrophysiological response unless exogenous channels are introduced. The cystic fibrosis transmembrane regulator (CFTR) contains a chloride channel that can be activated by receptors that couple to Gs or Gi.[5] Expression of CRLR in oocytes alone or with CFTR produced no new responses to CGRP and we were forced to conclude that CRLR did not encode a CGRP receptor.

During these studies we became convinced that some batches of oocytes had an endogenous CGRP receptor. This was detected via the CFTR as a small inward current after CGRP application (23 ± 15 nA at -60 mV, n=15) and it could be found whether the follicle cells around the oocyte were completely removed or not.[6] An electrophysiological readout of CGRP receptor activation would facilitate expression cloning. Very few activated receptors would be needed to produce a response and so large pools of cDNA could be screened with a reasonable chance that a cDNA encoding a CGRP receptor would be detected. Although it was clearly a risk to try and clone the receptor on

The CGRP Family: Calcitonin Gene-Related Peptide (CGRP), Amylin, and Adrenomedullin, edited by David Poyner. ©2000 EUREKAH.COM.

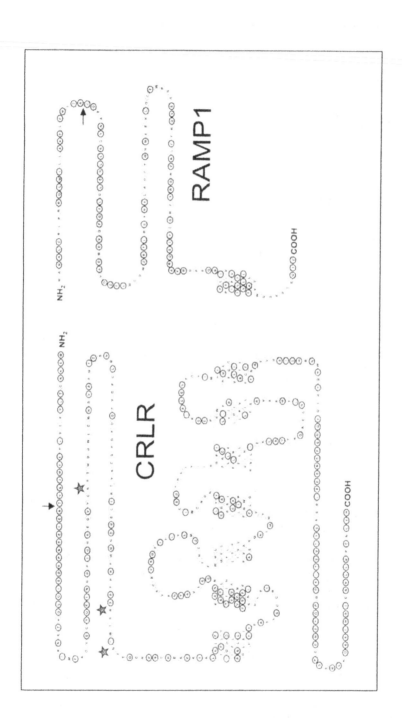

Fig. 6.1. A schematic representation of the amino acid sequences of CRLR and RAMP1. 6.1A. The amino acid sequence of human CRLR. Potential glycosylation sites are indicated *. Residues unique to CRLR over the human calcitonin receptor are indicated in bold circles. 6.1B. The amino acid sequence of human RAMP1. The residues unique to RAMP1 over RAMP2 and RAMP3 are indicated in bold circles.

top of an endogenous background the decision points were at the outset of the project, if we were unable to detect a signal above background we could not proceed. We were also encouraged by the observation that for each batch of oocytes the endogenous receptor was relatively constant.

6.2. RAMPS and CRLR

6.2.1. Cloning of RAMP1

We chose to clone from a cDNA library from SK-N-MC cells. These derive from a human neuroblastoma, bind $[^{125}I]\alpha$CGRP and respond to CGRP with an increase in intracellular cAMP.[7] The cDNA library was transcribed in vitro and pools of cRNA injected into oocytes along with cRNA encoding CFTR. Each oocyte received 50-100 ng of CFTR RNA and 100-400 ng of SK-N-MC pool RNA and was tested for CGRP responses after 3-14 days, under two electrode voltage clamp. In this format the membrane potential across the plasma membrane of the oocyte is detected with one electrode and maintained by the injection of current via a second. The endogenous oocyte β-adrenergic receptor controlled for CFTR expression and ensured any increased responses were specific for CGRP. A single pool of about 0.7 million clones was identified that induced CGRP responses that were significantly greater than background (Fig. 6.2). Six further subdivisions were made according to CGRP responses and corresponding to pool sizes of ~50,000, ~20,000, 4000 (gridded), 168, 40 and 2 clones respectively. A single cDNA was responsible for the novel CGRP responses and it encoded a 148 amino acid protein that we called 'Receptor Activity Modifying Protein' or RAMP1 (Fig 6.1B). In oocytes expressing RAMP1 large, dose-dependent responses to CGRP (1330 ± 315 nA, n=15) were recorded. We believe that RAMP1 potentiates the oocytes endogenous CGRP receptor rather that representing a CGRP receptor itself for a number of reasons.[8]

1. The amino acid sequence of RAMP1 has no homology to known G protein coupled receptors. The hydrophobicity plot of the RAMP1 protein is consistent with an N-terminal signal sequence and single putative transmembrane domain close to the C-terminus.

2. The EC_{50} of 10nM for CGRP in RAMP1 expressing oocytes is at least 10-fold lower than that for the CGRP receptor in SK-N-MC cells.

3. Expression of RAMP1 in mammalian cells (HEK293T, COS7 or Swiss3T3) failed to induce cAMP responses to CGRP or specific $[^{125}I]$CGRP1 binding.

4. RAMP1 enabled CRLR to confer CGRP responses and binding in mammalian cells but only when coexpressed with RAMP1 and CRLR increased the CGRP responses by oocytes expressing RAMP1.

The CGRP receptor reconstituted by coexpression of CRLR with RAMP1 (either transiently or through the construction of stable cell lines) displayed virtually identical pharmacology to the CGRP receptor in SK-N-MC cells and the CRLR-expressing cell line reported by Aiyar et al.[9] The requirement for CRLR and RAMP1 to reconstitute a CGRP receptor can explain its resistance to expression cloning, its failure to function in oocytes when expressed by itself and the observation that CRLR can only function as a CGRP receptor in certain cell lines (which presumably express RAMP1). [3,4,8-10]

6.2.2. RAMPS 2 and 3

The human RAMP1 sequence was partially represented by a number of Expressed Sequence Tags (ESTs) in public databases. These searches also led to the discovery of the partial sequences of two further RAMP1-like proteins, which we have termed RAMP2 and RAMP3. RAMP2 was cloned from the SK-N-MC cell cDNA library and RAMP3 by conventional RACE PCR from human spleen mRNA. The

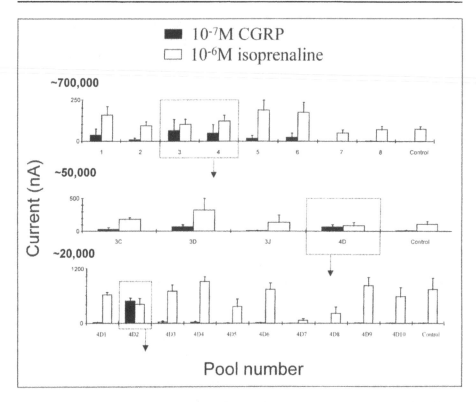

Fig. 6.2. The cloning scheme used for the isolation of RAMP1. Capped cRNA was injected into stage V-VI de-folliculated oocytes. Each oocyte received 50-100 ng of CFTR RNA and 1-20 ng of cloned RNA's or 100-400 ng of SK-N-MC pool RNA and was tested for CGRP responses after 3-14 days, under two electrode voltage clamp. The endogenous oocyte β-adrenergic receptor controlled for CFTR expression and ensured any increased responses were specific for CGRP. A pool of about 0.7 million clones produced a CFTR mediated response to CGRP. Six further subdivisions were made corresponding to pool sizes of ~50,000, ~20,000, 4000 (gridded), 168, 40 and 2 clones respectively. Oocytes were superfused with a Ca^{2+} free ND96 solution (96mM NaCl, 2mM KCl, 1mM $MgCl_2$, 5mM HEPES, pH 7.5 at 25°C) at a flow rate of 2ml/min. Recording electrodes had resistance of 0.5-1.0 MΩ when filled with 3M KCl.

RAMPs are about 31% identical and about 56% similar when compared to each other. We have identified rodent homologues of RAMPs (Fig. 6.3), but have been unable to find homologues from invertebrates. The hydrophobicity plots of the RAMP amino acid sequences are very similar and consistent with type I transmembrane proteins with the N-terminus extracellular and the C-terminus cytoplasmic (Fig. 6.4). This orientation has been confirmed for RAMP1, which, when immunotagged at the amino terminus, could be detected on the cell surface when co-expressed with CRLR. Otherwise epitope tagged RAMP1 was detected within the endoplasmic reticulum.

RAMP1 potentiates the CGRP receptor endogenous to *Xenopus* oocytes but RAMP2 or RAMP3 could not. None of the RAMPs could alter the endogenous oocyte responses to adenosine, VIP or β-adrenergic agonists (all of which will activate the CFTR) (Fig. 6.5).

Neither RAMP2 nor RAMP3 could enable CRLR to function as a CGRP receptor in HEK293T cells. However, by FACS

```
        1                 10                  20                  30                  40                  50
        |                 |                   |                   |                   |                   |
RAMP1   - - - - - - - M A P G C G R L P R R G L W L - - - - - - - - - - L L A H H L F M V T A C R D - - - -
RAMP1   - - - - - - - M A P G C G X X X X X G L W L - - - - - - - - - - L L A H H L F M V T A C R D - - - -
RAMP1   - - - - - - - M A R A L C R L P R R G L W L - - - - - - - - - - L L A H H L F M T T A C Q E - - - -
RAMP2   M A S L R V E R A - G G P R L P R T R V G R - - - - - - P A A V R L L L L L G A V L N P H E A L A Q
RAMP2   M A P L R V E R A P G G S R L G V T R A H A X X X X X X P L L L L L L L L L D A V I T S P E S L N Q
RAMP3   - - - - - - - M E T G A L R R P Q - - - - - - - - - - - - - - - L L P L L L L L C G G C P R A G G -
        <------------------------------- Signal Peptide --------------------------------->

        51                60                  70                  80                  90                  100
        |                 |                   |                   |                   |                   |
RAMP1   - - P D Y G - - - - - - - - - - - - - - - - - - - T L I Q E L C L S R F K E D M E T I G K T L W C D
RAMP1   - - P D Y G - - - - - - - - - - - - - - - - - - - T L I Q E L C L S R F K E N M E T I G K T L W C D
RAMP1   - - A N Y G - - - - - - - - - - - - - - - - - - - A L L R E L C L T Q F Q V D M E A V G E T L W C D
RAMP2   P L P T T G - - - - - T P G S E G - - G T V K N Y E T A V Q F C W N H Y K D Q M D P I E K - D W C D
RAMP2   S L P E S Q N Q S H P T E D S L V S K G K M E D Y E T H V L P C W Y E Y K S C M D S V K - - D W C N
RAMP3   - C N E T G - - - - - - - - - - - - - - - - - - M L E R L P L C G K A F A D M M G K V D V W K W C N
                                                                      *                                 *

        101               110                 120                 130                 140                 150
        |                 |                   |                   |                   |                   |
RAMP1   W G K T I G S Y G E L T D C T K L V A - - - - - - - - - - - - - - - - - - - - - - - - - - - - - - -
RAMP1   W G K T I Q S Y G E L T Y C T K H V A H T I G C F W P N P E V D R F F I A V H H R Y F S K C P I S G
RAMP1   W G R T I R S Y R E L A D C T W H M A E K L G C F W P N A E V D R F F L A V H G R Y F R S C P I S G
RAMP2   W T L I S R H Y S D L Q N C L E Y N A D K F G L G F P N P L A E N I I L E A H L I H F A N C S L V Q
RAMP2   W A M I S R P Y S T L R D C L E H F A E L F D L G F P N P L A E R I I F E T H Q I H F A N C S L V Q
RAMP3   L S E F I V Y Y E S F T N C T E M E A N V V G C Y W P N P L A Q G F I T G I H R Q F F S N C T V D R
                                  *                                                             *

        151               160                 170                 180                 190
        |                 |                   |                   |                   |
RAMP1   - - - - - - - - - - - - - - - - - - - - - - - - - - - - - - - - - - - - - - - -
RAMP1   R A V R D P P N S I L C P F I A L P I T V T L L M T A L V V W R S K R T E G I V
RAMP1   R A V R D P P G S I L Y P F I V V P I T V T L L V T A L V V W Q S K R T E G I V
RAMP2   P T F S D P P E D V L L A M I I A P I C L I P F L V T L V V W R S K D S D A Q A
RAMP2   P T F S D P P E D V L L A M I I A P I C L I P F L I T L V V W R S K D S E A Q A
RAMP3   V H L E D P P D E V L I P L I V I P V V L T V A M A G L V V W R S K R T D T L L
```

Fig. 6.3. An alignment of the amino acid sequences of human and murine RAMP1 and RAMP2 and human RAMP3. The alignments were initially constructed using the GeneWorks protocol and modified to favour common DNA homology when ambiguities arose. Murine RAMP1 and RAMP2 sequences were not determined experimentally but derived from the following ESTs. RAMP1 = ESTs W16467/W16468, H40103, H33975 (rat homologue) and W11529 (mouse homologue). RAMP2 = ESTs N93656, T93708, T55035, L77606/L77604 (localising the gene to chromosome 17q close to the BRCA1 locus), R56654, T54870, T27173 and W24378 (mouse homologue). RAMP3 = W69533, T79681 and AA039413

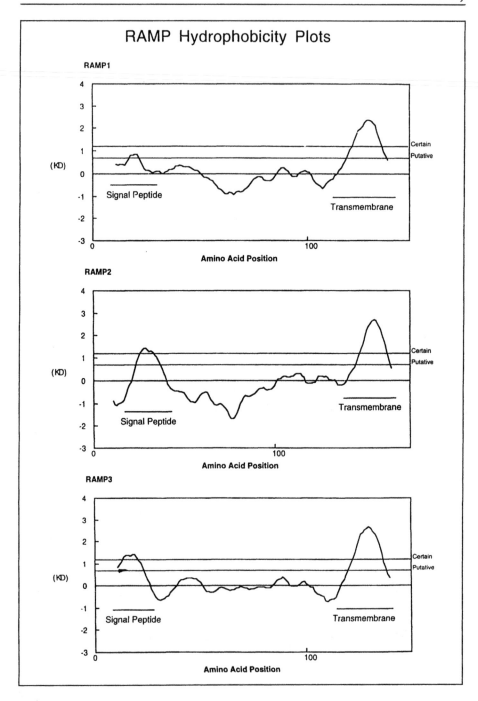

Fig. 6.4. The Kyte Doolittle Hydrophobicity plots of the RAMP protein sequences. The putative signal and transmembrane sequences are indicated.

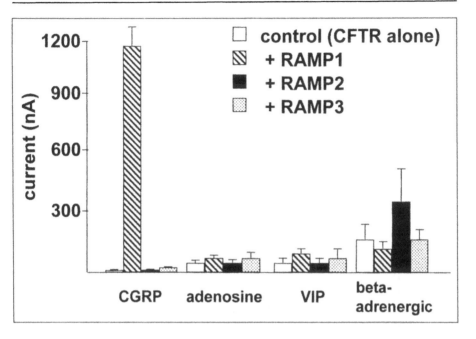

Fig. 6.5. RAMP2 and 3 have no activity against the endogenous responses to CGRP, adenosine, VIP or isoprenaline in *Xenopus* oocytes detected via the CFTR. The histograms represent mean +/- s.e.m. currents elicited from oocytes expressing CFTR +/- RAMP1, 2 or 3 in responses to 100 nM CGRP, 100 uM adenosine, 100 nM VIP and 10 uM isoprenaline applied at concentrations predicted to give maximal responses. Each response contains data from more than one donor and is controlled for differences between donors for each ligand between RAMP1 +ve and -ve oocytes. N> 10, 5, 5 and 12 respectively.

analysis we were able to demonstrate that RAMP2 and 3 were as effective at transporting epitope tagged CRLR to the cell surface as RAMP1. We speculated that RAMP2 and RAMP3 were unlikely to transport CRLR to the cell surface as a non-functional receptor and so we tested oocytes expressing RAMP2 and CRLR with ADM and found that they responded in a dose dependent manner to concentrations as low as 0.3nM (Fig. 6.6). The pharmacology of the ADM receptor created by the coexpression of CRLR with either RAMP2 or RAMP3 is similar to that native to NG108-15 cells.[11]

In oocytes expression of RAMP2 failed to generate ADM responses without CRLR. This may be because RAMP2 is unable to interact with the *Xenopus* homologue of CRLR or because the *Xenopus* receptor coexpressed with RAMP2 does not recognise human ADM. Oocytes might therefore express significant numbers of CRLR homologues that are committed to a '*Xenopus* ADM-like' phenotype. The expression of RAMP1 would change the phenotype of these receptors to that of a CGRP receptor which can be activated (by human CGRP). This would explain why the activity of RAMP1 in oocytes is so marked (it creates responses that are comparable to the expression of cRNA encoding an exogenous receptor). In either case, the novel responses to ADM appear to result entirely from exogenous RAMP2 and CRLR, a result that shows that ADM receptors have an intrinsic ability to respond to CGRP. This contrasts with the CGRP receptor generated by the expression of CRLR and RAMP1 which does not respond significantly to ADM.

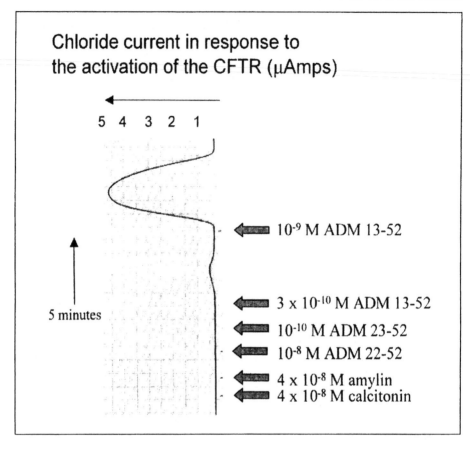

Fig. 6.6. Responses of a single oocyte to ADM. Oocytes were injected with cRNA encoding CFTR, CRLR and RAMP2. Recordings were made 4 days post-injection and at a holding potential of -60mV. A single representitive trace is shown but this sensitivity to ADM was observed routinely. All peptides represented human sequence.

6.2.3. Mechanism Of RAMP Action

Each of the RAMPs is able to transport CRLR to the surface of HEK293T cells yet the receptor pharmacology observed is that of a CGRP receptor for the RAMP1/CRLR combination and an ADM receptor for the RAMP2/CRLR and RAMP3/CRLR combinations. CRLR that is transported by RAMP1 runs on SDS PAGE as a glycoprotein of Mr 66KDa. This is the size of the protein to which $[^{125}I]\alpha$CGRP cross-links in membrane preparations from tissues that express the CGRP receptor.[12] Similar results were obtained when $[^{125}I]\alpha$CGRP1 was crosslinked to the surface of intact cells

expressing RAMP1 and CRLR. In contrast CRLR that is transported by RAMP2 or RAMP3 runs on SDS PAGE as a 58kDa glycoprotein.[8] We have considered two possible explanations for the different pharmacologies displayed when CRLR is transported by the different RAMPs.

1. RAMPs participate in the binding of CGRP or ADM and so define whether CRLR is a CGRP or ADM receptor.

2. The sugar residues present on the two forms of CRLR (as transported by RAMP1 or RAMP2 for example) determine the binding of CGRP and ADM.

We cannot distinguish between these possibilities with any certainty at the moment. However in support of the first hypothesis RAMP1 is expressed at the plasma membrane when it is expressed with CRLR and in close association with the CGRP binding site so it is possible that it could contribute directly to CGRP receptor function.[8]

6.3. RAMPS and Other Receptors/Accessory Factors

6.3.1. Non-CGRP/ADM Receptors

We have found that RAMP1 and RAMP2 are expressed in HEK293T and Kelly cells as well as in SK-N-MC cells. Neither HEK293T nor Kelly cells express CRLR (Fig. 6.7) or CGRP receptors. This might be interpreted as suggesting that RAMPs have a role beyond the creation of CGRP or ADM receptors however it is also possible that HEK293T and Kelly cells did once express CRLR and have lost the ability to do so through extended periods of culture. There is considerable interest in whether RAMPs regulate other receptor systems. A wide variety of receptors do not express well in commonly used host cell lines: GABAb, Eotaxin (CKR3), thrombin (PAR3) and a number of olfactory receptors represent a few examples.

6.3.2. RDC1 and G10d

At the outset of our studies two related members of the 'Family A' class of GPCRs, RDC1 and G10D, were reported to be the receptors for CGRP and ADM respectively.[13,14] We expressed both in oocytes but were unable to detect any novel responses to either CGRP or ADM. Similarly their expression in HEK293T cells also failed to confer CGRP or ADM receptors. We did not control for the expression of either RDC1 or G10D however the clones were those used in the original demonstration of their CGRP and ADM receptor activity (provided by Dr. Adrian Clark). The data in these studies was convincing and yet we have been unable to reproduce it in experimental systems that

clearly are able to express both CGRP and ADM receptors. It is possible that RDC1 and G10D express in COS1 cells but not in HEK293T cells. An alternative explanation is that expression of RDC1 and G10D can lead to the expression of RAMP1 or RAMP2/3 respectively.

6.3.3. Receptor Component Protein

Another observation that remains to be integrated into our scheme of CGRP and ADM receptor regulation is the activity of the Receptor Component Factor (RCF) (See Rosenblatt et al, this volume). RCF is a cytosolic protein that was expression cloned for its ability to induce CGRP responses in *Xenopus laevis* oocytes.[15] RCF has no homology to RAMPs. In our experiments CRLR and RAMPs will reconstitute CGRP or ADM receptors alone. On this basis RCF would be more likely to be a regulatory protein rather than a component of the receptor.

RAMPs suggest a novel mechanism whereby cells/tissues could change their responsiveness from one neuropeptide to another. CGRP and ADM levels alter in a number of physiological and patholophysiological states although their levels have not been measured concurrently.[16] In terms of the consequences of activating CGRP and ADM receptors in the studies reported so far, CGRP and ADM appear to activate the same second messengers and have similar bioactivities. Therefore it remains unclear what the physiological importance of switching receptors might be.

6.4. Future Directions

Are there any more CGRP or ADM receptors to be discovered? Subtle pharmacologies may result from the probable coexpression of CGRP and ADM receptors and these might be misinterpreted as indicating novel receptors.[17] However we have not been able to reconstitute the CGRP receptors defined by insensitivity to $CGRP_{8-37}$. Moreover CGRP receptors do exist in tissues that do not express significant levels of CRLR, notably the cerebellum.

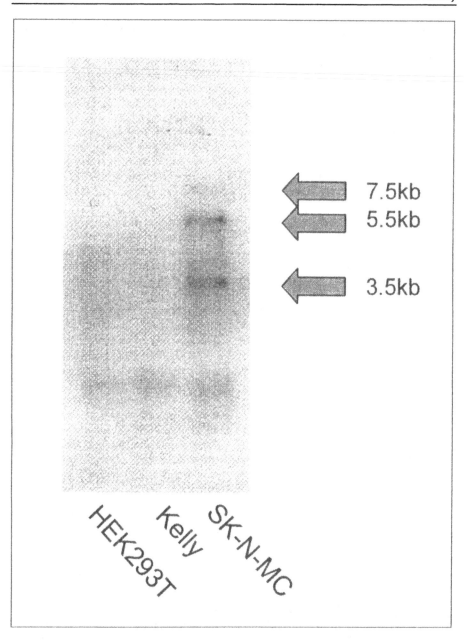

Fig. 6.7. Northern blot analysis of CRLR expression in 10 ug poly A+ mRNA isolated from SK-N-MC, Kelly and HEK293T cells. The expression of RAMPs in these samples has been published previously. [8]

The most rewarding scientific discoveries tend to be those that provide satisfactory answers to long standing questions yet also raise more questions of their own. We hope that the discovery of RAMPs qualifies on both counts.

References

1. Lin, HY, Harris TL, Flannery MS et al. Expression cloning of an adenylate cyclase-coupled calcitonin receptor. Science 1991; 254:1022-1024.
2. Njuki F, Nicholl CG, Howard A et al. A new calcitonin-receptor-like sequence in rat pulmonary blood vessels. Clin Sci 1993; 85:385-388.
3. Chang C-P, Pearse RV, O'Connell S et al. Identification of a seven transmembrane helix receptor for corticotropin-releasing factor and sauvagine in mammalian brain. Neuron 1993;11:1187-1195.
4. Fluhmann B, Muff R, Hunziker W et al. A human orphan calcitonin receptor-like structure. Biochem Biophys Res Commun 1995; 206:341-347.
5. Uezono Y, Bradley J, Min C et al. Receptors that couple to 2 classes of G proteins increase cAMP and activate CFTR expressed in Xenopus oocytes. Receptors & Channels 1993; 1:233-241.
6. Guillemare E, Lazdunski, M, Honore E. CGRP-induced activation of K_{ATP} channels in follicular *Xenopus* oocytes. Pflügers Arch 1994; 428:604-609.
7. Zimmermann U, Fischer JA, Muff R et al. Calcitonin gene-related peptide (CGRP) receptors are linked to cyclic adenosine monophosphate production in SK-N-MC human neuroblastoma cells. Neuroscience Letts 1990; 119:195-198.
8. McLatchie LM, Fraser NJ, Main MJ et al. RAMPs regulate the transport and ligand specificity of the calcitonin-receptor-like receptor. Nature 1998; 393:333-339.
9. Aiyar N, Rand K, Elshourbagy NA et al. A cDNA encoding the calcitonin gene-related peptide type 1 receptor. J Biol Chem 1996; 271:11325-11329.
10. Han Z-Q, Coppock HA, Smith DM et al. The interaction of CGRP and adreno-medullin with a receptor expressed in the rat pulmonary vascular endothelium. J Mol Endocrinol 1997; 18:267-272.
11. Zimmermann U, Fischer JA, Frei K et al. Identification of adrenomedullin receptors in cultured rat astrocytes and in neuroblastoma x glioma hybrid cells (NG108-15). Brain Res 1996; 724: 238-245.
12. Stangl D, Born W, Fischer J. Characterisation and photoaffinity labeling of a calcitonin gene-related peptide receptor solubilised from human cerebellum. Biochemistry 1991; 30:8605-8611.
13. Kapas S, Catt KJ, Clark AJ. Cloning and expression of cDNA encoding a rat adrenomedullin receptor. J Biol Chem 1995; 270:25344-25347.
14. Kapas S, Clark AJL. Identification of an orphan receptor gene as a type 1 calcitonin gene-related receptor. Biochem Biophys Res Commun 1995; 217:832-838.
15. Luebke AE, Dahl GP, Roos BA et al. Identification of a protein that confers calcitonin gene-related peptide responsiveness to oocytes by using a cystic fibrosis transmembrane conductance regulator assay. Proc Natl Acad Sci USA 1996; 93:3455-3460.
16. Wimalawansa SJ. Calcitonin gene-related peptide and its receptors: Molecular genetics, physiology, pathophysiology, and therapeutic potentials. Endocrine Reviews 1996; 17:533-585.
17. Poyner DR. Calcitonin gene-related peptide: Multiple actions, multiple receptors. Pharmacol Ther 1992; 56:23-51.

CHAPTER 7

The CGRP-Receptor Component Protein: A Novel Signal Transduction Protein

M.I. Rosenblatt, B.N. Evans, M. Nagashpour, L.O. Mnayer, A.E. Luebke, G.P. Dahl and I.M. Dickerson

7.1. Introduction

We have identified a novel protein required for signal transduction at CGRP receptors. This protein does not appear to be a receptor itself, but is required for CGRP receptor activation. We hypothesize that this protein is part of a complex of proteins that together constitute a CGRP receptor, and have named this protein the CGRP-receptor component protein (RCP). RCP expression is regulated in parallel with CGRP biological activity, and preliminary data suggest that RCP is a peripheral membrane protein required for CGRP-mediated signal transduction in *Xenopus laevis* oocytes and in NIH3T3 cells. These data suggest that in oocytes and NIH3T3 cells, RCP works in conjunction with a membrane-spanning CGRP-binding protein to form functional CGRP receptors.

7.2. Discovery of the CGRP- Receptor Component Protein (RCP)

We initially set out to clone the CGRP receptor from the cochlea. Vestibular (semi-circular canal) and cochlear (organ of Corti) hair cells are innervated by efferent neurons, which contain CGRP and acetylcholine.[1-5] A cDNA library made from the guinea pig organ of Corti (a gift from Ed Wilcox, NIH);[6] was obtained, and was screened using an expression-cloning strategy based on an assay described by Uezono et al.[7] This strategy used the cystic fibrosis transmembrane conductance regulator (CFTR) as a sensor for cAMP levels when expressed in *Xenopus* oocytes.[7,8] The CFTR is a protein kinase A (PKA)-activated chloride channel. Thus, increased intracellular cAMP, which activates PKA, results in a CFTR-mediated Cl⁻ current in oocytes expressing the CFTR (Fig. 7.1). The specificity of the oocyte-CFTR assay is demonstrated in Figure 7.2, using the calcitonin receptor as a control G protein-coupled receptor.[9] The CFTR cDNA and the calcitonin receptor cDNA were linearized and transcribed in vitro, and injected into oocytes either alone or in combination. Following a 48 hr incubation, oocytes were incubated with 100 nM calcitonin. No current was observed in response to calcitonin in oocytes injected with either CFTR cRNA alone or calcitonin receptor cRNA alone (Fig. 7.2a and b), while a robust signal (800 nA) was observed in oocytes coinjected with cRNA for CFTR and calcitonin receptor together (Fig. 7.2c).

The organ of Corti cDNA library was screened in pools of 5000 clones, which were

The CGRP Family: Calcitonin Gene-Related Peptide (CGRP), Amylin, and Adrenomedullin, edited by David Poyner. ©2000 EUREKAH.COM.

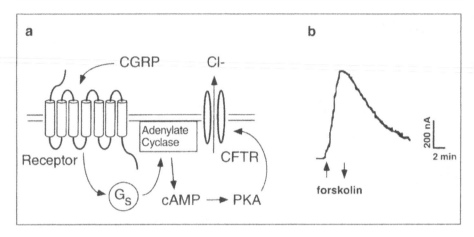

Fig. 7.1. Oocyte-CFTR assay. a) Addition of ligand results in activation of G-protein and adenylate cyclase, resulting in increased production of cAMP, and phosphorylation and activation of cystic fibrosis transmembrane conductance regulator (CFTR). b) Activation of CFTR upon application of 20 mM forskolin.

linearized and transcribed in vitro, and the resulting capped cRNA was coinjected with CFTR cRNA into oocytes. Pools that demonstrated a CGRP-induced membrane current (Fig. 7.2d) were further subdivided and retested, until a candidate RCP cDNA was identified.[10] For initial rounds of library screening, oocytes were incubated for 72 hr after injection to allow for synthesis and transport of library and CFTR protein. The RCP cDNA was enriched during successive rounds of purification, and the CGRP-induced CFTR currents increased accordingly. As a result, it was only necessary to incubate oocytes for 24 hr in later rounds of purification to detect the CGRP-induced currents. In the final round of purification with only 24 hr post-injection incubation, application of CGRP resulted in an average current greater than 1500 nA. An EC_{50} of 14 nM was obtained in these experiments for CGRP (Fig. 7.3), which was in agreement with previous in vivo studies on the effect of CGRP on beat rate in guinea pig left atria.[11,12]

RCP is specific for CGRP receptor activation in the oocyte-CFTR assay. In oocytes coinjected with RCP and CFTR cRNA, Cl- currents were observed only upon addition of CGRP. Application of 1 nM CGRP produced large currents, while no detectable current was observed upon subsequent application of 100 nM calcitonin (rat or salmon), amylin, vasoactive intestinal peptide (VIP), neuropeptide Y (NPY), or β-endorphin. No CGRP-dependent currents were observed in uninjected oocytes or in oocytes injected with CFTR alone,[10] or in oocytes coinjected with cRNA for calcitonin receptor and CFTR.[9] The oocyte-CFTR assay also detected receptor activity in mRNA isolated from tissues reported to contain CGRP binding sites.[9,10,13] To confirm the requirement for RCP, antisense oligonucleotides were made to the RCP cDNA. When injected into oocytes, these antisense oligonucleotides completely inhibited CGRP responsiveness in oocytes coinjected with CFTR cRNA and cerebellar mRNA, indicating that the cloned sequence was present in the cerebellar mRNA, and was required for CGRP receptor activation. RCP does not appear to be a transcription factor for an endogenous oocyte CGRP receptor, as neither enucleation (which functionally remove the endogenous transcriptional machinery), or coinjection of α-amanitin (an inhibitor of RNA polymerase II transcription)[14,15] diminished CGRP-induced Cl- currents in oocytes coinjected with RCP

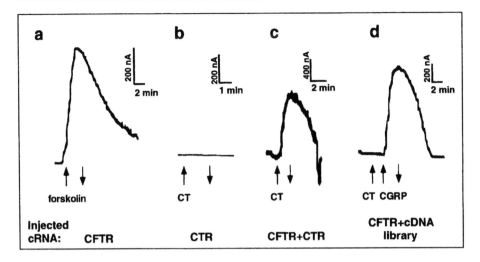

Fig. 7.2. Characterization of oocyte-CFTR assay. Oocytes injected with a) CFTR cRNA alone incubated with forskolin, b) calcitonin receptor cRNA alone incubated with 100 nM calcitonin, c) CFTR and calcitonin receptor cRNA incubated with 100 nM calcitonin, d) CFTR cRNA and cochlear cDNA library cRNA incubated sequentially with 100 nM calcitonin and then 100 nM CGRP.

and CFTR cRNA (Dahl and Dickerson, unpublished).

The guinea pig RCP cDNA encodes a 146 amino acid protein, which is largely hydrophilic and lacks the size and hydrophobic transmembrane domains characteristic of G protein-coupled receptors. The small size and hydrophilic nature of RCP suggests that it is not a membrane-spanning protein, and is therefore not a CGRP receptor. We hypothesize that RCP works with a separate membrane-spanning protein to form a functional CGRP receptor. In addition to guinea pig, we have also cloned RCP from mouse[13] and human (Luebke, Dahl and Dickerson; Genbank Accession #AF073792). The RCP protein is >80% identical between species and does not contain any obvious functional domains when searched against GenBank and Prosite databases. There are no conserved glycosylation sites in RCP, but there are several conserved sites for phosphorylation (Fig. 7.4), raising the possibility that RCP activity may be regulated by phosphorylation in vivo.

7.3. Expression of RCP in Cells and Tissues

RCP expression was detected specifically in cells previously reported to contain CGRP binding sites, supporting the role of RCP in CGRP receptor function. For example, using immunohistochemistry RCP has been detected in the rabbit eye, where RCP expression was limited to the ciliary epithelium of the iris, and the walls of blood vessels that course through the iris (Rosenblatt and Dickerson, submitted), which have been previously described to contain CGRP binding sites.[16-18] Using in situ hybridization we have detected RCP expression in the cochlea that is limited to the outer hair cells of the basal (high frequency) regions of the cochlea, where it is colocalized with CGRP-immunoreactive efferent neurons.[10] This is in agreement with previous studies of CGRP's presence on nerve fibers that terminate on outer hair cells of the guinea pig and rat cochleas.[3,19]

In cell culture, RCP was detected by Western blot in cell lines which express CGRP binding sites, such as SK-N-MC and

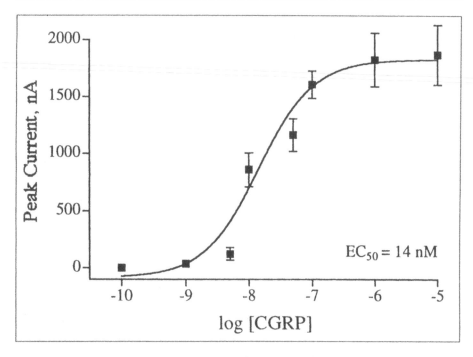

Fig. 7.3. Dose-response curve in oocytes coinjected with RCP and CFTR cRNA. CFTR currents in response to increasing doses of CGRP.

NIH3T3,[20-22] and was absent in cells which lack CGRP binding sites, such as COS-7 (Fig. 7.5a). However, RCP mRNA was detected in COS cells by RT-PCR (Fig. 7.5b), even though COS cells do not make RCP protein, indicating that RCP expression can be regulated post-transcriptionally, and that analysis of RCP mRNA may give an exaggerated estimate of RCP expression. RCP was also detected by Western blot in membranes from cerebellum and atria (Fig. 7.5a), two tissues previously reported to contain CGRP binding sites.[11,12,23-25] In these studies, RCP was detected as a monomer and as a dimer. It is not known if the dimer is present in vivo or if it results from analysis by SDS-PAGE, but the dimer is observed in purified preparations of recombinant RCP, so the higher molecular weight form is unlikely to represent a separate but homologous protein in the membrane preparations.

RCP appears to be a peripheral membrane protein. Proteins that use ionic interactions to associate with membranes can often be removed from membrane preparations by incubation with 1M NaCl, while integral membrane proteins remain in the membrane fraction. Ionic interactions also imply association with membrane-bound proteins (such as receptors), since lipid-attached proteins (prenylation, palmitylation) such as G-proteins are usually not removed by salt wash treatments. SK-N-MC cells express CGRP receptors,[22] and were used to test if RCP could be removed from membrane preparations by salt wash. SK-N-MC cells membranes were prepared by scraping cells from plates, and the 100,000 x g membrane pellet was resuspended in 50 mM Hepes pH 8.0/5 mM $MgCl_2$/5 mM EDTA, as described.[9] SK-N-MC membranes were then incubated in 0.1M Na_2CO_3 or 1 M NaCl for 30 minutes on ice, and the samples were centrifuged at 100,000 x g for 60 minutes. The pellet (integral membrane proteins) and supernatant (peripheral membrane proteins) were analyzed by Western blot with antibody R82, using conditions

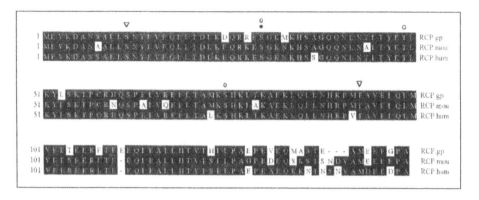

Fig. 7.4. Alignment of deduced amino acid sequences of RCP isolated from guinea pig, mouse, and human. Conserved sites are marked for potential phosphorylation by cAMP or cGMP protein kinase (●), protein kinase C (O), or casein kinase II (∇).

previously determined by our lab.[13] A dimer form of RCP was detected in the membranes before salt wash, and was detected in the salt wash supernatant after centrifugation, indicating that in SK-N-MC cells RCP appears to be a peripheral and not an integral membrane protein (Fig. 7.5b).

7.4. RCP Function

RCP expression correlates with CGRP efficacy in vivo. During pregnancy, CGRP inhibits acetylcholine-induced myometrial contractions in the uterus.[26,27] We measured the inhibitory effect of CGRP on acetylcholine-induced contraction during pregnancy, and compared the inhibitory effect of CGRP with expression of RCP protein.[13] RCP protein expression declined at embryonic day E12 and at parturition (Fig. 7.6a). Importantly, expression of RCP protein paralleled the biological potency of CGRP, which showed a similar decline at day E12 and at parturition, strongly implicating RCP in mediation of CGRP signal transduction in the myometrium (Fig. 7.6b).

More direct evidence for the role of RCP in CGRP-mediated signal transduction was obtained by creating cell lines expressing antisense constructs of RCP mRNA. Antisense strategies have evolved into an effective method of inhibiting protein expression in cell culture,[28-30] where expression of antisense RNA results in loss of protein expression, either by increased turnover of mRNA or block of protein translation.[31,32] NIH3T3 cells express CGRP receptors and are a well-characterized system in which to study receptor-mediated signal transduction.[21,33]

RCP-antisense cells were made by cloning the full-length RCP cDNA from NIH3T3 cells into the eukaryotic expression plasmid pcDNA3 (Invitrogen) in the antisense orientation. Following transfection into NIH3T3 cells, stable cell lines were selected by incubation with the antibiotic G418, and multiple independent cell lines were obtained as performed previously in our laboratory.[34] All the lines tested were indistinguishable from wild type for growth rate and cell morphology. Cell lines were first tested for the loss of RCP expression by Western blot (Fig. 7.7a), and then tested for cAMP response to incubation with CGRP (Fig. 7.7b). Cells were incubated overnight in serum-free medium supplemented with 2 μCi/ml ^{3}H-adenine, which was incorporated into [^{3}H]-ATP. Cells were then incubated with IBMX and varying concentrations of CGRP for 20 minutes, and extracted in trichloroacetic acid. [^{3}H]-ATP and [^{3}H]-cAMP were separated by sequential chromatography through dowex and alumina columns and percent conversion of

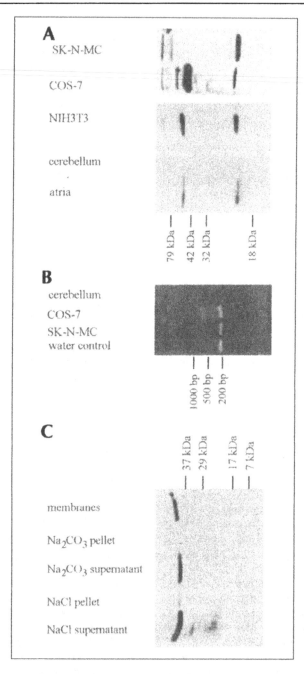

Fig. 7.5. RCP expression in CGRP-responsive cells. a) RCP protein expression is limited to CGRP responsive cell lines (SK-N-MC and NIH3T3) and tissue (cerebellum and atria) when analyzed by Western blot. RCP is present as monomer and dimer. b) In contrast, RCP mRNA is detected in non-responsive cells (COS) as well as CGRP-responsive sources (cerebellum and SK-N-MC cells) when analyzed by RT-PCR. c) RCP in SK-N-MC cell membranes is removed by incubation with salt, indicating that RCP is a peripheral membrane protein.

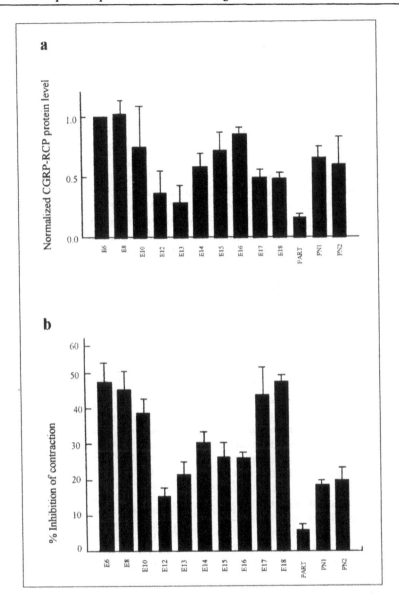

Fig. 7.6. RCP protein expression correlates with the physiological efficacy of CGRP during gestation in mouse. a) Quantitative analysis of RCP expression during gestation by Western blot (results averaged from three Western blots, three separate animals per data point). b) Quantitative analysis of CGRP inhibition on acetylcholine-induced contraction of uterine smooth muscle from pregnant and post-partum mice (three animals per data point, experiment represents results averaged from three separate experiments). Decreases in RCP protein at days E12 and partuition is correlated with diminished ability of CGRP to inhibit ACh-induced contractions, but not with RCP mRNA levels. Reproduced from Nagashpour, M., Rosenblatt, M.I., Dickerson, and G.P. Dahl. (1997) "The inhibitory effect of CGRP on myometrial contractility is diminished at partuition." Endocrinol 138:4207-4214, with permission from the Endocrine Society.

ATP to cAMP was monitored as described.[9] In Fig. 7.7b each data point represents triplicate values, and this experiment has been performed at least two times with similar results. As shown in Fig. 7.7, RCP-antisense cells had no RCP detectable by Western blot (panel a), and the CGRP-induced cAMP response was greatly reduced (panel b).

7.5. Conclusion

We have discovered a novel protein required for signal transduction at CGRP receptors in *Xenopus* oocytes. This receptor component protein (RCP) does not appear to be a ligand-binding protein, and represents one of the first examples of a receptor-associated protein required for function of a G protein-coupled receptor. Preliminary data indicate that RCP is required for CGRP-mediated signal transduction in *Xenopus* oocytes, that RCP is expressed in cells reported to contain CGRP receptors, and that RCP is likely to work in conjunction with a membrane-spanning protein to form a functional CGRP receptor. We do not yet know how RCP exerts its effects, but possible roles might include altering receptor conformation, coupling of receptor to G-proteins, or enhancing receptor trafficking to the cell surface. Two candidate receptors that might require RCP for function are the CGRP receptors RDC1 and CRLR,[35,36] neither of which consistently confers CGRP receptor activity when transfected in COS cells. However, we have not been able to reconstitute CGRP receptor activity in cotransfection experiments with RCP and either RDC1 or CRLR, implicating the requirement for a still unidentified receptor that works with RCP (Rosenblatt and Dickerson, unpublished).

Recently, an accessory protein for the CRLR receptor has been identified, named the Receptor Activity Modulator Protein (RAMP1: see Foord et al., this volume and reference 37). Cotransfection of CRLR with RAMP1 into COS cells resulted in expression of high affinity CGRP receptors. RCP and RAMP1 do not share amino acid homology, and appear to act via distinct CGRP receptors: 1) RAMP1 forms functional receptors with CRLR, while RCP does not, and 2) RCP and RAMP1 confer distinct CGRP receptor kinetics in oocytes; the RAMP1 currents have characteristic symmetrical peaks,[37] whereas RCP currents have a much faster on-rate and a characteristic biphasic peak.[10] These differences imply that RCP and RAMP1 are working with different receptors in the *Xenopus* oocyte. The CGRP receptor subfamily appears to use cofactors or receptor associated proteins (RAMP1, RAMP2, RCP) to form functional receptor complexes, and receptor diversity is increased by combinatorial use of transmembrane receptors and receptor-associated proteins.

References

1. Wackym PA. Ultrastructural organization of calcitonin gene-related peptide immunoreactive efferent axons and terminals in the vestibular periphery. Am J Otol 1993; 14:41-50.
2. Ishiyama A, Lopez I, Wackym PA. Subcellular innervation patterns of the calcitonin gene-related peptidergic efferent terminals in the chinchilla vestibular periphery. Otolaryngol Head Neck Surg 1994; 111:385-95.
3. Sliwinska-Kowalska M, Parakkal, M, Schneider et al. CGRP-like immunoreactivity in the guinea pig organ of Corti: A light and electron microscopy study. Hear Res 1989; 42:83-95.
4. Ohno K, Takeda N, Tanaka-Tsuji, M et al. Calcitonin gene-related peptide in the efferent system of the inner ear. A review. Acta Otolaryngol Suppl (Stockh) 1993; 501:16-20.
5. Safieddine S, Eybalin M. Triple immunofluorescence evidence for the coexistence of acetylcholine, enkephalins and calcitonin gene-related peptide within efferent (olivocochlear) neurons of rats and guinea-pigs. Eur J Neurosci 1992; 4:981-92,
6. Wilcox ER, Fex J. Construction of a cDNA library from microdissected guinea pig organ of Corti. Hearing Research 1992; 62:124-6.

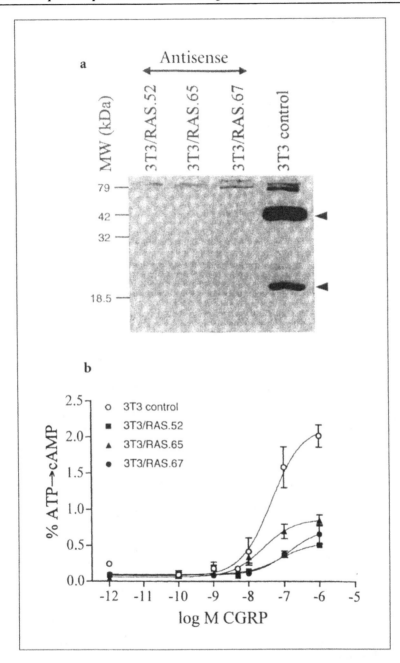

Fig. 7.7. RCP is required for CGRP-mediated signal transduction. a) Western blot of membranes (25 µg/lane) from three independant NIH3T3 RCP/antisense cell lines (3T3/RAS) and control (non-transfected) NIH3T3 cells. Monomer (20 kD) and dimer (42 kD) of RCP indicated by arrowheads. b) CGRP-induced cAMP production from RCP antisense (3T3/RAS) and control NIH3T3 cells.

7. Uezono Y, Bradley J, Min C et al. Receptors that couple to 2 classes of G proteins increase cAMP and activate CFTR expressed in Xenopus oocytes. Receptors Channels 1993; 1:233-421.

8. Birnbaum AK, Wotta DR, Law PY et al. Functional expression of adrenergic and opioid receptors in Xenopus oocytes: Interaction between alpha 2- and beta 2-adrenergic receptors. Brain Res Mol Brain Res 1995; 28:72-80.

9. Sarkar A, Dickerson IM. Cloning, characterization, and expression of a calcitonin receptor from guinea pig brain. J Neurochem 1997; 69:455-464.

10. Luebke AE, Dahl GP, Roos BA et al. Identification of a protein that confers calcitonin gene-related peptide responsiveness to oocytes by using a cystic fibrosis transmembrane conductance regulator assay. Proc Natl Acad Sci USA 1996; 93:3455-60.

11. Dennis T, Fournier A, St Pierre S et al. Structure-activity profile of calcitonin gene-related peptide in peripheral and brain tissues. Evidence for receptor multiplicity. J Pharmacol Exp Ther 1989; 251:718-725.

12. Dennis T, Fournier A, Cadieux A et al. hCGRP8-37, a calcitonin gene-related peptide antagonist revealing calcitonin gene-related peptide receptor heterogeneity in brain and periphery. J Pharmacol Exp Ther 1990; 254:123-128.

13. Naghashpour M, Rosenblatt MI, Dickerson IM et al. Inhibitory effect of calcitonin gene-related peptide on myometrial contractility is diminished at parturition. Endocrinology 1997; 138:4207-4214.

14. Bentley DL, Brown, WL, Groudine M. Accurate, TATA box-dependent polymerase III transcription from promoters of the c-myc gene in injected Xenopus oocytes. Genes Dev 1989; 3:1179-89.

15. Voellmy R, Rungger D. Transcription of a Drosophila heat shock gene is heat-induced in Xenopus oocytes. Proc Natl Acad Sci U S A 1982;79:1776-80.

16. Franco-Cereceda A, Gennari C, Nami R et al. Cardiovascular effects of calcitonin gene-related peptides I and II in man. Circ Res 1987;60:393-7.

17. Crook RB, Yabu JM. Calcitonin gene-related peptide stimulates intracellular cAMP via a protein kinase C-controlled mechanism in human ocular ciliary epithelial cells. Biochem Biophys Res Commun 1992; 188:662-70.

18. Malminiemi OI, Malminiemi KH. [125I] calcitonin gene-related peptide binding in membranes of the ciliary body-iris block. Curr Eye Res 1992; 11:1079-85.

19. Vetter, DE, Adams JC., Mugnaini E. Chemically distinct rat olivocochlear neurons. Synapse 1991; 7:21-43.

20. Aiyar N, Nambi P, Griffin E et al. Identification and characterization of calcitonin gene-related peptide receptors in porcine renal medullary membranes. Endocrinology 1991; 129:965-9.

21. Sakuta H, Inaba K, Muramatsu S. Calcitonin gene-related peptide enhances cytokine-induced IL-6 production by fibroblasts. Cell Immunol 1995; 165:20-5.

22. Van Valen F, Piechot G, Jurgens H. Calcitonin gene-related peptide (CGRP) receptors are linked to cyclic adenosine monophosphate production in SK-N-MC human neuroblastoma cells. Neurosci Lett 1990; 119:195-8.

23. Wimalawansa SJ, MacIntyre I. Calcitonin gene-related peptide and its specific binding sites in the cardiovascular system of rat. Int J Cardiol 1988; 20:29-37.

24. Chatterjee TK, Moy JA, Cai JJ et al. Solubilization and characterization of a guanine nucleotide-sensitive form of the calcitonin gene-related peptide receptor. Mol Pharmacol 1993; 43:167-75.

25. Chatterjee TK, Fisher RA. Multiple affinity forms of the calcitonin gene-related peptide receptor in rat cerebellum. Mol Pharmacol 1991; 39:798-804.

26. Samuelson UE, Dalsgaard CJ, Lundberg JM et al. Calcitonin gene-related peptide inhibits spontaneous contractions in human uterus and fallopian tube. Neurosci Lett 1985; 62:225-30.

27. Shew RL, Papka RE, McNeill DL. Calcitonin gene-related peptide in the rat uterus: Presence in nerves and effects on uterine contraction. Peptides 1990; 11:583-589.

28. Bloomquist BT, Eipper BA, Mains RE. Prohormone-converting enzymes: Regulation and evaluation of function using antisense RNA. Mol Endocrinol 1991; 5:2014-24.

29. Khokha R, Waterhouse P, Yagel S et al. Antisense RNA-induced reduction in

murine TIMP levels confers oncogenicity on Swiss 3T3 cells. Science 1989; 243:947-50.

30. Johanning K, Mathis JP, Lindberg I. Role of PC2 in proenkephalin processing: Antisense and overexpression studies. J Neurochem 1996; 66:898-907.

31. Green, PJ, Pines O, Inouye M. The role of antisense RNA in gene regulation. Annu Rev Biochem 1986; 55:569-97.

32. Eguchi Y, Itoh T, Tomizawa J. Antisense RNA. Annu Rev Biochem 1991; 60:631-52.

33. Aiyar N, Mattern MR, Hofmann GA et al. Down-regulation of CGRP-mediated cAMP accumulation in ras-transformed 3T3 fibroblasts. Ann N Y Acad Sci 1992; 657:449-451.

34. Rosenblatt MI, Dickerson IM. Endoproteolysis at tetrabasic amino acid sites in procalcitonin gene-related peptide by pituitary cell lines. Peptides 1997; 18:567-576.

35. Kapas S, Clark AJ. Identification of an orphan receptor gene as a type 1 calcitonin gene- related peptide receptor. Biochem Biophys Res Commun 1995; 217:832-838.

36. Aiyar N, Rand K, Elshourbagy NA et al. A cDNA encoding the calcitonin gene-related peptide type 1 receptor. J Biol Chem 1996; 271:11325-11329.

37. McLatchie LM, Fraser NJ, Main MJ et al. RAMPs regulate the transport and ligand specificity of the calcitonin- receptor-like receptor. Nature 1998; 393:333-339

CHAPTER 8

What Makes an Amylin Receptor?

Patrick M. Sexton, Katie J. Perry, George Christopoulos, Maria Morfis and Nanda Tilakaratne

8.1. Introduction

Amylin, also known as diabetes associated protein or islet amyloid polypeptide (IAPP), is a 37 amino acid peptide that is co-secreted with insulin from pancreatic beta cells following nutrient ingestion.[1] Amylin acts to inhibit insulin secretion from the pancreas[2] but its most characterized actions are in skeletal muscle where amylin acts to both decrease insulin-stimulated incorporation of glucose into glycogen and to promote glycogen breakdown.[3,4] A decrease in glycogen synthase activity and an increase in glycogen phosphorylase activity, at least in part, mediate these actions. Thus, amylin is thought to act as a partner to insulin in metabolic regulation. Evidence is also emerging for a prominent role for amylin outside of skeletal muscle including the gastrointestinal tract, where it is a potent inhibitor of gastric emptying,[5] and the kidney where it has diverse actions including proliferative effects on tubule epithelium,[6] modulation of Ca^{++} excretion and thiazide receptor levels,[7] as well as increasing renin activity.[6]

Amylin receptors are also widely expressed in the central nervous system where administered peptide engenders multiple potent actions.[8] These include decreased appetite and gastric acid secretion, adipsia, hyperthermia and reduction of growth hormone releasing hormone induced release of growth hormone.[8] Central injection of amylin may also affect memory and the extrapyramidal motor system.[8,9]

8.2. The Amylin, Calcitonin, CGRP, Adrenomedullin Family of Peptides

Alignment of amino acid sequences of amylin peptides from different species reveals a high level of conservation with the exception of residues 24-29, which is amyloidogenic in humans.[10-12] Amylin presents with 43-46% amino acid identity with the CGRPs and ~30% (with a gapped alignment) identity with teleost/avian calcitonins (CTs). However, only ~18% identity is displayed with rat/human CT (Fig. 8.1). Adrenomedullin shares ~22% identity with amylin across its C-terminal 38 amino acids. Notwithstanding differences in primary sequence, the peptides also share additional secondary structural features which include a cysteine disulfide-bridged loop of 6 or 7 amino acids at the amino terminus, a predicted region of amphipathic α-helical secondary structure from residues 8-18 (8-22 for salmon CT) and a C-terminally amidated aromatic amino acid at the carboxy terminal end of the peptide.

In concord with the conservation of structure among the peptides, amylin can induce CGRP-like actions in vasculature and CT-like effects in osteoclasts.[13-15] However, for these actions amylin is much weaker than either CGRP or CT suggesting that

The CGRP Family: Calcitonin Gene-Related Peptide (CGRP), Amylin, and Adrenomedullin, edited by David Poyner. ©2000 EUREKAH.COM.

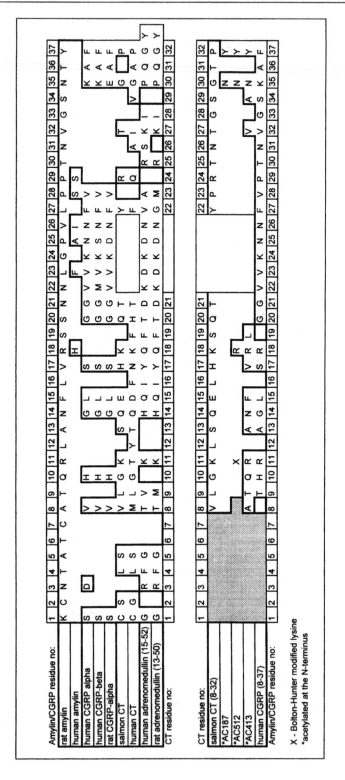

Fig. 8.1. Alignment of amylin amino acids with those of calcitonin, CGRP and adrenomedullin peptides. Agonist peptides are depicted in the upper panel. Antagonist peptides are depicted in the lower panel.

amylin acts through classical CGRP or CT receptors.

8.3. Amylin Receptors

8.3.1. Introduction

Specific, high affinity amylin binding sites were first identified in the rat central nervous system by Beaumont and colleagues[16] and these correspond to the C3 binding site originally described by our group as novel CT-sensitive CGRP binding sites.[17] Analysis of the specificity of agonist peptide interaction with the C3-type amylin receptor, revealed high affinity for rat amylin and also for salmon CT (sCT), moderate to high affinity for the CGRPs but low affinity for rat/human CT, and this is similar to that described for amylin-mediated physiological responses in skeletal muscle.[18,19] Thus, the C3-amylin receptor is likely to mediate the metabolic actions of amylin.

Removal of the N-terminal disulfide-bridged loop region of salmon CT or CGRP produces peptides which act as antagonists at CT or CGRP receptors, respectively.[20,21] Although amylin8-37 is also an antagonist of amylin effects in muscle, it is only very weak in this action, being less potent than CGRP8-37.[22,23] Amylin8-37 is also a weak competitor of amylin binding to the accumbens nucleus (Sexton PM, unpublished data) consistent with its lack of effectiveness in antagonizing amylin's actions. Both sCT8-32 and CGRP8-37 are effective competitors of amylin binding to accumbens nucleus, with sCT8-32 exhibiting ~10-fold higher affinity.[18,24] Analogues of sCT8-32 substituted with C-terminal amylin residues (eg. AC187, AC512, AC413; Figure 1), display improved affinity for C3-amylin binding sites over the unmodified peptide.[18] These peptides, and in particular sCT8-32, have very little interaction with CGRP receptors and thus have been effectively used to discriminate between C3-amylin receptor-mediated and CGRP-receptor mediated actions of amylin.[18] CGRP8-37, while a weak antagonist of amylin receptors in the

rat, is a relatively selective antagonist of type-1 CGRP receptors.

The greater relative efficacy of sCT (and its truncated peptides), compared with CGRP at C3-amylin binding sites is indicative of an important role for both peptide secondary structure and, in particular, the C-terminal region of amylin and sCT in binding site recognition. Despite only ~30% amino acid identity (in gap alignment) between amylin and sCT, 6/10 residues in the C-terminus are common between the two peptides, while amylin and CGRP share 5/10 in this region. Adrenomedullin is only a very weak competitor for amylin binding to C3-amylin sites,[25,26] consistent with its lack of C-terminal homology with amylin and sCT.

8.3.2. Comparative Receptor Pharmacology

High affinity amylin binding sites also occur in rat kidney and these sites display comparable ligand specificity to brain receptors,[6] with sCT, amylin and rat αCGRP all demonstrating high affinity. Similarly, sCT8-32 and AC413 also displayed moderate to high affinity while CGRP8-37 was a relatively weak competitor of amylin binding.

Relatively little information is available on the profile of amylin receptor specificity outside the rat, however, a limited number of studies are coming to light which indicate that phenotype of the amylin receptor may vary either between species or due to alterations in cellular environment. In 1995, Hannah and colleagues[2] identified a C3-like amylin receptor in the mouse α-TSH thyrotroph cell line. Further, analysis of this receptor revealed moderate to high affinity binding for sCT, amylin and CGRP, with only poor competition of either ^{125}I-amylin or ^{125}I-CGRP binding by rat CT in both whole cells and cell membrane preparations.[28] However, the relative specificity of amylin was not identical between the two preparations with amylin being ~10-fold less potent in competition for [^{125}I]CGRP binding in whole cells versus membranes.

This difference in relative specificity appeared to be due to the effect of G-protein coupling as the higher affinity amylin binding seen in membranes could be reduced to the level in whole cells upon treatment with GTPγs. Interestingly only the profile of amylin binding appeared to be affected in these cells, with the affinity of salmon CT and CGRP being essentially unchanged. This may reflect the greater sensitivity of amylin to the effects of G-protein coupling, with [^{125}I]amylin binding reduced by 52% by GTPγs, while [^{125}I]CGRP binding was reduced by 39% and [^{125}I]sCT binding was unchanged.

We have recently analyzed the distribution and specificity of [^{125}I]amylin binding to kidney[29] and brain of monkey (Paxinos G, Chai SY, Chrisopoulos G et al, submitted. The pattern of receptor distribution in both tissues paralleled strongly the distribution of C3-amylin receptors identified in the rat, indicating that binding occurred to the primate homologue of the rat amylin receptor. Nonetheless, significant differences in the specificity of ligand recognition occurred between the monkey and rat receptors. In particular the mammalian CTs demonstrated increased efficacy, with human CT being at least as potent as rat amylin.

While not specifically designed to test for multiple affinity states, analysis of amylin competition binding in the monkey kidney sections was suggestive of both a high and a low affinity state, with the lower affinity state being more abundant. This may indicate a relatively low level of G-protein dependent high affinity sites in the monkey tissue sections. If this is the case then the apparent efficacy of agonist peptides in competition for ^{125}I-amylin binding may be dependent on their relative sensitivity to G-protein coupling.

Analysis of human amylin receptors has been problematic, due to lack of cell lines expressing specific amylin binding or difficulty in obtaining human tissues and problems with postmortem delay on viability of binding sites (PM Sexton, unpublished

data). However, high affinity amylin binding sites have recently been identified in the human breast carcinoma cell MCF-7.[30,31] Characterization of these sites in MCF-7 cell membranes revealed a comparable profile of interaction to C3-amylin binding sites in rat with high affinity for rat amylin, moderate to high affinity for CGRP but low affinity for rat and human CT.[30] Similarly, the profile of antagonist binding was also consistent with the rat C3-amylin receptor with sCT8-32, AC512 and AC413 all exhibiting high affinity in competition for [^{125}I]amylin binding. The greatest discrepancy between specificity profiles for the MCF-7 cell membranes and rat receptors was the relatively low affinity of MCF-7 amylin binding sites for CGRP8-37. In this, the MCF-7 receptors resembled the amylin binding sites in monkey tissues which also had relatively poor interaction with CGRP8-37 (Paxinos et al, submitted; see also ref. 29).

8.3.3. *Other Receptors*

While the majority of amylin receptors, whether identified by radioligand binding or functional studies, exhibit a phenotype consistent with a C3-like amylin receptor, amylin receptors with alternative pharmacologies have also been described. Thus, in rat acinar AR42J cells amylin induced production of inositol trisphosphate and mobilization of intracellular calcium to stimulate secretion of amylase and cholesterol esterase, but this action was not mimicked by either CGRP or sCT.[32] Similarly, the action of amylin to induce bone mineralization and proliferation of osteoblasts in mice appears to be via a receptor independent of the C3-amylin receptor.[33]

8.4. Second Messengers

Inhibition of G-protein coupling by co-incubation with the non-hydrolyzable GTP analogue, GTPγS, reduces specific ^{125}I-amylin binding to renal cortex by ~90%, indicating that the amylin receptor is G-protein coupled.[6] Similarly, GTPγS treatment reduced binding of [^{125}I]amylin and

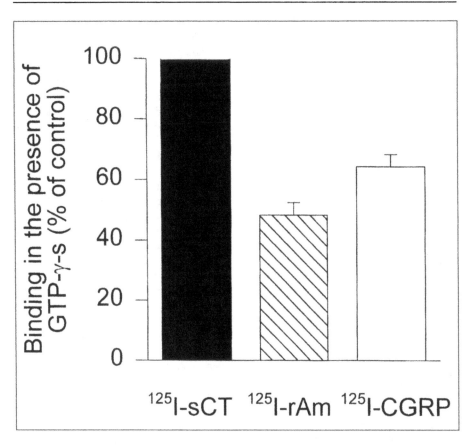

Fig. 8.2. Effect of GTPγs (20 μM) on specific binding of ^{125}I-sCT, ^{125}I-rat amylin and ^{125}I-rat CGRP to α-TSH cell membranes. Adapted from Perry KJ, Quiza M, Myers DE et al. Characterization of amylin and calcitonin receptor binding in the mouse α-thyroid-stimulating hormone thyrotroph cell line. Endocrinology 1997: 138:3486-3496.

[125]I]CGRP to C3-like receptors in mouse α-TSH cell membranes (Fig. 8.2).[27,28]

Relatively little information is available on the second messenger(s) utilized by amylin in mediating its actions. In rat soleus muscle, amylin stimulates production of cAMP. sCT is equipotent with amylin for this action, with the relative efficacy of CGRP and rat CT consistent with a C3-receptor phenotype.[19] There was a direct correlation between the specificity and efficacy of peptides in stimulating cAMP and the potency of peptides in inhibiting [14C]glycogen incorporation indicating that cAMP is likely to underlie the metabolic action of amylin in this tissue.[19] This contrasts with earlier data of Kreuttner et al[34]

who failed to detect changes in cAMP levels following amylin treatment. The basis of this discrepancy is unclear.

In rat osteoclasts, the decrease in motility (or quiescence) mediated by amylin and related peptides, also appears to be dependent upon stimulation of adenylate cyclase.[35,36] In contrast, amylin's inhibition of GIP-stimulated insulin release in the pancreas is prevented by pretreatment with pertusis toxin.[37] Similarly, amylin decreases isopreterenol- or forskolin-stimulated insulin release from Rin m5F rat insulinoma cells with a concomitant decrease in cAMP levels.[38] This action was also blocked by pertusis toxin pretreatment indicating that the action of amylin to inhibit insulin secretion is mediated by pertusis toxin-sensitive G_i proteins. In the mouse α-TSH thyrotroph cell line, amylin failed to mobilize intracellular calcium and only poorly activated adenylate cyclase, suggesting that the C3-amylin receptor, in this cell line, does not couple to significantly to either G_s or G_q.[28] Nonetheless, binding of [^{125}I]amylin to α-TSH membranes was highly sensitive to co-incubation with GTPγS, indicating that the amylin receptor is likely to couple to alternative G-proteins such as G_i.

Coupling of receptors to multiple G-proteins is a common feature of class II (family B) G-protein coupled receptors such as the CT receptor,[39] and is consistent with the C3-amylin receptor being a member of this receptor family. Nonetheless, the molecular identity of this has been remains unclear. The high potency of the lower vertebrate CTs, along with the strong colocalization of amylin receptors with CT receptors and the difficulty in isolating an independent cDNA for this receptor have led to speculation that the amylin receptor phenotype may arise from either post-translational modification to or alternate coupling of associating proteins (such as G-proteins) to the CT receptor.[8] Evidence in support of this hypothesis is presented below.

8.5. Relationship Between Amylin and Calcitonin Receptors

Calcitonin receptors have been cloned from a number of species including pig,[40] rat[41,42] and human[43,44] and demonstrate ~70% amino acid identity across species. Classically CT receptors have highest affinity for CTs, with relatively weak interaction with the related peptides CGRP and amylin. We have termed CT receptors C1[17,45] and this forms the basis of the C1a, C1b nomenclature used for the rat CT receptor isoforms.[41,42]

8.5.1. Porcine CT Receptor

Unlike cloned receptors from rat or human, porcine CT (pCT) receptors transfected into either HEK-293 or COS cells demonstrated moderate to high affinity binding for amylin, being equipotent with the presumed endogenous ligand for this receptor, pCT[46] and more potent than rat/human CT. In cAMP accumulation studies, amylin, sCT and pCT were essentially equipotent, being about two orders of magnitude more potent than human CT. Intriguingly, transfection of the pCT receptor into CHO-K1 cells yielded a specificity profile in cAMP accumulation studies that more closely resembled that of the rat C1a CT receptor,[47] indicating that pCT receptor phenotype (in particular its interaction with amylin) can vary depending upon the cellular background in which it is expressed.

8.5.2. Human CT Receptor

Recently, Chen and colleagues[30] demonstrated low levels of high affinity [^{125}I]amylin binding to cloned human CT receptors in either transiently transfected COS-7 cells (~1.5% of sCT binding) or stably transfected HEK-293 cells (~8% of sCT binding). Characterization of these sites revealed both similarities and differences with amylin binding sites characterized in either human MCF-7 cells or rat accumbens nucleus membranes. Consistent with the MCF-7 cells, the receptors retained high affinity for amylin, but also for CGRP and the amylin/CT chimeric antagonist peptide AC413, and in this specificity differed from sites labeled with [^{125}I]human CT. However, unlike C3-like amylin receptors in the accumbens nucleus and MCF-7 cells, human CT also maintained a high affinity for the amylin binding sites, being at least equipotent with amylin in competing for [^{125}I]amylin binding. While superficially this distinguishes the high affinity amylin binding sites in human CT receptor transfected cells from C3-like receptors in the MCF-7 cells, this profile of peptide competition is similar to that observed in sections of monkey brain (Paxinos et al., submitted).

As described above, although the monkey brain amylin receptors exhibited higher affinity for CT peptides than their rat counterparts, these receptors had an equivalent distribution, indicating that they are the species homologue of the rat brain C3-amylin receptor. It is possible, as speculated by Chen and colleagues[30] that the level and type of G-protein present modulate the observed amylin receptor phenotype. Curiously, no specific, high affinity amylin binding was observed when human CT receptors were transfected into *Ti ni* insect cells,[30] despite similar levels of $[^{125}I]sCT$ and $[^{125}I]$human CT binding to that seen in COS-7 cells. This may imply the absence, in insect cells, of additional proteins required for expression of an amylin receptor phenotype such as receptor activity modifying proteins (see below) or particular G-proteins.

8.5.3. α-TSH Cells

As described above, mouse α-TSH cells express a C3-amylin receptor phenotype. Analysis of the receptor binding for $[^{125}I]$ amylin and $[^{125}I]sCT$, in cross-linking and deglycosylation studies, revealed different apparent molecular weights for the major amylin and sCT binding proteins.[28] Thus, $[^{125}I]$amylin bound a protein of ~80kDa while the major sCT binding protein was ~70kDa. This difference in size was due to increased glycosylation of the amylin receptor and appeared to be associated with a site on the receptor accessible to aqueous deglycosylation. Full deglycosylation of the proteins under denaturing conditions indicated that the native binding proteins for both ligands were of similar molecular weight (~50kDa) and this was consistent with the cDNA-predicted protein size of unmodified CT receptor.[28] However, despite this difference in receptor biochemistry both proteins were immunoprecipitated by anti-rodent CT receptor antibodies, although anti-human CT receptor antibodies recognized neither receptor (Fig. 8.3). These antibodies were raised against the hypervariable C-terminus of the receptor and indicated

that, if not the same receptor, the α-TSH amylin and CT binding proteins are highly homologous. Indeed, more homologous than the human–rodent species homologues of the CT receptor.[28]

8.6. Receptor Associating Proteins

Recently, a family of novel proteins termed receptor activity modifying proteins (RAMPs), were identified (Foord et al, Ch. 6,this volume). The first of these proteins, RAMP1, was cloned during an attempt to expression clone the receptor for CGRP,[48] and coexpression of RAMP1 was subsequently demonstrated to engender a CGRP receptor phenotype from the calcitonin receptor-like receptor (CRLR). For the CGRP receptor phenotype, RAMP1 induced increased receptor glycosylation, but also remained associated with the receptor on the cell surface, suggesting that it may have multiple modes of action. Intriguingly, coexpression of RAMP2 with the CRLR induced an adrenomedullin receptor phenotype, indicating that a single receptor protein could demonstrate at least two distinct phenotypes through interaction with associating proteins.[48]

Cloning of the RAMPs has raised the tantalizing possibility that association of a RAMP-like protein may also contribute to the expression of an amylin receptor phenotype from the calcitonin receptor gene. We have recently demonstrated that at least two distinct amylin receptor phenotypes can be generated through RAMP association with the 'calcitonin' receptor, which differ in their interaction with CGRP. The RAMP1-CT receptor phenotype has moderate to high affinity interaction with sCT, amylin and CGRP but low affinity for human CT. The RAMP3-CT receptor phenotype, however, displays markedly reduced affinity for CGRP, while the affinity for sCT, amylin and human CT are similar to that seen with the RAMP1-CT receptor phenotype.[49]

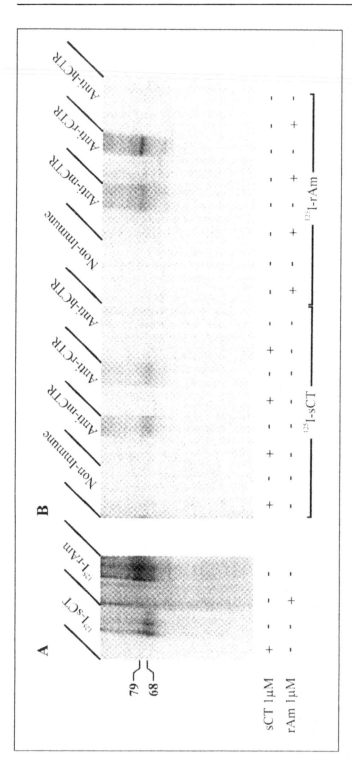

Fig. 8.3. Immunoprecipitation of amylin and CT binding proteins from α-TSH cells with anti-CT receptor antibodies. A.) Autoradiography of ^{125}I-sCT and ^{125}I-rat amylin (^{125}I-rAm) cross-linked to α-TSH membranes. Samples were incubated in the absence (total binding) or presence (nonspecific binding) of 1 μM unlabeled homologous peptide. B.) Autoradiography of cross-linked membranes (from A) immunoprecipitated with antibodies raised against the C-terminus of the mouse CT receptor (anti-mCTR), the rat CT receptor (anti-rCTR), the human CT receptor (anti-hCTR) or nonimmune sera. The relative molecular mass of the cross-linked proteins is indicated on the left of panel A (x1000). The data is shown from an individual experiment (n ≥ 3). Reproduced with permission from Perry KJ, Quiza M, Myers DE, Christopoulos, G and Sexton, PM, Characterization of amylin and calcitonin receptor binding in the mouse α-thyroid-stimulating hormone thyrotroph cell line. Endocrinology 1997; 138:3486-3496.

8.7. Conclusion

The sensitivity of amylin binding to guanine nucleotides such as GTPγs indicate that the C3-amylin receptor is a member of the superfamily of G-protein coupled receptors, while second messenger studies indiCTcate signaling via multiple G-proteins. Accumulating evidence indicates that the amylin receptor derives from the CT receptor gene product, through association with receptor activity modifying proteins. However, the observed receptor phenotype is likely also to be influenced by other proteins including the level and type of G-protein present.

References

1. Cooper GJS. Amylin compared with calcitonin gene-related peptide: Structure, biology, and relevance to metabolic disease. Endocr Rev 1994; 15:162-201.
2. Silvestre RA, Salas M, Rodriguez-Gallardo J et al. Effect of (8-32) salmon calcitonin, an amylin antagonist, on insulin, glucagon and somatostatin release: Study in the perfused pancreas of the rat. Br J Pharmacol 1996; 117:347-350.
3. Pittner RA, Albrandt K, Beaumont K et al. Molecular physiology of amylin. J Cell Biochem 1994; 55S:19-28.
4. Young AA, Mott DM, Stone K et al. Amylin activates glycogen phosphorylase in the isolated soleus muscle of the rat. FEBS Lett 1991; 281:149-151.
5. Young A. Role of amylin in nutrient intake–animal studies. Diabet Med 1997; 14:S14-S18.
6. Wookey PJ, Tikellis C, Du H-C et al. Amylin binding in rat renal cortex, stimulation of adenyl cyclase, and activation of plasma renin. Am J Physiol 1996; 270:F289-F294.
7. Blakely P, Vaughn DA, Fanestil DD. Amylin, calcitonin gene-related peptide, and adrenomedullin: Effects on thiazide receptor and calcium. Am J Physiol 1997; 272:F410-F415.
8. Sexton PM, Perry KJ. Amylin receptors in the central nervous system. Recent Res Devel in Neurochem 1996; 1:157-166

9. Clementi G, Valerio C, Emmi I et al. Behavioral effects of amylin injected intracerebroventricularly in the rat. Peptides 1996; 17:589-591.
10. Nishi M, Chan SJ, Nagamatsu S et al. Conservation of the sequence of islet amyloid polypeptide in five mammals is consistent with its putative role as an islet hormone. Proc Natl Acad Sci USA 1989; 86:5738-5742.
11. Nishi M, Bell GI, Steiner DF. Sequence of a cDNA encoding Syrian hamster islet amyloid polypeptide precursor. Nucleic Acids Res 1990; 18:6726.
12. Nishi M, Steiner DF. Cloning of complementary DNAs encoding islet amyloid polypeptide, insulin, and glucagon precursors from a New World rodent, the degu, Octodon degus. Mol Endocrinol 1990; 4:1192-1198.
13. Young AA, Rink TJ, Wang MW. Dose-response characteristics for the hyperglycemic, hyperlactemic, hypotensive and hypocalcemic actions of amylin and calcitonin gene-related peptide-I (CGRPα) in the fasted anaesthetized rat. Life Sci 1993; 52:1717-1726.
14. Brain SD, Wimalawansa SJ, MacIntyre I et al. The demonstration of vasodilator activity of pancreatic amylin amide in the rabbit. Am J Pathol 1990; 136:487-490.
15. Alam ASMT, Moonga BS, Bevis PJR et al. Amylin inhibits bone resorption by a direct effect on the motility of rat osteoclasts. Exp Physiol 1993; 78:183-196.
16. Beaumont K, Kenney MA, Young AA et al. High affinity amylin binding sites in rat brain. Mol Pharmacol 1993; 44:493-497.
17. Sexton PM, McKenzie JS, Mendelsohn FAO. Evidence for a new subclass of calcitonin/calcitonin gene-related peptide binding site in rat brain. Neurochem Int 1988; 12:232-335.
18. Beaumont K, Moore CX, Pittner RA et al. Differential antagonism of amylin's metabolic and vascular actions with amylin receptor antagonists. Can J Physiol Pharmacol 1995; 1025-1029.
19. Pittner RA, Wolfe-Lopez D, Young AA et al. Different pharmacological characteristics in L₆ and C₂C₁₂ muscle cells and intact rat skeletal muscle for amylin, CGRP and calcitonin. Br J Pharmacol 1996; 117:847-852.

20. Feyen JH, Cardinaux F, Gamse R et al. N-terminal truncation of salmon calcitonin leads to calcitonin antagonists. Structure activity relationship of N-terminally truncated salmon calcitonin fragments in vitro and in vivo. Biochem Biophys Res Commun 1992; 187:8-13.

21. Chiba T, Yamaguchi A, Yamatani T et al. Calcitonin gene-related peptide receptor antagonist human CGRP-(8-37). Am J Physiol. 1989; 256:E331-E335.

22. Deems RO, Cardinaux F, Deacon RW et al. Amylin or CGRP (8-37) fragments reverse amylin-induced inhibition of 14C-glycogen accumulation. Biochem Biophys Res Commun 1991; 181:116-120.

23. Wang MW, Young AA, Rink TJ et al. 8-37h-CGRP antagonizes actions of amylin on carbohydrate metabolism in vitro and in vivo. FEBS Lett. 1991; 291:195-198.

24. Sexton PM, Paxinos G, Kenney MA et al. In vitro autoradiographic localization of amylin binding sites in rat brain. Neuroscience 1994; 62:553-567.

25. Van Rossum D, Menard DP, Chang JK et al. Comparative affinities of human adrenomedullin for [125I]-labelled human alpha calcitonin gene related peptide ([125I]hCGRP alpha) and [125I]-labelled Bolton-Hunter rat amylin ([125I]BHrAMY) specific binding sites in the rat brain. Can J Physiol Pharmacol 1995; 73:1084-1088.

26. Vine W, Beaumont K, Gedulin B et al. Comparison of the in vitro and in vivo pharmacology of adrenomedullin, calcitonin gene-related peptide and amylin in rats. Eur J Pharmacol 1996; 314:115-121.

27. Hanna FWF, Smith DM, Johnstone CF et al. Expression of a novel receptor for the calcitonin peptide family and a salmon calcitonin-like peptide in the α-thyrotropin thyrotroph cell line. Endocrinology 1995; 136:2377-2382.

28. Perry KJ, Quiza M, Myers DE et al. Characterization of amylin and calcitonin receptor binding in the mouse α-thyroid-stimulating hormone thyrotroph cell line. Endocrinology 1997; 138:3486-3496.

29. Chai SY, Christopoulos C, Cooper ME et al. Characterization of binding sites for amylin, calcitonin, and CGRP in primate kidney. Am J Physiol 1998; 273:F51-F62.

30. Chen W-J, Armour S, Way J et al. Expression cloning and receptor pharmacology of human calcitonin receptors from MCF-7 cells and their relationship to amylin receptors. Mol Pharmacol 1997; 52:1164-1165.

31. Zimmermann U, Fluehmann B, Born W et al. Coexistence of novel amylin-binding sites with calcitonin receptors in human breast carcinoma MCF-7 cells. J Endocrinol 1997; 155:423-431.

32. Huang Y, Fischer JE, Balasubramaniam A. Amylin mobilizes [Ca2+]i and stimulates the release of pancreatic digestive enzymes from rat acinar AR42J cells: Evidence for an exclusive receptor system of amylin. Peptides 1996; 17:497-502.

33. Cornish J, Callon KE, Coope GJ et al. Amylin stimulates osteoblast proliferation and increases mineralized bone volume in adult mice. Biochem Biophys Res Commun 1995; 207:133-139.

34. Kreuttner DK, Orena SJ, Torchia AJ et al. Amylin and CGRP induce insulin resistance via a receptor distinct from cAMP coupled CGRP receptor. Am J Physiol 1993; 264:E606-E613.

35. Alam AS, Moonga BS, Bevis PJ et al. Selective antagonism of calcitonin-induced osteoclastic quiescence (Q effect) by human calcitonin gene-related peptide-(Val8Phe37). Biochem Biophys Res Commun 1991; 179:134-139

36. Alam AS, Bax CM, Shankar VS et al. Further studies on the mode of action of calcitonin on isolated rat osteoclasts: pharmacological evidence for a second site mediating intracellular Ca2+ mobilization and cell retraction. J Endocrinol 1993; 136:7-15.

37. Silvestre RA, Salas M, Garcia-Hermida O et al. Amylin (islet amyloid polypeptide) inhibition of insulin release in the perfused rat pancreas: implication of the adenylate cyclase/cAMP system. Regul Peptides 1994; 50:193-199.

38. Suzuki S, Murakami M, Abe S et al. The effects of amylin on insulin secretion from Rin m5F cells and glycogen synthesis and lipogenesis in rat primary cultured hepatocytes. Diabetes Res Clin Pract 1992; 15:77-84.

39. Peroutka SJ. Handbook of receptors and channels. 1994; Vol. 1, CRC Press, Florida.

40. Lin HY, Harris TL, Flannery MS et al. Expression cloning of an adenylate cyclase-coupled calcitonin receptor. Science 1991; 254:1022-1024

41. Sexton PM, Houssami S, Hilton JM et al. Identification of brain isoforms of the rat calcitonin receptor. Mol Endocrinol 1993; 7:815-821.

42. Albrandt K, Mull E, Brady EMG et al. Molecular cloning of two receptors from rat brain with high affinity for salmon calcitonin. FEBS Lett 1993; 325:225-230.

43. Gorn AH, Lin HY, Yamin M et al. Cloning, characterization, and expression of a human calcitonin receptor from an ovarian carcinoma cell line. J Clin Invest 1992; 90:1726-1735.

44. Kuestner RE, Elrod R, Grant FJ et al. Cloning and characterization of an abundant subtype of the human calcitonin receptor. Mol Pharmacol 1994; 46:246-255

45. Sexton PM. Central nervous system binding sites for calcitonin and calcitonin gene-related peptide. Mol Neurobiol 1991; 5:251-273.

46. Sexton PM, Houssami S, Brady CL et al. Amylin is an agonist of the renal porcine calcitonin receptor. Endocrinology 1994; 134:2103-2107.

47. Christmanson L, Westermark P, Betsholtz C. Islet amyloid polypeptide stimulates cyclic AMP accumulation via the porcine calcitonin receptor. Biochem Biophys Res Commun 1994; 205:1226-1235.

48. McLatchie LM, Fraser NJ, Main MJ et al. RAMPs: A family of proteins that regulate the transport and ligand specificity of the Calcitonin Receptor Like Receptor. Nature 1998; 393:333-339.

49. Christopolous F, Perry KJ, Morfis M et al. Multiple amylin receptor phenotypes arise from RAMP interaction witht he calcitonin receptor gene product. Mol Pharmacol 1999; 56:235-242.

CHAPTER 9

Neuroendocrine Actions of Amylin

Andrew Young, Candace Moore, John Herich and Kevin Beaumont

9.1. Introduction

9.1.1. The Amylin Molecule

Amylin is a 37-amino acid peptide that is localized with, and secreted with, insulin from the pancreatic beta cell. In humans, amylin circulates at basal plasma levels of 4 to 8 pM, rising to levels of 20 to 25 pM post-prandially.[1,2] Amylin-like immunoreactivity is also present in the amygdala and other brain regions.[3] Amylin shows some structural similarity to calcitonin gene-related peptide (CGRP),[4] the calcitonins (especially teleost calcitonins)[5] and adrenomedullin.

9.1.2. Survey of Amylin Actions

Nearly 60 different effects have been reported in various experiments using amylin (see Table 9.1). Some of these are likely to represent artifacts present only via non-relevant receptors at pharmacological concentrations. Advances in the understanding of the role of amylin by identifying actions which occur at physiological concentrations have been comparatively recent, and have included studies where the response to administration of selective antagonists was used to infer the effect of endogenous amylin. Other indications of physiologically-relevant effects have been obtained in dose-response studies where responses have been observed at doses that result in changes in plasma concentration that are comparable to those observed with endogenous peptide.

The first effect of amylin to be interpreted as physiological by the use of selective antagonists was the inhibition of insulin secretion.[7, 8] Administration of antagonist alone resulted in exaggerated insulin responses to nutrient stimuli, a result replicated by others in different preparations.[9] The inhibition by amylin of β-cell secretion thus appears similar to the situation in other neuroendocrine tissues where secreted products feed back to limit secretion.

9.1.3. Amylin, Insulin Resistance and Syndrome-X

Earlier hypotheses on the possible physiological role of amylin were guided by the historical order in which biological actions were discovered. A potent effect to inhibit insulin- stimulated glycogen formation in isolated skeletal muscle in rats[10,11] but not impair insulin action in fat,[12] as well an effect to increase lactate turnover,[13] blunt first-phase insulin secretion,[14] and stimulate the renin-angiotensin system,[15] mirrored many of the features of insulin-resistance and contributed to the hypotheses that amylin may be involved in the pathogenesis of insulin- resistance[16] and syndrome-X.[17] Effects in rat muscle were identified as due to cAMP-mediated activation of glycogen phosphorylase,[8] resulting in release of lactate into plasma[19, 20] and substrate-mediated gluconeogenesis.[21]

The CGRP Family: Calcitonin Gene-Related Peptide (CGRP), Amylin, and Adrenomedullin, edited by David Poyner. ©2000 EUREKAH.COM.

Table 9.1. Reported actions of amylin

Action	Reference
Inhibition of muscle glycogen synthesis	Leighton, Cooper. Nature 1998; 333:632-635.
Activation of muscle glycogenolysis	Young et al. FEBs Lett 1991; 281:149-151.
Inhibition of peripheral glucose uptake	Young et al. Amer J Physiol 1990; 259:E457-E461.
Increase in plasma glucose	Young et al. FEBS Lett 1991; 291:101-104.
Increase in plasma lactate	Young et al. FEBS Lett 1991; 291:101-104.
Stimulation of endogenous glucose production	Molina et al. Diabetes 1990; 39:260-265.
Stimulation of glucose release from muscle	Young et al. FEBS Lett 1993; 334:317-321.
Increase in liver glycogen content	Young et al. FEBS Lett 1991; 287:203-205.
Increase in Cori cycling	Young et al. J Cell Biochem1991; 15 Part B:68.
Inhibition of insulin secretion	Dégano et al. Regul Peptides 1993; 43:91-96.
Inhibition of arginine-stimulated glucagon secretion	Gedulin et al. Metabolism 1997; 46:67-70.
Stimulation of exocrine pancreatic metabolism	Iwamoto et al. Diabetes 1991; 40:67A
Inhibition of CCK-stimulated pancreatic secretion	Gedulin et al. Diabetes 1998; 47:280A
Inhibition of gastric emptying	Young et al. Diabetologia 1995; 38:642-648.
Reduction of post-prandial hyperglycemia	Kolterman et al. Clin Res 1994; 42:87A
Inhibition of gastric acid secretion	Guidobono et al. Peptides 1994; 15:699-702.
Stimulation of gastrin secretion	Funakoshi et al. Regul Pept 1992; 38:135-143.
Reduction of antral gastrin	Makhlouf et al. Digestive Disease Week 1996; A-230.
Stimulation of somatostatin in fundus	Zaki et al. Digestive Disease Week 1996; A-2.
Inhibition of ethanol-induced gastritis	Jodka et al. Diabetes 1997; 40:365A.
Inhibition of indomethacin-induced gastritis	Guidobono et al. Br J Pharmacol 1997; 120:581-586.
Gut relaxation	Mulder et al. Peptides 1997; 771-783.
Reduction of plasma calcium	MacIntyre. Lancet 1989; 2:1026-1027.
Reduction of plasma potassium	Young et al. Drug Develop Res 1996: 37:231-248.
Inhibition of osteoclasts	Alam et al. Exp Physiol 1993;78:183-196.
Stimulation of osteoblasts	Romero et al. J Bone Min Res 1993; 8:2369.
Stimulation of calciuria	Miles et al. Calcified Tissue Int 1994; 55:265-273.
Central inhibition of food intake	Chance et al. Brain Res 1991; 539:352-354.
Peripheral inhibition of food intake	Morley, Flood. Peptides 1991; 12:865-869.
Increase in water intake	Rauch et al. Pfleugers Arch-Eur J Physiol 1997; 433:619.
Centrally, increase in body temperature	Chance et al. Surg Forum 1991; 42:84-86.
Modulation of learning/memory	Flood, Morley. Peptides 1992; 13:57-580.
Peripherally, analgesic	Young. US Patent 1997; 5,677,279.
Centrally, decrease in locomotion	Bouali et al. Regul Peptides 1995; 56:167-174.
Stimulation of renin secretion	Young et al. J Hypertension 1994; 12:S152.
Stimulation of aldosterone secretion	Nuttal et al. Am J Hypertens 1995; 8:108A.
Increase in tubular sodium excretion	Harris et al. Amer J Physiol 1997; 41:F13-F21.
Increase in urine volume	Vine et al. Hormone Metab Res 1998; 30:518-522.
Increase in urinary sodium excretion	Vine et al. Hormone Metab Res 1998; 30:518-522.
Stimulation of cutaneous vasodilation	Brain et al. Am J Pathol 1990; 136:487-490.
Stimulation of pulmonary vasodilation	Dewitt et al. Eur J Pharmacol 1994; 257:303-306.
Stimulation of tracheal mucus secretion	Wagner et al. Res Exp Med 1995; 195:289-296.
Relaxation of airway smooth muscle	Bhogal et al. Peptides 1994; 15:1243-1247.
Reduction in blood pressure	Young et al. Life Sci 1993; 52:1717-1726.
Umbilical venous endothelial proliferation	Datta et al. Biochem Soc Trans 1990; 18:1276.
Stimulation of cardiocyte growth	Bell et al. J Mol Cell Cardiol 1195; 27:2433-2443.
Increases renal thiazide receptor	Blakely et al. Amer J Physiol 1997; 41:F410-F415.
Stimulation of CNS tyrosine and tryptophan transport	Balasubramaniam et al. Soc Neuroscie Abstr1991; 17:976.
Anti-inflammatory action	Clementi et al. Life Sci 1995; 57:PL193-PL197.

Table 9.1., cont. Reported actions of amylin

Action	Reference
Amplification of eosinophiol responses	Hom et al. J Leukocyte Biol 1995; 58:526-532.
Reduction of plasma fructosamine	Thompson et al. Diabetes Care 1998; 21:987-993.
Reduction in plasma HbA1c	Thompson et al. Diabetes Care 1998; 21:987-993.
Reduction in plasma lipids and LDL/HDL	Thompson et al. Diabetes Care 1998; 21:987-993.
Increase in cardiocyte contractility	Bell, McDermott. Regul Peptides 1995; 60:125-133.
Inhibition of growth hormone release	Netti et al. Neuroendocrinology 1995; 62:313-318.
Apoptosis in cultured nerve cells	May et al. J Neurochem 1993; 61:2330-2333.
Apoptosis in cultured β-cells	Lorenzo et al. Nature 1994; 368:756-760.
Protective effect in islets	Mulder et al. Didabetologia 1997; 40:135A.
Growth factor in kidney	Wookey et al. Kidney Int 1998; 53:25-30.

There was some indication that effects related to muscle glycogen metabolism and lactate release could occur at physiological amylin concentrations in the rat.[8, 20, 22] But this sequence of events appeared eventually not to be a prominent feature of amylin action in humans; amylin antagonists amplified insulin secretion but did not otherwise alter insulin action in clinical studies.[23]

While the effects of amylin to stimulate the renin-angiotensin-aldosterone system in rats and man[24,25] appeared potent enough to prevail at pathophysiological amylin concentrations,[26] chronic administration of amylin antagonist did not ameliorate essential hypertension in humans (Moyses et al, unpublished) or in fat-fed dogs (Hall et al., unpublished). Conversely, chronic administration of the human amylin analog, pramlintide, to people with diabetes did not result in changes in blood pressure (Kolterman et al, unpublished). It is possible that effects on sodium retention and blood pressure predicted by amylin's stimulation of the renin-angiotensin-aldosterone system are opposed by effects to increase urine flow and urinary sodium excretion.[27]

9.2. Amylin and Regulation of Nutrient Load

9.2.1. Effects on Gastric Emptying

Amylin's role in metabolic regulation was further elucidated following identification of a potent effect of amylin to inhibit gastric emptying[28] that was more potent than that of other known endogenous peptides.[29] A similar potency has been observed in man with the human amylin analog, pramlintide.[30] A determination that the ability of amylin to inhibit gastric emptying is a physiological effect is supported by experiments showing that gastric emptying was accelerated in [amylin-deficient] BB rats[28,31] and by prior administration of the amylin antagoinst, AC187.[32]

9.2.2. Effects on Food Intake

Several studies have documented an effect of amylin to inhibit food intake in rodents.[33-35] The potency of amylin's inhibition of food intake is comparable to that of its gastric effect, and is increased in the presence of cholecystokinin,[36] a situation that normally follows the ingestion of a mixed meal. The proposal that endogenous amylin may contribute to the control of food intake is supported by an increase in food intake after administration of the selective antagonist, AC187, alone (unpublished results), and by an increase in body weight in amylin gene knockout mice.[37, 38] Decreases in body weight in human subjects chronically treated with pramlintide at doses which mimick endogenous plasma amylin levels,[39] and anecdotal reports of fullness, could be also consistent with amylin functioning as a peripheral satiety agent in man.

9.2.3. Effects on Pancreatic Digestive Enzyme Secretion

We have recently identified an effect of amylin to inhibit cholecystokinin-stimulated secretion of amylase and lipase from the exocrine pancreas in the anesthetized rat.[40] The dose-response of this effect indicates a potency similar to that of amylin's gastric actions (inhibition of gastric emptying, inhibition of acid secretion, and mucosal protection), implying that it too is a physiological action. The magnitude of exocrine pancreatic secretion (up to a 70% inhibition of digestive enzyme secretion) is one which would be expected to affect rates of nutrient digestion and absorption. Whether this effect of amylin exists in humans is presently unknown. If it does exist, it may have therapeutic utility, because modulation of digestive enzyme activity (for example, with α-glucosidase inhibitors[41,42] and lipase inhibitors[43]) is reported to be useful in diabetes and obesity.

9.2.4. Effects on Nutrient-Stimulated Glucagon Secretion

Diabetic patients, who typically exhibit an absolute or relative deficiency of amylin,[2] have been reported to show exaggerated glucagon secretion in response to protein meals.[44,45] Relative hyperglucagonemia is proposed by some to be responsible for the excessive glucose production and fasting hyperglycemia present during diabetes,[46] and appears necessary for ketoacidosis during loss of metabolic control.[47] In rats[48] and in clinical studies,[49,50] amylin or pramlintide potently inhibited arginine- or meal-stimulated secretion of glucagon. The dose-response for this effect in both species suggests that it too is a physiological response. Additional support for the interpretation that endogenous amylin normally inhibits glucagon secretion is derived from the observation that administration of neutralizing anti-amylin antibody and AC187 can increase circulating glucagon concentrations.[51]

9.2.5. Amylin and the Integrated Control of Nutrient Influx

The four potent actions of amylin just described (the inhibition of food intake, inhibition of gastric emptying, inhibition of digestive enzyme secretion, and inhibition of glucagon secretion) share a common outcome—they each restrict nutrient load, the rate of appearance (Ra) of nutrient in blood. In an integrated picture, summarized in Figure 9.1, we propose that amylin physiologically orchestrates, via several parallel processes, the rate of entry of nutrient into the circulation. In this way, amylin's function may be viewed as complemetary to that insulin (secreted from the same pancreatic β-cells), which orchestrates the exit of nutrient from blood and its storage in peripheral tissues.

The following discussion addresses the emerging picture that, although amylin is co-secreted with an endocrine hormone from endocrine tissue (the pancreatic islets), the target for its most potent and physiologically relevant effects appears to be the central nervous system. Amylin may thus be primarily regarded as a neuroendocrine hormone.

9.2.6. Amylin's Effects on Gastric Motility: Site of Action

The major brain site regulating gastric motility is the dorsal vagal complex of the brainstem, composed of the nucleus tractus solitarius (NTS), dorsal motor nucleus of the vagus (DMV), and area postrema. This region receives information from visceral afferents supplying the NTS and area postrema, and integrates this information to regulate efferent nerve activity to the stomach originating in the DMV.[52] The area postrema, which is one of the few brain

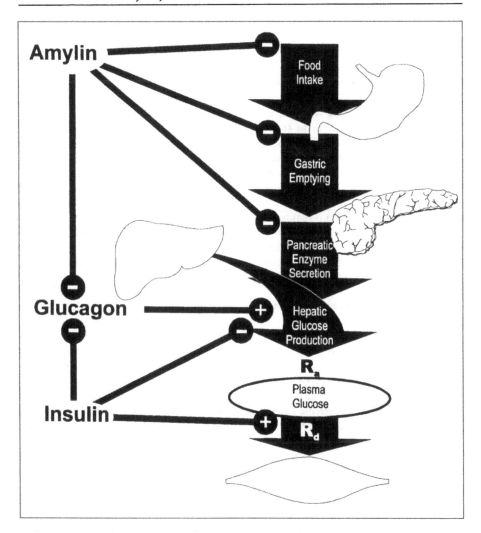

Fig. 9.1. Proposed integrated control of amylin of processes that collectively determine the rate at which nutrients enter the circulation (R_a). Insulin, cosecreted with amylin from pancreatic b-cells, is the prime controller of processes that govern the rate at which nutrients leave the circulation (R_d).

areas lacking a blood-brain barrier, samples circulating hormones and relays this information to the NTS and higher brain sites.

Recent studies of amylin's effects on gastric emptying in rats support the idea that these effects are mediated by the central nervous system. Firstly, an intact vagus nerve is necessary for amylin to slow gastric emptying.[53] In Sprague-Dawley rats that

underwent a subdiaphragmatic vagotomy, subcutaneously injected amylin had no effect on the rate of radiolabeled glucose appearance in the blood following an oral gavage. The same doses of amylin in sham-operated control animals produced a prolonged delay in glucose appearance in the blood, reflecting delayed gastric emptying.

Secondly, the potency of amylin to inhibit gastric emptying depended upon the intracranial location of injection.[54] Injection of 0.1 to 1µg of amylin into the lateral ventricle was little more effective at delaying gastric emptying than equal subcutaneous doses of amylin. However, amylin was approximately 10-fold more potent at inhibiting gastric emptying when injected through cannulae implanted in the 4th ventricle. These results are consistent with a central site of action for amylin, particularly a brainstem site located near the 4th ventricle.

Thirdly, recent studies demonstrate that an intact area postrema is required for amylin's effect on gastric emptying. In rats that underwent surgical ablation of the area postrema and then were allowed to recover, amylin was no longer effective at inhibiting gastric emptying.[55]

Thus, evidence from physiological studies in rodents indicate that:
1. amylin does not influence gastric emptying in vagally- lesioned animals,
2. injection of amylin in the 4th ventricle of the brain, directly adjacent to the area postrema and dorsal vagal complex, potently inhibits gastric emptying, and
3. ablation of the area postrema blocks the ability of amylin administered systemically to slow gastric emptying.

Furthermore, as described in the following sections, receptors that bind amylin with high affinity are concentrated in the area postrema. These results are consistent with the idea that amylin acts at receptors in the area postrema, thereby regulating gastric motility via the vagal neural input to the stomach

9.2.7. Amylin Receptors: Binding and Localization

High affinity amylin binding sites, identified by the specific binding of radioiodinated rat amylin at low picomolar concentrations, are unevenly distributed in brain.[56,57] The regions with highest binding densities in rat brain as determined in autoradiographic studies include nucleus accumbens and fundus striati, area postrema, subfornical organ, vascular organ of the lamina terminalis, and locus coeruleus.[58] The distribution of amylin receptors is similar in monkey brain, with high densities in area postrema, NTS, locus coeruleus and dorsal raphe.[59] However, relative to the rat, the density of amylin binding was lower in monkey nucleus accumbens, and highest binding densities were present in monkey hypothalamus.

The pharmacological specificity of these binding sites was initially determined for receptors in rat nucleus accumbens membranes. Amylin had a binding affinity (Kd) of approximately 30 pM for these sites, which matches well with circulating concentrations. The related peptides βCGRP and αCGRP, which share approximately 45% amino acid sequence identity with amylin, had 3- to 12- fold lower affinity than amylin in competitive binding studies. Both rat and human calcitonins, which have 15% amino acid sequence identity with amylin, had quite low affinities. However, salmon calcitonin, which has 30% amino acid sequence identity with amylin, was equipotent with amylin in competitive binding studies.[56]

These studies show that receptors with a high binding affinity for amylin, consistent with its low picomolar circulating concentrations, are present in the brain. These receptors have a characteristic binding specificity and distribution in several species. Amylin binding sites are highly localized to circumventricular organs, notably the area postrema, which have access to amylin circulating in the blood.

9.2.8. Amylin Receptors in Area Postrema

We have recently characterized the binding of radioiodinated amylin to membranes from porcine area postrema. Autoradiographic studies have shown that binding sites for [^{125}I]- Bolton-Hunter-labeled rat amylin ([^{125}I]-BH- amylin) are concentrated in the area postrema of rat brain.[58] However, due to the limited size of this region in rats, the pharmacological characteristics of these binding sites were not determined. For the studies described here, amylin binding was determined using area postrema tissue dissected from freshly collected pig brains and frozen on dry ice (ABS Inc., Maryland). Tissue was thawed, homogenized in ice-cold 20 mM HEPES, pH 7.4, and area postrema membranes were collected using three cycles of washing in fresh buffer followed by centrifugation for 15 minutes at 48,000 x g. The final membrane pellet was resuspended in 20 mM HEPES buffer containing 0.2 mM phenylmethylsulfonyl fluoride (PMSF). Radioligand binding and data analysis were done by previously described methods.[56]

Saturation binding isotherms (see Figure 9.2) demonstrated that area postrema contained receptors with high binding affinity for [^{125}I]- BH-amylin (Kd = 25 ± 4 pM), present at high density (87 ± 20 fmol/mg protein). Lower affinity binding was also observed, but was not saturable over the concentrations of radioligand used. The ability of several structurally-related peptides to compete for [^{125}I]-BH-amylin binding to porcine area postrema membranes was determined. Salmon calcitonin (sCT), and rat and human amylin, were the most potent peptides at competing for [^{125}I]-BH-amylin binding (see Table 9.2). CGRPs had lower potencies, as did pig calcitonin, while rat and human calcitonin were relatively inactive. Thus, receptors in porcine area postrema were similar to receptors in rat brain[56] and in porcine nucleus accumbens[60] in their affinity for amylin and their relative potency for structurally-related peptides.

The competitive binding potencies of N-terminally-truncated peptides and the analog AC187,[8] which are antagonists of amylin action, were also determined. Salmon calcitonin (8-32) and AC187 effectively competed for [^{125}I]-BH-amylin binding at concentrations lower than 100 pM, while CGRP(8-37) was less active (see Table 9.2). Previous studies have shown that

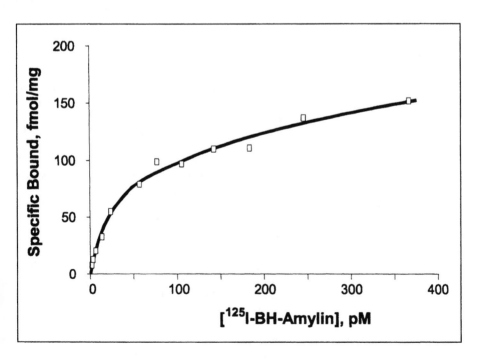

Fig. 9.2. [^{125}I]-BH-amylin saturation isotherm in porcine area postrema membranes.

Table 9.2. Binding IC50s for amylin-related peptides and antagonists

Peptide	IC_{50}, nM
Salmon calcitonin	0.022 ± 0.009
Rat amylin	0.31 ± 0.22
Human amylin	0.32 ± 0.20
Human βCGRP	3.52 ± 0.46
Pig calcitonin	37.2 ± 8.56
Human αCGRP	99.7 ± 38.8
Rat calcitonin	>1000
Human calcitonin	>1000
Salmon calcitonin (8-32)	17.1 ± 15.5
AC187	70.6 ± 52.4
Human CGRP (8-37)	~1000

mean±sd; n = 3-5 separate experiments

intravenous AC187 administration blocks the ability of rat amylin to slow gastric emptying in rats.[61]

9.2.9. Amylin Receptors: Molecular and Biochemical Characterization

Although the molecular identity of the amylin receptor has not been reported, several studies suggest a close relationship of this receptor to products of the calcitonin receptor gene. High affinity amylin binding sites have been detected in MCF-7 human breast carcinoma cells.[62,63] Expression cloning experiments using a peptide antagonist of amylin and calcitonin receptors identified two isoforms of the calcitonin receptor gene in MCF-7 cells. Expression of one of these calcitonin receptor isoforms in cell lines generated a typical calcitonin recep-

tor, as well as much smaller amounts of a high affinity amylin binding site. However, generation of the amylin binding site was dependent on the cell background in which the gene was expressed. These studies demonstrated that factors varying with the cell background may determine whether expression of the calcitonin receptor gene results in a receptor able to bind amylin.

High affinity amylin binding sites have also been studied in mouse TSH thyrotroph cells.[64,65] Like the MCF-7 cells, these cells express at least two isoforms of the calcitonin receptor gene, and contain typical calcitonin receptors in addition to high affinity amylin sites. Receptors covalently labeled with radioiodinated amylin in TSH cell membranes were immunoprecipated with antibodies to the calcitonin receptor, suggesting a shared peptide backbone in the antigenic region. However, amylin receptors differed biochemically from the major calcitonin receptor present in TSH cells in their molecular size and extent of glycosylation.

The recent discovery of a family of accessory proteins (RAMPs) required for expression of CGRP and adrenomedullin receptors raises the possibility that these proteins are involved in amylin receptor expression.[66] RAMPs (Receptor Activity Modifying Proteins) are single-trans-membrane proteins which are required to transport an otherwise inactive 7-transmembrane protein (the calcitonin-receptor-like receptor, CRLR) to the cell surface (see chapters by Foord et al, Legon and Aiyar et al, this volume). Three distinct RAMPs have been identified. Coexpression of RAMP1 with the CRLR gene generates a functional CGRP receptor, while coexpression of RAMP2 with the CRLR gene generates a receptor for adrenomedullin. Thus, a single gene product (CRLR) is converted into either CGRP or adrenomedullin receptor depending on the nature of the associated RAMP. This novel mechanism for generating receptor specificity may apply to other members of the amylin/CGRP/calcitonin/adrenomedullin peptide group. Indeed, cellular coexpression of RAMP1 with a calcitonin receptor isoform generates high affinity amylin binding sites (see Sexton et al in this monograph).

9.3. Conclusion

Results of lesioning and local injection studies in rodents suggest that amylin's effects on gastric emptying require an intact vagus, are blocked by area postrema ablation, and are potently reproduced by amylin injection in the 4th ventricle of the brain. Along with the presence of high densities of high affinity amylin receptors in the area postrema, these results are consistent with the idea that amylin slows gastric emptying by acting at the area postrema region of the brainstem's dorsal vagal complex, thereby influencing the neural output to the gastrointestinal tract.

References

1. Percy AJ, Trainor DA, Rittenhouse J et al. Development of sensitive immunoassays to detect amylin and amylin-like peptides in unextracted plasma. Clin Chem 1996; 42 (4):576-585.
2. Koda JE, Fineman M, Rink TJ et al. Amylin concentrations and glucose control. Lancet 1992; 339(8802):1179-1180.
3. Dilts RP, Phelps J, Koda J et al. Comparative distribution of amylin and calcitonin gene related peptide (CGRP). Soc Neurosci Abstr 1995; 21:1116.
4. Cooper GJS, Willis AC, Clark A et al. Purification and characterization of a peptide from amyloid-rich pancreases of type 2 diabetic patients. Proc Natl Acad Sci USA 1987; 84(23):8628-8632.
5. Young AA, Wang MW, Gedulin B et al. Diabetogenic effects of salmon calcitonin are attributable to amylin-like activity. Metabolism 1995; 44(12):1581-1589.
6. Kitamura K, Kangawa K, Kawamoto M et al. Adrenomedullin: a novel hypotensive peptide isolated from human pheochromocytoma. Biochem Biophys Res Commun 1993; 192(2):553-560.
7. Young AA, Carlo P, Rink TJ et al. 8-37hCGRP, an amylin receptor antagonist, enhances the insulin response and perturbs the glucose response to infused arginine in anesthetized rats. Mol Cell Endocrinol 1992; 84(1-2):R1-R5.
8. Young AA, Gedulin B, Gaeta LSL et al. Selective amylin antagonist suppresses rise in plasma lactate after intravenous glucose in the rat - Evidence for a metabolic role of endogenous amylin. FEBS Lett 1994; 343(3):237-241.
9. Silvestre RA, Salas M, Rodriguez-Gallardo J et al. Effect of (8-32) salmon calcitonin, an amylin antagonist, on insulin, glucagon and somatostatin release: Study in the perfused pancreas of the rat. Br J Pharmacol 1996; 117(2):347-350.
10. Leighton B, Cooper GJS. Pancreatic amylin and calcitonin gene-related peptide cause resistance to insulin in skeletal muscle in vitro. Nature 1988; 335(6191):632-635.
11. Young AA, Gedulin B, Wolfe-Lopez D et al. Amylin and insulin in rat soleus muscle: Dose responses for cosecreted noncompetitive antagonists. Amer J Physiol 1992; 263(2 Pt 1):E274-281.
12. Lupien JR, Young AA. No measurable effect of amylin on lipolysis in either white or brown isolated adipocytes from rats. Diabet Nutr Metab 1993; 6(1):13-18.
13. Young AA. Amylin and its effects on the Cori cycle. Taking Control in Diabetes. Surrey, UK: Synergy Medical Education, 1993: p. 16-20.
14. Dégano P, Silvestre RA, Salas M et al. Amylin inhibits glucose-induced insulin secretion in a dose-dependent manner—study in the perfused rat pancreas. Regul Peptides 1993; 43(1-2):91-6.
15. Young AA, Vine W, Carlo P et al. Amylin stimulation of renin activity in rats: a possible link between insulin resistance and hypertension. J Hypertension 1994; 12(Suppl 3):S152.
16. Cooper GJS, Leighton B, Dimitriadis GD et al. Amylin found in amyloid deposits in human type 2 diabetes mellitus may be a hormone that regulates glycogen metabolism in skeletal muscle. Proc Natl Acad Sci USA 1988; 85(20):7763-7766.
17. Young AA, Rink TJ, Vine W et al. Amylin and syndrome-X. Drug Develop Res 1994; 32(2):90-99.
18. Pittner R, Beaumont K, Young A et al. Dose- dependent elevation of cyclic AMP, activation of glycogen phosphorylase, and release of lactate by amylin in rat skeletal muscle. BBA-Mol Cell Res 1995; 1267(2-3):75-82.

19. Young AA, Wang M-W, Cooper GJS. Amylin injection causes elevated plasma lactate and glucose in the rat. FEBS Lett 1991; 291(1):101-104.

20. Vine W, Smith P, Lachappell R et al. Lactate production from the rat hindlimb is increased after glucose administration and is suppressed by a selective amylin antagonist: Evidence for action of endogenous amylin in skeletal muscle. Biochem Biophy Res Commun 1995; 216(2):554-559.

21. Young AA, Cooper GJS, Carlo P et al. Response to intravenous injections of amylin and glucagon in fasted, fed, and hypoglycemic rats. Amer J Physiol 1993; 264(6 Part 1):E943-E950.

22. Vine W, Percy A, Gedulin B et al. Evidence for metabolic action of endogenous amylin from effects of a neutralizing monoclonal antibody in rats. Diabetes 1995; 44(Suppl 1):251A.

23. Leaming R, Johnson A, Hook G et al. Amylin modulates insulin secretion in humans. Studies with an amylin antagonist. Diabetologia 1995; 38(Suppl 1):A113.

24. Cooper ME, Mcnally PG, Phillips PA et al. Amylin stimulates plasma renin concentration in humans. Hypertension 1995; 26(3):460-464.

25. Vine W, Smith P, Percy A et al. Concentration-response for amylin stimulation of plasma renin activity in rats. J Amer Soc Nephrol 1995; 6(3):750.

26. Young A, Nuttall A, Moyses C et al. Amylin stimulates the renin-angiotensin-aldosterone axis in rats and man. Diabetologia 1995; 38(Suppl 1):A225.

27. Vine W, Smith P, LaChappell R et al. Effects of rat amylin on renal function in the rat. Hormone Metab Res 1998; 30(8):518-22.

28. Young AA, Gedulin B, Vine W et al. Gastric emptying is accelerated in diabetic BB rats and is slowed by subcutaneous injections of amylin. Diabetologia 1995; 38(6):642-8.

29. Young AA, Gedulin BR, Rink TJ. Dose-responses for the slowing of gastric emptying in a rodent model by glucagon-like peptide (7-36)NH2, amylin, cholecystokinin, and other possible regulators of nutrient uptake. Metabolism 1996; 45(1):1-3.

30. Kong MF, Stubbs TA, King P et al. The effect of single doses of pramlintide on gastric emptying of two meals in men with IDDM. Diabetologia 1998; 41(5):577-83.

31. Nowak TV, Roza AM, Weisbruch JP et al. Accelerated gastric emptying in diabetic rodents: Effect of insulin treatment and pancreas transplantation. J Lab Clin Med 1994; 123(1):110-116.

32. Gedulin B, Jodka C, Green D et al. Effect of the Amylin receptor antagonist, AC187, on gastric emptying of glucose in corpulent LA/N rats. Program and Abstracts, the American Diabetes Association 29th Research Symposium 1994:40.

33. Chance WT, Balasubramaniam A, Zhang FS et al. Anorexia following the intrahypothalamic administration of amylin. Brain Res 1991; 539:352-354.

34. Lutz TA, Delprete E, Scharrer E. Reduction of food intake in rats by intraperitoneal injection of low doses of amylin. Physiol Behav 1994; 55(5):891-895.

35. Morley JE, Flood JF. Amylin decreases food intake in mice. Peptides 1991; 12(4):865-869.

36. Bhavsar S, Watkins J, Young A. Synergy between amylin and cholecystokinin for inhibition of food intake in mice. Physiol Behav 1998; 64:557-561.

37. Gebre-Medhin S, Mulder H, Pekny M et al. IAPP (amylin) null mutant mice; plasma levels of insulin and glucose, body weight and pain responses. Diabetologia 1997; 40(Suppl 1):A26.

38. Devine E, Young AA. Weight gain in male and female mice with amylin gene knockout. Diabetes 1998; 47(Suppl 1):A317.

39. Whitehouse F, Ratner R, Rosenstock J et al. Pramlintide showed positive effects on body weight in type 1 and type 2 diabetes. Diabetes 1998; 47(Suppl):A9.

40. Gedulin B, Jodka L, Lawler R et al. Amylin inhibits lipase and amylase secretion from the exocrine pancreas in rats. Diabetes 1998; 47(Suppl 1):A280.

41. Balfour JA, McTavish D. Acarbose: an update of its pharmacology and therapeutic use in diabetes mellitus. Drugs 1993; 46(6):1025-1054.

42. Coniff RF, Shapiro JA, Seaton TB et al. A double-blind placebo-controlled trial evaluating the safety and efficacy of acarbose for the treatment of patients with insulin-requiring type II diabetes. Diabetes Care 1995; 18(7): 928-932.

43. James WP, Avenell A, Broom J et al. A one- year trial to assess the value of orlistat in the management of obesity. Int J Obes Relat Metab Disord 1997; 21(Suppl 3):S24-30.

44. Samols E, Bonner-Weir S, Weir GC. Intra-islet insulin-glucagon-somatostatin relationships. Clin Endocrinol Metab 1986; 15(1):33-58.

45. Unger RH, Aguilar-Parada E, Muller WA et al. Studies of pancreatic alpha cell function in normal and diabetic subjects. J Clin Invest 1970; 49(4):837-848.

46. Unger RH, Orci L. The role of glucagon in the endogenous hyperglycemia of diabetes mellitus. Annu Rev Med 1977; 28:119-130.

47. Foster DW, McGarry JD. The regulation of ketogenesis. Ciba Found Symp 1982; 87:120-131.

48. Gedulin BR, Rink TJ, Young AA. Dose-response for glucagonostatic effect of amylin in rats. Metabolism 1997; 46(1):67-70.

49. Fineman M, Koda J, Kolterman O. Subcutaneous administration of a human amylin analogue suppresses postprandial plasma glucagon concentrations in type I diabetic patients. Diabetes 1998; 47(Suppl 1):A89.

50. Fineman M, Organ K, Kolterman O. The human amylin analogue pramlintide suppressed glucagon secretion in patients with type 2 diabetes. Diabetologia 1998; 41(Suppl 1):A167.

51. Gedulin B, Jodka C, Percy A et al. Neutralizing antibody and the antagonist AC187 may inhibit glucagon secretion in rats. Diabetes 1997; 40(Suppl 1):238A.

52. Rogers RC, McTigue DM, Hermann GE. Vagal control of digestion: modulation by central neural and peripheral endocrine factors. Neurosci Biobehav Rev 1996; 20(1):57-66.

53. Jodka C, Green D, Young A et al. Amylin modulation of gastric emptying in rats depends upon an intact vagus nerve. Diabetes 1996; 45(Suppl 2):235A.

54. Dilts R, Gedulin B, Jodka C et al. Central infusion of amylin delays gastric emptying in the rat. The Pharmacologist 1997; 39(1):32.

55. Edwards GL, Gedulin B, Jodka C et al. Area postrema (AP)-lesions block the regulation of gastric emptying by amylin. Neurogastroenterol Motil 1998; 10(4):365.

56. Beaumont K, Kenney MA, Young AA et al. High affinity amylin binding sites in rat brain. Mol Pharmacol 1993; 44(3):493-497.

57. Van Rossum D, Menard DP, Fournier A et al. Autoradiographic distribution and receptor binding profile of [I-125] Bolton Hunter-rat amylin binding sites in the rat brain. J Pharmacol Exp Ther 1994; 270(2):779-787.

58. Sexton PM, Paxinos G, Kenney MA et al. In vitro autoradiographic localization of amylin binding sites in rat brain. Neuroscience 1994; 62(2):553-567.

59. Christopoulos G, Paxinos G, Huang XF et al. Comparative distribution of receptors for amylin and the related peptides calcitonin gene related peptide and calcitonin in rat and monkey brain. Can J Physiol Pharmacol 1995; 73(7):1037-1041.

60. Aiyar N, Baker E, Martin J et al. Differential calcitonin gene-related peptide (CGRP) and amylin binding sites in nucleus accumbens and lung: Potential models for studying CGRP/amylin receptor subtypes. J Neurochem 1995; 65(3):1131-1138.

61. Gedulin B, Jodka C, Lawler R et al. Effect of amylin blockade on gastric emptying in normal HSD and LA/N corpulent rats. Program and Abstracts: 79th Annual Meeting of The Endocrine Society 1997:194.

62. Chen WJ, Armour S, Way J et al. Expression cloning and receptor pharmacology of human calcitonin receptors from MCF-7 cells and their relationship to amylin receptors. Mol Pharmacol 1997; 52(6):1164-1175.

63. Zimmermann U, Fluehmann B, Born W et al. Coexistence of novel amylin-binding sites with calcitonin receptors in human breast carcinoma MCF-7 cells. J Endocrinol 1997; 155(3):423-431.

64. Hanna FWF, Smith DM, Johnston CF et al. Expression of a novel receptor for the calcitonin peptide family and a salmon calcitonin-like peptide in the alpha-thyrotropin thyrotroph cell line. Endocrinology 1995; 136(6):2377-2382.

65. Perry KJ, Quiza M, Myers DE et al. Characterization of amylin and calcitonin receptor binding in the mouse alpha-thyroid-stimulating hormone thyrotroph cell line. Endocrinology 1997; 138(8): 3486-3496.

66. McLatchie LM, Fraser NJ, Main MJ et al. RAMPs regulate the transport and ligand specificity of the calcitonin-receptor-like receptor. Nature 1998; 393(6683):333-339.

CHAPTER 10

Characterization of Adrenomedullin Receptor Binding in Rat Tissues and Cell Lines

David M. Smith, Hedley A. Coppock, Ali A. Owji, Gillian M. Taylor, Arif A. Gharzi, Mohammad A. Ghatei and Stephen R. Bloom

10.1. Introduction

Adrenomedullin, originally isolated from human phaeochromocytomas in 1993,[1] is the most recently discovered member of the calcitonin family of peptides.[2] It has some sequence and structural homology with amylin and CGRP including an N-terminal six membered ring formed by an intramolecular disulphide bridge.[1] Adrenomedullin peptide and mRNA are widely distributed with high concentrations in adrenal medulla, lung, heart, spleen and kidney.[3,4] Particularly high mRNA levels are present in endothelial and vascular smooth muscle cells[5,6] but the peptide is also found in the CNS.[7] Many functions, mainly cardiovascular, have been proposed for adrenomedullin. The most important of these is probably a powerful long-lasting vasodilator response, similar to CGRP.[2,8] Adrenomedullin also stimulates or inhibits cell growth depending on cell type[9] and has a number of effects on salt and water balance.[10] The main aims of our study were; a) to determine the distribution of adrenomedullin binding sites in tissues and different mammals and to compare these with CGRP receptor distribution, b) to ascertain whether specific adrenomedullin receptors exist and to compare their pharmacology with CGRP receptors. The results of these studies showed the presence of adreno - medullin receptors in skeletal muscle and in this chapter we will also present some data on the function and mechanism of action of adrenomedullin in this tissue.

10.2. Distribution of Adrenomedullin Receptors

We studied the binding of [^{125}I]rat adrenomedullin to crude tissue membranes. Assays were performed as previously described in Owji et al.[11] Briefly, rat adrenomedullin (Peninsula, St Helens UK) was iodinated by the iodogen method and membranes prepared by differential centrifugation. Binding was performed at 4°C for 30 mins in the presence of 100 pM [^{125}I]rat adrenomedullin with bound and free ligand separated by centrifugation at 15,000xg. Adrenomedullin receptor binding (single ligand concentration only) in rat peripheral tissues and CNS is shown in figures 10.1 and 10.2 respectively. High levels of binding were observed in lung, heart, spleen and liver, which correlates well with adrenomedullin mRNA distribution.[3] These tissues are highly vascular so this observation is not unexpected for a cardiovascular peptide. A similar distribution for CGRP binding has been shown.[12] The exception to this was

---

The CGRP Family: Calcitonin Gene-Related Peptide (CGRP), Amylin, and Adrenomedullin, edited by David Poyner. ©2000 EUREKAH.COM.

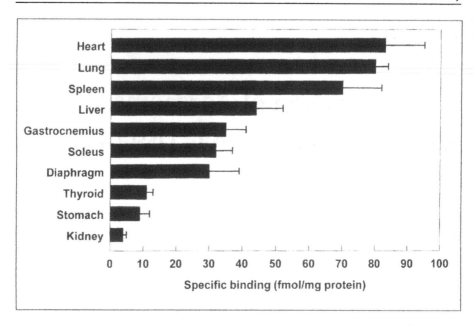

Fig. 10.1. Binding of ^{125}I rat adrenomedullin to rat peripheral tissue membranes. Binding in fmol/mg of membrane protein was measured at a single ligand concentration (100 pM). Data is the mean ± s. e. mean of triplicate assays performed on a minimum of three membrane preparations from each tissue.

kidney which showed low binding but high adrenomedullin immunoreactivity and mRNA.[3,4] We used whole kidney for our membrane preparations so it is possible that a high density of receptors are present but localised to a specific structure or area. Interestingly, high to moderate binding was found in the muscle tissues, diaphragm, soleus and gastrocnemius. This again may be vascular but is much higher than the binding for CGRP in muscle.[12] Moderate to low levels of binding were found throughout the brain but with the highest levels in spinal cord. Similar results have been shown in the human brain.[13] Overall this distribution was not very different from the CGRP receptor distribution[12] and CGRP receptors have been proposed as mediating adrenomedullin effects.[8] Hence, we next compared the affinity of CGRP and adrenomedullin receptors for the respective peptides.

10.3. Affinities of Adrenomedullin and CGRP at Their Respective Receptors

CGRP receptor binding was measured using ^{125}I]Tyr⁰-rat αCGRP as previously described.[14] Table 10.1 shows a comparison of ^{125}I]Tyr⁰-rat αCGRP and ^{125}I]rat adrenomedullin binding in a number of different rat tissue membranes with competition by rαCGRP and rat adrenomedullin. These data are pooled from a number of publications by our group[11,15-20] and this table is not meant to be a definitive review of the literature but a comparison of affinities (mainly K_D and K_i) using identical binding assays and membrane preparations. Both CGRP and adrenomedullin receptors show high affinity for the homologous ligands with affinities commonly below 1 nM. The clear difference is seen with the lack of competition for ^{125}I]adrenomedullin by CGRP, even at 1 μM, compared with adrenomedullin competing with ^{125}I]CGRP

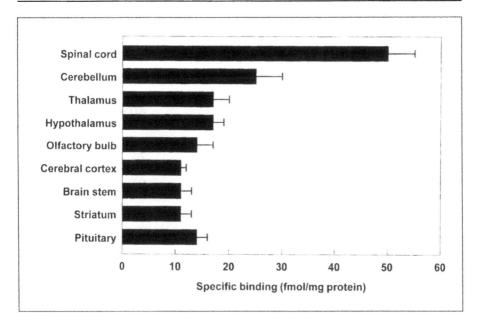

Fig. 10.2. Binding of ^{125}I rat adrenomedullin to rat CNS tissue membranes. Binding in fmol/ mg of membrane protein was measured at a single ligand concentration (100 pM). Data is the mean ± s. e. mean of triplicate assays performed on a minimum of three membrane preparations from each tissue.

at low concentrations (around 5 nM). Thus, although the affinity of adrenomedullin for CGRP receptors is much less than CGRP itself there is still a high affinity interaction between the two. Similar results have been shown with adrenomedullin having high affinity binding to CGRP receptors in SK-N-MC neuroblastoma cells[21] and rat heart.[22] Specific adrenomedullin receptors which do not bind CGRP have also been demonstrated in mouse astrocytes[23] and vascular smooth muscle cells.[24] This means that adrenomedullin effects, particularly those measured in vitro with high concentrations of adrenomedullin could be mediated via either specific adrenomedullin receptors or CGRP receptors. Competition for ^{125}I]adrenomedullin by amylin was also poor and salmon calcitonin had no effect at 1 μM in any tissue.[11]

10.4. Comparison of Adrenomedullin and CGRP Binding in Mammalian Lung

CGRP receptors are abundant and widespread in tissues from mammals, including man.[12] Adrenomedullin binding is widespread in the rat but is it abundant in other mammals? We examined receptor concentration (B_{max}, fmol/mg membrane protein, Fig. 10.3) and affinity (IC_{50}, nM, Fig. 10.4) in lungs from rat, rabbit, guinea pig, pig, monkey (marmoset) and bull and compared them with CGRP receptors in the same membrane preparations. In these experiments we used only ^{125}I]rat adrenomedullin. It is possible that different results could be obtained using the radioligand of that particular species but mainly these sequences or peptides were not available. Tissues were obtained from the Imperial College School of Medicine biological services unit except for bull and pig which were

Table 10.1. *Competition for ^{125}I-rat adrenomedullin and ^{125}I-αCGRP binding by unlabelled adrenomedullin and CGRP in rat tissues and cell lines*

Label	[^{125}I]ADM		[^{125}I]CGRP	
Competitor (affinity nM)	ADM	CGRP	ADM	CGRP
Lung[a]	5.80	>1000	5.0	0.30
Heart[a]	0.20	240.0	0.8	0.05
Hypothalamus	0.54	>1000	4.6	0.10
Blood vessels	1.40	>1000	4.8	1.04
Spinal cord	0.45	>1000	34.6	0.18
Uterus	0.08	>1000	1.7	0.14
L6 cells	0.22	>1000	8.7	0.13
Rat-2 cells	0.20	>1000	-[b]	-[b]

Rat adrenomedullin and rat αCGRP were used in all cases except hypothalamus where human adrenomedullin was used. All affinities are derived by non-linear regression analysis of competition binding as K_D or K_i except for tissues marked superscript[a] where IC_{50} values are given. Data obtained from Owji et al[11] for lung and heart, Taylor et al[15] for hypothalamus, Nandha et al[16] for blood vessels, Owji et al[17] for spinal cord, Upton et al[18] for uterus, Coppock et al[19] for L6 cells and Owji et al[20] for Rat-2 cells. b: no detectable CGRP binding.

obtained from a local abattoir. Adrenomedullin receptors are far more numerous in rat lung (2800 fmol/mg) than CGRP receptors (320 fmol/mg), again providing evidence for two types of site. In the guinea pig both adrenomedullin (2500 fmol/mg) and CGRP (1600 fmol/mg) receptors are expressed at high levels. All the other mammals express both receptors but at lower levels. It should be noted that, for example, in bull where adrenomedullin receptors are expressed at 470 fmol/mg, this level of expression is only low relative to rat lung and would be considered generally high expression for a peptide receptor. The affinity of the two receptors in the lungs are shown in figure 10.4. In general affinities are nanomolar with CGRP receptors ... wing

a higher affinity for the homologous ligand than adrenomedullin receptors. It is also clear that the IC_{50} for adrenomedullin at the adrenomedullin receptor in the mammalian lung is higher than other tissues (compare Fig. 10.4 and table 10.1). This could be due to the comparison of IC_{50}s in the lung (Fig. 10.4 and rat lung in table 10.1), which may be higher when high levels of binding are expressed, with K_Ds, which are not affected by receptor concentration, in other tissues (table 10.1). Variations in affinity and receptor concentration have also been observed for CGRP[14] and amylin[25] receptors in lung in different species.

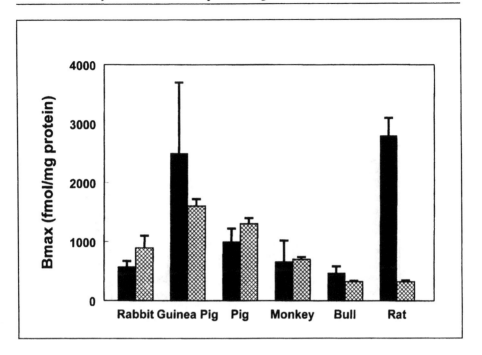

Fig. 10.3. Concentration of adrenomedullin and CGRP receptors in mammalian lung membranes. Receptor concentration (Bmax) is expressed as fmol/mg membrane protein (mean ± s. e. mean, n=3) and was calculated by non-linear regression analysis of equilibrium competition data. Labelled peptides were ^{125}I-rat adrenomedullin (100 pM, filled bars) and ^{125}I-Tyr$_0$-rat αCGRP (28 pM, cross-hatched bars) with unlabelled rat adrenomedullin and rat αCGRP, respectively, present from 1 pM to 1 μM.

10.5. Adrenomedullin and CGRP Binding in L6 Rat Skeletal Muscle Cells

In figure 10.1 we demonstrated binding sites for adrenomedullin in several muscles including heart, gastrocnemius, soleus and diaphragm, but it was not clear whether these sites were vascular or on the myocytes. This is of interest as direct effects of CGRP and amylin on myocyte glucose metabolism, particularly pathways stimulated by insulin, have been shown.[26] Adrenomedullin certainly shares the vasodilator effects of CGRP but little is known of its effects on muscle glucose metabolism. Therefore we examined ^{125}I]adrenomedullin binding in the rat skeletal muscle cell line L6, which has already been shown to exhibit CGRP binding sites.[27] Construction and

analysis of competition curves for CGRP and adrenomedullin against their corresponding ^{125}I-labelled peptide gave similar IC$_{50}$s (0.13 nM and 0.22 nM respectively, table 10.1) and B$_{max}$ values (0.83 and 0.95 pmol/mg of protein respectively). Therefore, to confirm that the CGRP and adrenomedullin binding sites were distinct from each other, competition curves were constructed for each of the ^{125}I-labelled peptides using CGRP and adrenomedullin (see table 10.1). [125]I]adrenomedullin binding was weakly inhibited by CGRP (IC$_{50}$>10 μM). [125]I]CGRP binding was inhibited by adrenomedullin with high affinity (IC$_{50}$=8.7 nM). The CGRP$_1$ receptor antagonist, CGRP$_{8-37}$, competed with high affinity at CGRP receptors (1.1 nM) but competed very weakly for ^{125}I]adrenomedullin

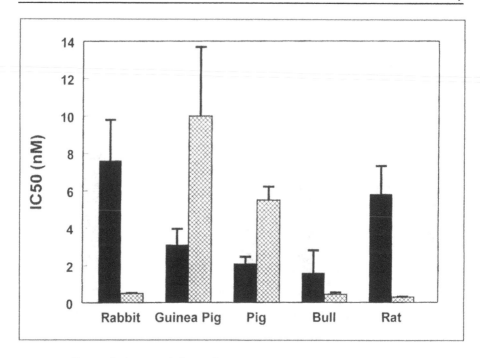

Fig. 10.4. Affinity of adrenomedullin and CGRP receptors in mammalian lung membranes. Receptor affinity (IC_{50}) is expressed as nM (mean ± s. e. mean, n=3) and was generated by non-linear regression analysis of equilibrium competition data. Labelled peptides were ^{125}I-rat adrenomedullin (100 pM, filled bars) and ^{125}I-Tyr$_0$-rat αCGRP (28 pM, cross hatched bars) as described in figure 3.

binding (601 nM).[19] These results indicate that in L6 cells, the adrenomedullin and CGRP binding sites are pharmacologically distinct from each other.

10.6. Adrenomedullin and CGRP Effects on Adenylyl Cyclase Activity in L6 Rat Skeletal Muscle Cells

Second messenger systems mediating the effects of adrenomedullin may involve intracellular Ca^{2+} or nitric oxide but most reports demonstrate an increase in cAMP mediated by G_s stimulated adenylyl cyclase (for a review see ref. 28). CGRP elevates intracellular cAMP in L6 cells.[26] Therefore we examined the ability of adrenomedullin to activate adenylyl cyclase in L6 membranes and compared this with CGRP actions. Stimulation with CGRP or adrenomedul-

lin increased adenylyl cyclase activity with similar dose dependencies (Fig. 10.5, approximate EC_{50}'s were between 1 and 10 nM for both peptides). CGRP$_{8-37}$ alone had no effect as expected. Concentrations of CGRP$_{8-37}$ exceeding 25 nM were sufficient to totally inhibit CGRP or adrenomedullin stimulated adenylyl cyclase activity (Fig. 10.6). At this low concentration of CGRP$_{8-37}$ there would be no expected interaction with adrenomedullin receptors strongly suggesting that adrenomedullin is activating adenylyl cyclase via the CGRP binding site. An elevation of intracellular cAMP is associated with increased phosphorylation of intracellular proteins by protein kinase A. CGRP and adrenomedullin increased protein kinase A activity with similar dose dependencies in L6 cells.[19]

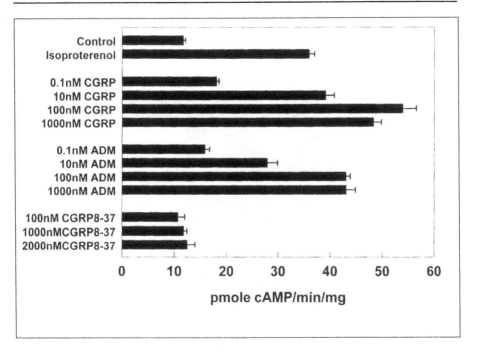

Fig. 10.5. Effect of adrenomedullin, CGRP and CGRP$_{8-37}$ on adenylyl cyclase activity in L6 membranes. Adenylyl cyclase activity was assayed as described in Coppock et al[19] in the presence of 2 mM isobutylmethylxanthine. Isoproterenol at 10 µM was included as a positive control. Data are expressed as pmoles of cyclic AMP generated per min per mg of membrane protein as means ± s. e. mean, n=6.

Eguchi et al suggested that adrenomedullin exerts its relaxant effects in rat vascular smooth muscle cells via specific adrenomedullin receptors coupled to increased intracellular cAMP.[29] Here, CGRP$_{8-37}$ did inhibit the increase in cAMP mediated by adrenomedullin albeit at high concentration (100 nM). This was not discussed by Eguchi et al and the absence of binding data for CGRP$_{8-37}$ makes the study difficult to interpret. Adrenomedullin is reported to increase $[Ca^{2+}]_i$ and inositol 1,4,5-trisphosphate by a cholera toxin-sensitive G-protein in bovine aortic endothelial cells.[30] Specific adrenomedullin receptors were demonstrated on the endothelial cells but the signalling mechanisms were not differentiated from those actions via CGRP receptors which are also present on endothelial cells. These studies highlight an important general point for CGRP and adrenomedullin receptor studies. To determine that the receptors mediating the effects of adrenomedullin are not CGRP receptors, binding studies using both [^{125}I]CGRP and [^{125}I]rat adrenomedullin competed by adrenomedullin, CGRP and CGRP$_{8-37}$ should be compared with inhibition of the agonist effect with low concentrations (to avoid interaction with adrenomedullin receptors) of CGRP$_{8-37}$. To unambiguously define the signal transduction pathways of specific adrenomedullin receptors we need to use cell lines expressing only specific adrenomedullin receptors or to use cell transfected with specific receptors. We have used Swiss 3T3 cells[31] and Rat-2 cells,[20] which express only [^{125}I]adrenomedullin binding, and in both cell lines a powerful activation of adenylyl cyclase was observed.

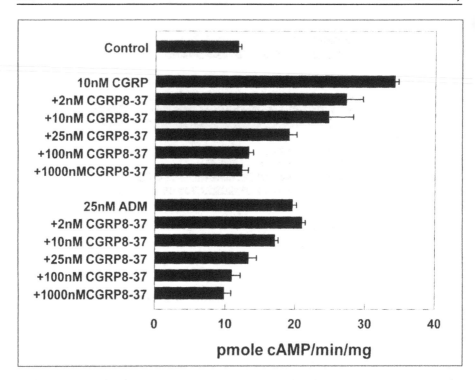

Fig. 10.6. Effect of $CGRP_{8-37}$ on adrenomedullin and CGRP stimulated adenylyl cyclase activity in L6 membranes. Adenylyl cyclase activity was assayed as described in figure 10.5. Data are expressed as pmoles of cyclic AMP generated per min per mg of membrane protein as means ± s. e. mean, n=6.

Similarly, using expression of the L1[32] and calcitonin-receptor-like receptor (CRLR)[33] clones in cell lines adrenomedullin has been shown to stimulate cAMP. The effect of adrenomedullin on CRLR may prove important as although CRLR was initially thought to be a CGRP receptor it now seems that in the presence of receptor-activity modifying protein 2 or 3 vthis receptor may present the pharmacology of a specific adrenomedullin receptor.[34]

10.7. Adrenomedullin Receptor Binding in Rat Soleus Muscle

L6 cells are a good model of skeletal muscle myocytes but caution should be observed in transferring findings from any cell line to normal tissue. Therefore, to compare the specificity of the ^{125}I-adrenomedullin binding sites in the L6 cell line with that in

rat soleus muscle, we constructed soleus muscle membrane competition curves for ^{125}I]adrenomedullin, $CGRP_{8-37}$ and rat adrenomedullin$_{17-50}$ (rat adrenomedullin without the N-terminal ring, equivalent to $CGRP_{8-37}$). Soleus membranes showed a high affinity adrenomedullin binding site with an IC_{50} for adrenomedullin of 0.12±0.02 nM. $CGRP_{8-37}$ (IC_{50}=58.4 nM) and adrenomedullin$_{17-50}$ (IC_{50}=18.1 nM) both competed with adrenomedullin (Fig. 10.7). CGRP was unable to compete at the adrenomedullin binding site. Binding of ^{125}I]CGRP was not detectable in our hands, despite having well characterised inhibitory effects on insulin stimulated glucose uptake.[26] The adrenomedullin binding site in the L6 cell-line was compared with that of the soleus muscle and can be seen to be similar (IC_{50} for adrenomedullin in L6=0.22 nM

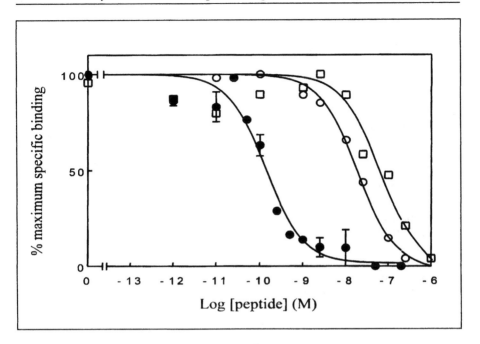

Fig. 10.7. Equilibrium competition binding of ^{125}I-rat adrenomedullin in rat soleus membranes. Effect of rat adrenomedullin (filled circles), rat adrenomedullin$_{17-50}$ (open circles) and CGRP$_{8-37}$ (open squares). Values are expressed as a percentage of the maximal specific binding. Each point is the mean ± s. e. mean of three separate assays, performed in triplicate.

and for soleus 0.125 nM), thereby decreasing the possibility that the cell-line binding properties are due to the non-physiological conditions of cell culture.

10.8. The Effects of Adrenomedullin and CGRP on Insulin Stimulated Glucose Uptake in Isolated Rat Soleus Muscles

Adrenomedullin receptors are present on both L6 cells and soleus muscle. To establish whether adrenomedullin has effects on muscle glucose metabolism we measured glucose uptake in the isolated rat soleus muscle preparation. Both soleus muscles were removed from ten male Wistar rats (150-200 g body weight), killed by asphyxiation and cervical dislocation, for each study. Soleus muscles were quickly washed, blotted dry, weighed then incubated in 10 ml Ehrlenmeyer flasks for 30-50 minutes in 5 ml of gassed Krebs-Henseleit bicarbonate buffered saline (4.7 mM KCl, 118.5 mM NaCl, 25 mM NaHCO$_3$, 1.19 mM KH$_2$PO$_4$, 1.19 mM MgSO$_4$.7H$_2$O, 2 mM pyruvate) in a shaking water bath at 37˚C. After the first incubation, the buffer was replaced with fresh Krebs containing one of the following; 10 nM porcine insulin, 100 nM CGRP or adrenomedullin, 10 nM insulin and 100 nM CGRP or adrenomedullin, or Krebs alone (control). Muscles were incubated in the second buffer for 15 minutes to allow peptide binding. For the third and final incubation, the buffer in each flask was replaced with fresh Krebs containing the same hormones as the second incubation, plus 0.5 μCi of 2-deoxy-D-[U-^{14}C] glucose, 1 mM 2-deoxy-D-glucose and 2.5 μCi of [^{3}H]inulin and the muscles were incubated for 30 minutes. After the final incubation muscles were removed and blotted dry, then dissolved in 1 M NaOH and 2-deoxy-D-[U-^{14}C] glucose uptake quantified by liquid scintillation counting. The extracellular volume of muscles was calculated based on the specific activity of the membrane impermeable polysaccharide [^{3}H]inulin. This allowed a value for extracellular 2-deoxy-D-[U-^{14}C] glucose to be calculated and subtracted from the total 2-deoxy-0-[U-^{14}C] glucose counts. Finally the rate of 2-deoxy-D-[U-^{14}C] glucose uptake was expressed as nmoles/30 minutes/mg wet weight.

For CGRP experiments (Fig. 10.8) the basal rate of glucose uptake in soleus muscles was 0.25 ±0.02 nmoles/30min/mg wet weight. Treatment with 10 nM insulin increased the rate of uptake to 0.62±0.03 nmoles/ 30min/ mg wet weight. CGRP alone (100 nM) had no effect on uptake (0.23±0.01 nmoles/ 30min/ mg wet weight), but in combination with insulin, CGRP decreased stimulated uptake to 0.48±0.05 nmoles/ 30min/mg wet weight (p<0.05 vs insulin alone, n=5 by ANOVA). For adrenomedullin experiments (Fig. 10.9) the basal rate of glucose uptake in soleus muscles was 0.28±0.02 nmoles/30min/mg wet weight, 10 nM insulin elevated uptake to 0.55±0.06 nmoles/30min/mg wet weight. Adrenomedullin alone (100 nM) had no effect on uptake (0.31±0.02 nmoles/30min/mg wet weight). Simultaneous administration of adreno- medullin and insulin had no significant effect on insulin stimulated glucose uptake (0.55± 0.06 nmoles/30min/mg wet weight). Thus it appears that adrenomedullin has little or very weak (since we only examined one high concentration) effects on glucose uptake in a rat soleus muscle model where CGRP was effective as expected. In a very extensive study Vine et al[35] showed that adrenomedullin had no effect on insulin stimulated glycogen synthesis in rat soleus compared with potent effects of amylin and CGRP in the same system. The in vivo effect of adrenomedullin (injected subcutaneously or intravenously) on plasma lactate and glucose levels in the rat were also very weak compared to CGRP and amylin.[35] Thus it appears that adrenomedullin is a poor mimic

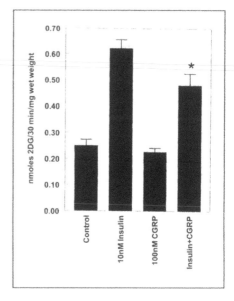

Fig. 10.8. Effect of CGRP (100 nM) on insulin stimulated glucose uptake in isolated rat soleus muscle. The uptake of 2-deoxyglucose (2DG) as nmoles/30 min per mg wet weight of muscle (mean ± s. e. mean) was measured as described in the text. Each group contained five separate muscles from different animals. * $P< 0.05$ vs 10 nM insulin alone (ANOVA).

of CGRP in terms of skeletal muscle glucose metabolism.

10.9. Conclusion

Thus we have demonstrated specific adrenomedullin receptors which do not bind CGRP and are inhibited by CGRP$_{8-37}$ only at very high concentrations. These receptors are present in many tissues of the rat and in the lungs of all the mammals examined. Specific adrenomedullin receptors are present on L6 skeletal muscle cells and in rat soleus but adrenomedullin stimulated cAMP in the L6 cells is mediated by CGRP$_1$ receptors. Adrenomedullin has no effect on glucose metabolism in the isolated rat soleus muscle. Hence, the function of adrenomedullin receptors on skeletal muscle remains a mystery.

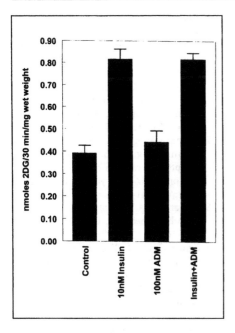

Fig. 10.9. Effect of adrenomedullin (ADM, 100 nM) on insulin stimulated glucose uptake in isolated rat soleus muscle. The uptake of 2-deoxyglucose (2DG) as nmoles/30 min per mg wet weight of muscle (mean ± s. e. mean) was measured as described in the text. Each group contained five separate muscles from different animals.

References

1. Kitamura K, Kangawa K, Kawamoto M et al. Adrenomedullin: A novel hypotensive peptide isolated from human phaeochromocytoma. Biochem Biophys Res Commun 1993; 192:553-560.
2. Poyner DR. Pharmacology of receptors for calcitonin gene-related peptide and amylin. Trends Pharmacol Sci 1995; 12:424-428.
3. Sakata J, Shimokubo T, Kitamura K et al. Molecular cloning and biological activities of rat adrenomedullin, a hypotensive peptide. Biochem Biophys Res Commun 1993; 195:921-927.
4. Sakata J, Shimokubo T, Kitamura K et al. Distribution and characterisation of immunoreactive rat adrenomedullin in tissue and plasma. FEBS Lett 1994; 352:105-108.
5. Sugo S, Minamino N, Kangawa K et al. Endothelial cells actively synthesize and secrete adrenomedullin. Biochem Biophys Res Commun 1994; 201:1160-1166.
6. Sugo S, Minamino N, Shoji H et al. Production and secretion of adrenomedullin from vascular smooth muscle cells: Augmented production by tumor necrosis factor-α. Biochem Biophys Res Commun 1994; 203:719-726.
7. Satoh F, Takahashi K, Murakami O et al. Adrenomedullin in human brain, adrenal glands and tumor tissues of pheochromocytoma, ganglioneuroblastoma and neuroblastoma. J Clin Endocrinol Metab 1995; 80:1750-1752.
8. Nuki C, Kawasaki H, Kitamura K et al. Vasodilator effect of adrenomedullin and calcitonin gene-related peptide receptors in rat mesenteric vascular beds. Biochem Biophys Res Commun 1993; 196:245-251.
9. Miller MJ, Martinez A, Unsworth EJ et al. Adrenomedullin expression in human tumor cell lines: Its potential role as an autocrine growth factor. J Biol Chem 1996; 271:23345-23351.
10. Schell DA, Vari RC and Samson WK. Adrenomedullin: Newly discovered hormone controlling fluid and electrolyte homeostasis. Trends Endocrinol Metab 1996; 7:7-13
11. Owji AA, Smith DM, Coppock HA et al. An abundant and specific binding site for the novel vasodilator adrenomedullin in the rat. Endocrinology 1995; 136:2127-2134.
12. Wimalawansa SJ. Calcitonin gene-related peptide and its receptors: Molecular genetics, physiology, pathophysiology, and therapeutic potentials. Endocrine Rev 1996; 17:533-585.
13. Sone M, Takahashi K, Satoh F et al. Specific adrenomedullin binding sites in human brain. Peptides 1997; 18:1125-1129.
14. Bhogal R, Smith DM, Purkiss P et al. Molecular identification of binding sites for calcitonin gene-related peptide (CGRP) and islet amyloid polypeptide (IAPP) in mammalian lung: Species variation and binding of truncated CGRP and IAPP. Endocrinology 1993; 133:2351-2361.

15. Taylor GM, Meeran K, O'Shea D et al. Adrenomedullin inhibits feeding in the rat by a mechanism involving calcitonin gene-related peptide receptors. Endocrinology 1996; 137:3260-3264.

16. Nandha KA, Taylor GM, Smith DM et al. Specific adrenomedullin binding sites and hypotension in the rat systemic vascular bed. Regul Pept 1996; 62:145-151.

17. Owji AA, Gardiner JV, Upton PD et al. Characterisation and molecular identification of adrenomedullin binding sites in the rat spinal cord: A comparison with calcitonin gene-related peptide receptors. J Neurochem 1996; 67:2172-2179.

18. Upton PD, Austin C, Taylor GM et al. Expression of adrenomedullin (ADM) and its binding sites in the rat uterus: Increased number of binding sites and ADM messenger ribonucleic acid in 20-day pregnant rats compared with nonpregnant rats. Endocrinology 1997; 138:2508-2514.

19. Coppock HA, Owji AA, Bloom SR et al. A rat skeletal muscle cell line (L6) expresses specific adrenomedullin binding sites but activates adenylate cyclase via calcitonin gene-related peptide receptors. Biochem J 1996; 318:241-245.

20. Owji AA, Smith DM, Coppock HA et al. Specific adrenomedullin receptors stimulate cAMP accumulation in Rat-2 cells. Regul Pept 1995;57:204.

21. Zimmermann U, Fischer JA, Muff R. Adrenomedullin and calcitonin gene-related peptide interact with the same receptor in cultured human neuroblastoma SK-N-MC cells. Peptides 1995; 16:421-424.

22. Enzeroth M, Doods HN, Weiland HA et al. Adrenomedullin mediates vasodilation via CGRP1 receptors. Life Sci 1995; 56:19-25.

23. Yeung VT, Ho SK, Nicholls MG et al. Adrenomedullin, a novel vasoactive hormone, binds to mouse astrocytes and stimulates cyclic AMP production. J Neurosci Res 1996; 46:330-335.

24. Ishizaka Y, Ishizaka Y, Tanaka M et al. Adrenomedullin stimulates cyclic AMP formation in rat vascular smooth muscle cells. Biochem Biophys Res Commun 1994; 200:642-646.

25. Bhogal R, Smith DM, Owji AA et al. Binding sites for islet amyloid polypeptide in mammalian lung: species variation and effects on adenylyl cyclase. Can J Physiol Pharmacol 1995; 73:1030-1036.

26. Leighton B, Cooper GJS. The role of amylin in the insulin resistance of non-insulin dependent diabetes mellitus. Trends Biochem Sci 1990; 15:295-299

27. Poyner DR, Andrew DP, Brown D et al. Pharmacological characterisation of a receptor for calcitonin gene-related peptide on rat, L6 myocytes. Br J Pharmacol 1992; 105:441-447

28. Richards MA, Nicholls MG, Lewis L et al. Adrenomedullin. Clin Sci Col 1996; 91:3-16.

29. Eguchi S, Hirata Y, Kano H, et al. Specific receptors for adrenomedullin in cultured rat vascular smooth muscle cells. FEBS Lett 1994; 340:226-230.

30. Shimekake Y, Nagata K, Ohta S et al. Adrenomedullin stimulates two signal transduction pathways, cAMP accumulation and Ca^{2+} mobilization, in bovine aortic endothelial cells. J Biol Chem 1995; 270:4412-4417.

31. Withers DJ, Coppock HA, Seufferlein T et al. Adrenomedullin stimulates DNA synthesis and cell proliferation via elevation of cAMP in Swiss 3T3 cells. FEBS Lett 1996; 378:83-87.

32. Kapas S, Catt KJ, Clark AJL. Cloning and expression of a cDNA encoding a rat adrenomedullin receptor. J Biol Chem 1995; 270:25344-25347.

33. Aiyar N, Rand K, Elshourbagy NA et al. A cDNA encoding the calcitonin gene-related peptide type 1 receptor. J Biol Chem 1996; 271:11325-11329.

34. McLatchie LM, Fraser NJ, Main MJ et al. RAMPs regulate the transport and ligand specificity of the calcitonin-receptor-like receptor. Nature 1998; 393:333-339.

35. Vine W, Beaumont K, Gedulin B et al. Comparison of the in vitro and in vivo pharmacology of adrenomedullin, calcitonin gene-related peptide and amylin in rats. Eur J Pharmacol 1196; 314:115-121.

CHAPTER 11

Cardiovascular Actions of Adrenomedullin

Philip J. Kadowitz, Albert L. Hyman, Bobby D. Nossaman, Trinity
J. Bivalacqua, William A. Murphy, David H. Coy, Dennis B. McNamara,
and Hunter C. Champion

11.1. Introduction

Adrenomedullin is a novel vasoactive peptide first isolated from human pheochromocytoma cells.[1-3] Human adrenomedullin (hADM) consists of 52 amino acids and a disulfide bond that forms a six-membered ring structure similar to the ring structure in calcitonin gene-related peptide (CGRP) and pancreatic amylin.[1-3] Although ADM shares slight (21%) sequence homology with CGRP, it is considered to be a member of the CGRP/amylin superfamily of peptides based largely on the presence of the six-membered ring structure and C-terminal amidation that is highly conserved in this family.[1-6] The most notable difference between ADM, CGRP and amylin is an additional 14-residue extension at the amino terminus of the ADM sequence that is not present in CGRP or amylin.[1-3] The amino acid sequence of human, rat (r) and porcine (p) ADM has been elucidated.[2,7-9] pADM and hADM differ by one amino acid substitution of glycine for asparagine at position 40 of the sequence.[1,8] The sequence of rADM differs from the human sequence in the deletion of two residues at positions 7 and 8 and six amino acid substitutions at positions 11, 14, 23, 40, 41 and 45 of the human peptide.[1,2] Although the six-membered ring structure sequence is conserved in hADM, rADM and pADM, these sequences are different from the sequences of the ring structures present in CGRP and amylin.[2,7-9]

Although ADM was first isolated from pheochromocytoma tissue, the peptide is localized in a variety of tissues.[10-14] ADM has been found to be expressed in the adrenal medulla, lung, kidney, ventricle, spinal cord, stomach, anterior pituitary, thalamus and hypothalamus.[10-14] It has also been reported that ADM is synthesized by endothelial cells, and that this release rate is similar to that of endothelin-1.[2] This observation may suggest that ADM is another endothelium-derived relaxing factor (EDRF).[2] Moreover, it has been reported that vascular smooth muscle cells produce and release ADM.[2] In addition to the presence of ADM message and peptide in tissues and ADM in plasma under physiological conditions, plasma levels of ADM have been shown to be increased in disease states, such as congestive heart failure, hypertension, renal failure, septic shock, primary aldosteronism, diabetes, myocardial infarction and pulmonary hypertension.[11-16] ADM has been shown to have actions in the CNS, intestine and kidney, in addition to having vasodilator activity in many species and vascular beds, including the hindlimb, mesenteric and pulmonary vascular beds.[17-36]

The CGRP Family: Calcitonin Gene-Related Peptide (CGRP), Amylin, and Adrenomedullin,
edited by David Poyner. ©2000 EUREKAH.COM.

Although ADM has marked effects on renal blood flow and electrolyte excretion in the dog, it has been shown that constant infusion of ADM in a dose that has minimal hypotensive activity increases renal plasma flow and urinary sodium excretion in the rat.[37]

The DNA sequence encoding the ADM precursor proadrenomeudllin has been determined in rat, porcine and human tissue.[2,7-9] Proadrenomedullin contains 185 amino acids, and cleavage of the signal peptide between amino acids Thr_{21} and Ala_{22} yields a shortened propeptide composed of 164 amino acids, which contains ADM.[9] Proadrenomedullin contains three paired basic amino acids that can be sites for enzymatic cleavage of the 20 amino acid peptide named proadrenomedullin NH_2-terminal 20 peptide (PAMP).[2,9,36] Like hADM, hPAMP is found in a number of organ systems, such as the adrenal medulla, heart, kidney and brain, and is detectable in plasma.[2,36]

11.2. Cardiovascular Responses to ADM

Human ADM_{1-52} and $PAMP_{1-20}$ are products of the hADM gene and have hypotensive activity.[1-3,18-26,31-36] Although hADM has been shown to decrease vascular resistance by a direct action on vascular smooth muscle or by releasing nitric oxide from the endothelium, hPAMP has been reported to induce vasodilation by inhibiting the release of norepinephrine from adrenergic nerve endings in the rat.[1-3,18-26,29,31-36] However, hPAMP has been shown to decrease vascular resistance in the regional vascular bed of the cat and rat in experiments in which the sympathetic nerves have been removed surgically or the adrenergic nerve endings have been depleted of transmitter with reserpine.[19,21,23] Recent data have shown that hPAMP possesses direct vasodilator activity that is mediated by a cAMP-dependent mechanism and not an inhibitory effect on the adrenergic nervous system or on the release of nitric oxide or by increases in cGMP, the release of vasodi-

lator prostaglandins or the opening of K^+_{ATP} channels in the mesenteric or hindquarters vascular beds of the cat.[19,21,23] In the rat, vasodilator responses to hPAMP in the hindquarters vascular bed were not dependent on the presence of the adrenergic nervous system and were correlated closely with the baseline level of vasoconstrictor tone in the vascular bed.[19,21,23] These data suggest that PAMP may have two, independent, yet complementary mechanisms for reducing peripheral vascular resistance. The effects of ADM and PAMP are compared in Fig. 11.1 where it can be seen that injection of ADM and PAMP induced dose-related decreases in hindquarters vascular resistance in the denervated vascular bed. When doses of the peptides are compared on a nmol basis, ADM was approximately 100-fold more potent than PAMP in decreasing hindquarters vascular resistance.

The receptor with which ADM interacts is uncertain and appears to depend on species and tissue studied. While it has been reported in a number of systems that ADM interacts with the CGRP receptor, it has also been shown that responses to ADM are not reduced after administration of the CGRP receptor antagonist $CGRP_{8-37}$.[2,3,19-21,23,24,27,28,35] Recently, it has been reported that ADM selectively binds a unique ADM receptor that is structurally similar to the CGRP receptors but does not bind CGRP with high affinity.[2,3] The mechanism mediating responses to ADM appears to vary with species and experimental preparation studied. It has been reported that ADM acts via a cAMP-dependent mechanism.[1,21,23,24,27,28,30] However, it has also been shown that inhibition of nitric oxide synthase attenuates responses to ADM, suggesting that an nitric oxide-mediated mechanism is involved in mediating vasodilator responses in the rat and cat.[29,31,32] The mechanism of action of CGRP is also uncertain, and it has been reported that vasorelaxant responses to CGRP are mediated by the opening of K^+_{ATP} channels, the release of nitric oxide or cAMP.[2-6,21,23,24,38]

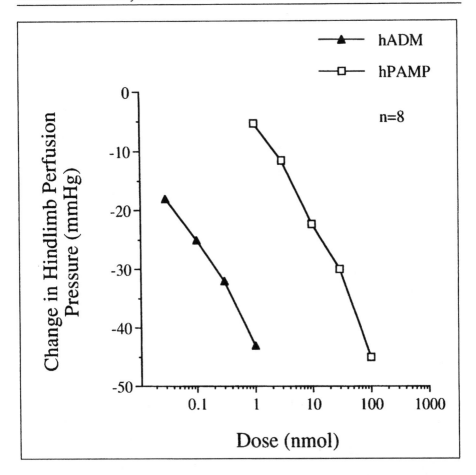

Fig. 11.1. Dose-response curves comparing decreases in hindquarters perfusion pressure in response to injections of hADM and hPAMP in the cat. Doses are expressed on a nmol basis and the vascular bed was denervated. n indicates number of experiments; and, on a nmol basis, hADM was approximately 100-fold more potent than hPAMP in decreasing regional vascular resistance.

The effects of ADM and the influence of the nitric oxide synthase inhibitor N^G-nitro-L-arginine methyl ester (L-NAME) on responses to ADM were compared in the pulmonary vascular bed of the intact-chest cat and rat, and these data are shown in Fig. 11.2. Injections of ADM into the pulmonary vascular bed of both species induced dose-related decreases in pulmonary vascular resistance (Fig. 11.2). However, L-NAME only inhibited vasodilator responses to ADM in the pulmonary vascular bed of the rat (Fig. 11.2). Pulmonary vasodilator responses to acetylcholine were reduced by L-NAME in both species whereas the nitric oxide synthase inhibitor did not decrease vasodilator

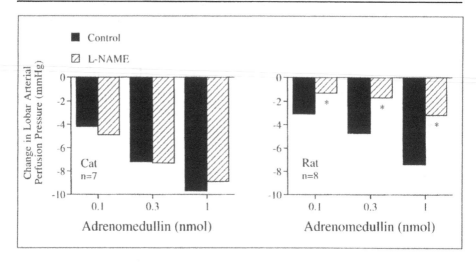

Fig. 11.2. Influence of L-NAME on decreases in lobar arterial perfusion pressure in the intact-chest cat and rat. Injections of hADM resulted in dose-related decreases in lobar arterial pressure in both species when baseline tone was increased with U-46619. The pulmonary vasodilator response to hADM was not changed by L-NAME in the cat. In contrast, the pulmonary vasodilator response to hADM in the rat was significantly decreased following treatment with the nitric oxide synthase inhibitor. n indicates number of experiments, and the asterisk indicates that the response is significantly different than control.

responses to nitroglycerin in either the cat or the rat.

11.3. Structure Activity Relationships for Adrenomedullin

11.3.1. Effect of N-Terminal Truncation

The most obvious structural difference between hADM, CGRP, and amylin is the presence of the 14 amino acid residue N-terminal extension in the ADM sequence. Since CGRP and amylin have full activity without this extension, it was hypothesized that residues 1-14 are not necessary for expression of full agonist activity of hADM.[2-6,18,24,25,27,33] Experiments have been conducted in the systemic, pulmonary and hindlimb vascular beds of the cat and rat using N-terminal truncated forms of hADM. Moreover, it has been shown that hADM$_{15-52}$ has vasodilator activity that is similar to hADM$_{1-52}$ in the pulmonary, me-

senteric and hindlimb vascular beds of the cat and rat.[18,24,25,27,33] These findings are supported by experiments in a number of other vascular models in a variety of species. These data suggest that the N-terminal amino acids located before the ring structure of hADM are not necessary for full expression of vasodilator activity. Vasodilator responses to ADM, ADM$_{15-52}$ and ADM$_{22-52}$ are compared in the hindquarters vascular bed of the cat, where it can be seen that ADM and ADM$_{15-52}$ have similar vasodilator activity whereas ADM$_{22-52}$ has no significant effect on hindquarters vascular resistance (Fig. 11.3).

11.3.2. Role of the Six-Membered Ring Structure

It has been shown with CGRP and amylin that removal of the six-membered ring structure results in the loss of activity. Fragments of ADM, such as hADM$_{22-52}$ and hADM$_{40-52}$, that do not possess the ring structure lack vasodilator activity.[18,24,25,27,33,34] In

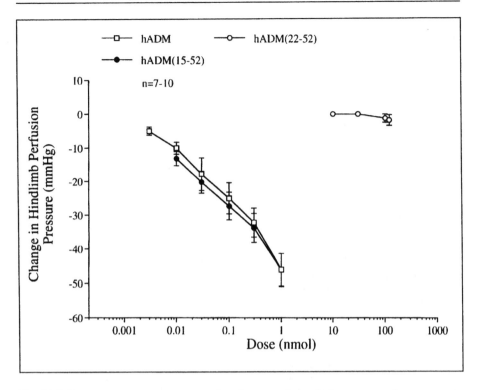

Fig. 11.3. Dose-response curves comparing decreases in perfusion pressure in response to injections of hADM, hADM$_{15-52}$, and hADM$_{22-52}$ into the perfusion circuit in the cat. When doses are expressed on a nmol basis, hADM and hADM$_{15-52}$ had similar vasodilator activity. However, hADM$_{22-52}$ had no significant vasodilator activity when injected in doses up to 300 nmol. n indicates number of experiments.

in vitro experiments with cultured vascular smooth muscle cells, it has been reported that hADM$_{22-52}$ binds to the ADM receptor and competitively inhibits the binding of ADM.[24] It has also been reported that hADM$_{22-52}$ inhibits ADM-induced cAMP formation in rat vascular smooth muscle cells in a dose-dependent manner.[24] It has recently been shown that hADM$_{22-52}$ possesses little or no activity in the hindlimb, mesenteric or pulmonary vascular beds of the cat and rat at doses up to 250 nmol.[24,25,33,34] Moreover, it has been shown that analogs of hADM and hCGRP lacking the six-membered ring structure selectively inhibit vasodilator responses to hCGRP in the hindlimb vascular bed of the cat.[24] In this study, however, responses to hADM and hADM$_{15-52}$

remained unchanged after administration of hADM$_{22-52}$.[24] These findings suggest that the ring structure is required for the expression of vasodilator activity of hADM and hCGRP and that the C-terminal fragment of ADM antagonizes vasodilator responses to hCGRP, but not hADM, in the hindlimb vascular bed of the cat.[24] The explanation for the inhibitory effect of hADM$_{22-52}$ on responses to hCGRP are uncertain but clearly involves the linear carboxy terminal portion of both peptides. The differential inhibitory effect of hADM$_{22-52}$ and hCGRP$_{8-37}$ on responses to hCGRP and hADM suggests that the two peptides do not act on the same receptor in the hindlimb vascular bed.[24] Current beliefs, however, suggest that the actions of ADM may be mediated by a group of receptors which bind

ADM, CGRP and amylin in a varied manner.

11.3.3. Effects of the ADM Ring Structure

It has been reported that N-terminal fragments of ADM, which contain the ring structure $hADM_{1-25}$ and $hADM_{16-21}$, have vasopressor activity in the systemic vascular bed of the rat.[39] Vasopressor responses to these peptides were attenuated by phenoxybenzamine, guanethidine and reserpine.[39] The observation that $hADM_{15-52}$, which comprises the ring structure of hADM, produces dose-dependent increases in systemic arterial pressure in the rat is in agreement with a previous study in which $hADM_{16-21}$, a smaller ring structure analog, was shown to increase systemic arterial pressure in the rat.[18,39] $hADM_{16-31}$ possesses the ability to increase systemic arterial pressure in the rat and was similar to $hADM_{15-22}$ (Fig. 11.4). These data extend the work of previous studies by demonstrating the role of a non-nicotinic mechanism in the release of catecholamines from the adrenal gland in mediating the pressor response to $hADM_{15-22}$ and $hADM_{16-31}$.[18,22] Although $hADM_{15-22}$ and $hADM_{16-312}$ had pressor activity in the rat, the peptides had little or no effect on systemic arterial pressure in the cat when injected in doses up to 1000 nmol/kg iv, $hADM_{1-52}$ and $hADM_{15-52}$ had similar activity in both species.[18,22] The results of experiments in the rat and cat show that there are marked species differences with regard to the effects of $hADM_{15-22}$ and $hADM_{16-31}$ on systemic arterial pressure.[18,22] These data provide evidence in support of the hypothesis that responses to the ADM ring structure analogs depend on species.[18,22]

Responses to the ADM ring structure peptides are compared in Fig. 11.4 where it can be seen that $hADM_{15-22}$ and $hADM_{16-31}$ produce similar increases in systemic arterial pressure in the rat, whereas the ring structure peptides for CGRP and amylin had no significant activity. Although the ring structure of hADM is similar to the ring structure of hCGRP and hamylin, the data

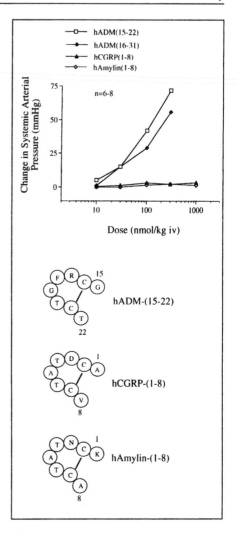

Fig. 11.4. Dose-response curves comparing changes in systemic arterial pressure in the response to iv injections of ring-structure peptides in the anesthetized rat. Intravenous injections of $hADM_{15-22}$ and $hADM_{16-31}$ caused dose-related increases in systemic arterial pressure. Injections of $hCGRP_{1-8}$ and $hamylin_{1-8}$ in doses up to 1000 nmol/kg iv had no significant effect on systemic arterial pressure. n indicates number of experiments.

from the present study show that the ring structure analogs of hCGRP, $hCGRP_{1-8}$ and hamylin, $hamylin_{1-8}$ possess no activity in

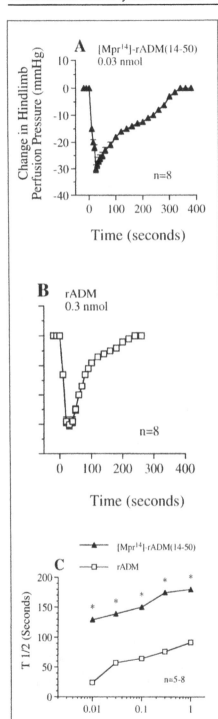

Fig. 11.5. (left) Panels A and B show the time-course of the decrease in hindquarters perfusion pressure in response to injections of $[Mpr_{14}]$-$rADM_{14-50}$ (0.3 nmol) and $rADM_{1-50}$ (0.3 nmol) in the cat. $[Mpr_{14}]$-$rADM_{14-50}$ is more potent than $rADM_{1-50}$ and the $t_{1/2}$ or recovery half-time of the vasodilator response to $[Mpr_{14}]$-$rADM_{14-50}$ is much longer than is the $t_{1/2}$ of the vasodilator response to $rADM_{1-50}$ in the hindquarters vascular bed of the cat. n indicates number of experiments, and the asterisk denotes that the response is significantly different than control.

the systemic vascular bed of the rat when injected in doses up to 300 nmol/kg iv. Furthermore, both hamylin and hamylin$_{1-8}$ had no effect on systemic arterial pressure in the rat or cat.[18] These data suggest that, although hCGRP, hamylin and hADM possess similar structural features, especially in the ring structure, the three truncated analogs have markedly different effects on systemic arterial pressure in the rat.[18]

The breakdown of hADM into inactive fragments by plasma endopeptidases may play a role in the inactivation or degradation of hADM. $[Mpr_{14}]$-$rADM_{14-50}$ is a novel analog of rADM that has a mercaptoproprionic acid (Mpr) residue replacing the cysteine in the 14 position.[20] Mercaptoproprionic acid forms a ring structure with the position 20 cysteine, thus removing the free amino acid terminal.[20] Recent studies in the hindlimb vascular bed of the cat provide evidence that $[Mpr_{14}]$-$rADM_{14-50}$ has vasodilator activity that is more potent than $hADM_{1-52}$ or $rADM_{1-50}$. Moreover, it has been shown that $[Mpr_{14}]$-$rADM_{14-50}$ has a longer duration of action when compared with the full-sequence form of the peptide. It has been suggested that this substitution of Mpr in the 14 position inhibits the degradation of the peptide.[20] The modified structure may decrease the inactivation of the analog, thus increasing the potency and duration of the half-life $[Mpr_{14}]$-$rADM_{14-50}$

in the hindlimb vascular bed of the cat.[20] Responses to $[Mpr_{14}]$-$rADM_{14-50}$ and $rADM_{1-50}$ are compared in Fig. 11.5 where it can be seen that the $t_{1/2}$ of the vasodilator response $[Mpr_{14}]$-$rADM_{14-50}$ is significantly longer than are responses to ADM_{1-50}.[20]

11.4. Conclusion

The data presented in this presentation indicate that ADM has significant vasodilator activity in all vascular beds surveyed. Although the peptide has vasodilator activity in all vascular beds studied, the mechanisms by which ADM decreases regional vascular resistance varies with species studied. In the hindlimb and pulmonary vascular beds of the rat and in the renal vascular bed of the dog, vasodilator responses to ADM are attenuated by inhibitors of nitric oxide synthase, suggesting that they are mediated at least in part by the release of nitric oxide from the endothelium. In this regard, in these vascular beds ADM acts as an endothelium-dependent vasodilator agent. However, in pulmonary and hindlimb vascular beds in the cat, vasodilator responses to ADM are not altered by inhibitors of nitric oxide synthase but appear to act by a cAMP-dependent mechanism. ADM and PAMP are products of the same gene and possess vasodilator activity. However, PAMP is 100-fold less active than ADM in its ability to decrease regional vascular resistance. It has been postulated that PAMP decreases vascular resistance by inhibiting the release of norepinephrine from adrenergic nerve endings whereas other studies have implicated a cAMP-dependent mechanism. It appears that both ADM and PAMP can decrease vascular resistance by different mechanisms and that mechanism may be species- or vascular bed-dependent.

References

1. Kitamura K, Kangawa K, Kawamoto M et al. Adrenomedullin: A novel hypotensive peptide isolated from human pheochromocytoma. Biochem Biophys Res Commun 1993; 192:553-560.

2. Kitamura K, Kangawa K, Matsuo H et al. Adrenomedullin: Implications for hypertension research. Drugs 1995; 49:485-495.

3. Schell DA, Vari RC, Samson WK. Adrenomedullin: A newly discovered hormone controlling fluid and electrolyte homeostasis. Trends Endocrinol Metab 1996; 7:7-13.

4. Brain SD, Cambridge H. Calcitonin gene-related peptide: Vasoactive effects and potential therapeutic role. Gen Pharmacol 1996; 27:607-611.

5. Poyner DR. Calcitonin gene-related peptide: Multiple actions, multiple receptors. Pharmacol Ther 1992; 56:23-51.

6. Poyner DR. Molecular pharmacology of receptors for calcitonin gene-related peptide, amylin, and adrenomedullin. Biochem Soc Trans 1997; 25:1032-1036.

7. Ishimitsu T, Kojima M, Kangawa K et al. Genomic structure of human adrenomedullin gene. Biochem Biophys Res Commun 1994; 203:631-639.

8. Kitamura K, Kangawa K, Kojima M et al. Complete amino acid sequence of porcine adrenomedullin and cloning of cDNA encoding its precursor. FEBS Lett 1994; 338:306-310.

9. Kitamura K, Sakata J, Kangawa K et al. Cloning and characterization of cDNA encoding a precursor for human adrenomedullin. Biochem Biophys Res Commun 1993; 194:720-725.

10. Ichiki Y, Kitamura K, Kangawa K et al. Distribution and characterization of immunoreactive adrenomedullin in human tissue and plasma. FEBS Lett 1994; 338:6-10.

11. Jougasaki M, Rodeheffer RJ, Redfield J et al. Cardiac secretion of adrenomedullin in human heart failure. J Clin Invest 1996; 97:2370-2376.

12. Jougasaki M, Wei C, McKinley LJ et al. Elevation of circulating and ventricular adrenomedullin in human congestive heart failure. Circulation 1995; 29:286-289.

13. Martinez A, Miller MJ, Unsworth EJ et al. Expression of adrenomedullin in normal human lung and pulmonary tumors. Endocrinology 1995; 136:4099-4105.

14. Sato K, Hirata Y, Imai T et al. Characterization of immunoreactive adrenomedullin in human plasma and urine. Life Sci 1995; 57:189-194.

15. Cheung B, Leung R. Elevated plasma levels of human adrenomedullin in cardiovascular, respiratory, hepatic and renal disorders. Clin Sci 1997; 92:59-62.

16. Ishimitsu T, Nishikimi T, Saito Y et al. Plasma levels of adrenomedullin, a newly identified hypotensive peptide, in patients with hypertension and renal failure. J Clin Invest 1994; 94:2158-2161.

17. Baskaya MK, Suzuki Y, Anzai M et al. Effects of adrenomedullin, calcitonin gene-related peptide, and amylin on cerebral circulation in dogs. J Cerebral Blood Flow Metab 1995; 15:827-834.

18. Champion HC, Fry R, Murphy WA et al. Catecholamine release mediates pressor effects of adrenomedullin-(15-22) in the rat. Hypertension 1996; 28:1041-1046.

19. Champion HC, Czapla MA, Friedman DE et al. Tone-dependent vasodilator responses to proadrenomedullin NH$_2$-terminal 20 peptide in the hindquarters vascular bed of the rat. Peptides 1997; 18:513-518.

20. Champion HC, Duperier CD, Fitzgerald WE et al. [Mpr^{14}]-rADM-(14-50), a novel analog of adrenomedullin, possesses potent vasodilator activity in the hindlimb vascular bed of the cat. Life Sci 1996; 59:PL1-PL7.

21. Champion HC, Erickson CC, Simoneaux ML et al. Proadrenomedullin NH$_2$-terminal 20 peptide has cAMP-mediated vasodilator activity in the mesenteric vascular bed of the cat. Peptides 1996; 17:1379-1387.

22. Champion HC, Friedman DE, Lambert DG et al. Adrenomedullin-(16-31) has pressor activity in the rat but not the cat. Peptides 1997; 18:133-136.

23. Champion HC, Murphy WA, Coy DH et al. Proadrenomedullin NH$_2$-terminal 20 peptide has direct vasodilator activity in the cat. Am J Physiol 1997; 272:R1047-R1054.

24. Champion HC, Santiago JA, Murphy WA et al. Adrenomedullin-(22-52) antagonizes vasodilator responses to calcitonin gene-related peptide but not adrenomedullin in the cat. Am J Physiol 1997; 272:R234-R242.

25. Cheng DY, DeWitt BJ, Wegmann MJ et al. Synthetic human adrenomedullin and ADM-(15-52) have potent short-lasting vasodilator activity in the pulmonary vascular bed of the cat. Life Sci 1994; 55:251-256

26. Cheng DY, DeWitt BJ, Wegmann MJ et al. Synthetic human adrenomedullin and ADM-(15-52) have potent short-lasting vasodilator activity in the pulmonary vascular bed of the cat. Life Sci 1994; 55:251-256.

27. DeWitt BJ, Cheng DY, Caminiti GN et al. Comparison of responses to adrenomedullin and calcitonin gene-related peptide in the pulmonary vascular bed of the cat. Eur J Pharmacol 1994; 257:303-306.

28. Eguchi S, Hirata Y, Iwasaki H et al. Structure-activity relationship of adrenomedullin, a novel vasodilator peptide, in cultured rat vascular smooth muscle cells. Endocrinology 1994; 135:2454-2458.

29. Entzeroth M, Doods HN, Wieland HA et al. Adrenomedullin mediates vasodilation via CGRP$_1$ receptors. Life Sci 1995; 56:PL-19-PL-25.

30. Feng CJ, Kang B, Kaye AD et al. L-NAME modulates responses to adrenomedullin in the hindquarters vascular bed of the rat. Life Sci 1994; 55:PL-433-PL-438.

31. Ishizaka Y, Tanaka M, Kitamura K et al. Adrenomedullin stimulates cyclic AMP formation in rat vascular smooth muscle cells. Biochem Biophys Res Commun 1994; 200:642-646.

32. Majid DSA, Kadowitz PJ, Coy DH et al. Renal responses to intra-arterial administration of adrenomedullin in dogs. Am J Physiol 1996; 270:F200-F205.

33. Nossaman BD, Feng CJ, Kaye AD et al. Pulmonary vasodilator responses to adrenomedullin are reduced by NOS inhibitors in rats but not in cats. Am J Physiol 1996; 270:L782-L789.

34. Santiago JA, Garrison EA, Purnell WL et al. Comparison of responses to adrenomedullin and adrenomedullin analogs in the mesenteric vascular bed of the cat. Eur J Pharmacol 1995; 272:115-118.

35. Santiago JA, Garrison EA, Ventura VL et al. Synthetic human adrenomedullin and adrenomedullin-(15-52) have potent short-lived vasodilator activity in the hindlimb vascular bed of the cat. Life Sci 1994; 55:PL-85-PL-90.

36. Shimosawa T, Ito Y, Ando K et al. Proadrenomedullin NH$_2$-terminal 20 peptide, a new product of the adrenomedullin gene, inhibits norepinephrine overflow from nerve endings. J Clin Invest 1995; 96:1672-1676.

37. Vari RC, Adkins SD, Samson WK. Renal effects of adrenomedullin in the rat. Proc Soc Exp Biol Med 1995; 211:178-183.

38. Marshall I. Mechanism of vascular relaxation by the calcitonin gene-related peptide. Ann NY Acad Sci 1992; 657:204-215.

39. Watanabe TX, Itahara Y, Inui T et al. Vasopressor activities of N-terminal fragments of adrenomedullin in anesthetized rat. Biochem Biophys Res Commun 1996; 219:59-63.

CHAPTER 12

Calcitonin Gene-Related Peptide in Gastrointestinal Homeostasis

Peter Holzer

12.1. Introduction

The localization of calcitonin gene-related peptide (CGRP) to distinct neurons in the gastrointestinal tract,[1] the cell-selective expression of CGRP receptors on gastrointestinal effector systems[2,3] and the peptide's biological actions in the gut[1] suggest that CGRP plays multiple roles in the regulation of digestive functions. This is particularly true for CGRP-positive primary afferent neurons which supply esophagus, stomach and intestine and whose functional implications are considerably better studied than those of enteric neurons containing CGRP. There is increasing evidence that CGRP released from the peripheral endings of extrinsic afferents regulates blood flow, secretory processes, motor activity and tissue homeostasis and that CGRP released from their central endings participates in visceral nociception. The subject of this article is to highlight some of the peptide's most important actions and functions in the gut and to delineate those areas in which disturbances of the CGRP system may contribute to gastrointestinal disorders and correction of these disturbances may be of therapeutic potential.

12.2. Expression and Release of CGRP in the Gastrointestinal Tract

The principal sources of CGRP in the digestive system are extrinsic primary afferent nerve fibers and intrinsic enteric neurons (Fig. 12.1). These two neuronal systems contain different molecular forms of the peptide, given that most of the CGRP expressed in extrinsic afferents is αCGRP, whereas the only form of CGRP in enteric neurons is βCGRP.[3,4] The density of the gastrointestinal innervation by CGRP-immunoreactive neurons varies greatly among different regions of the gut and different mammalian species as does the relative contribution of extrinsic afferent and intrinsic enteric neurons to the overall CGRP content of the gut.[1] Most of the CGRP found in the esophagus and stomach of small rodents (rat, mouse, guinea-pig) is derived from extrinsic afferent neurons while the small and large intestine contains a sizeable number of intrinsic enteric neurons expressing CGRP.[1,5,6,7] The CGRP-immunoreactive neurons of the myenteric and submucosal plexus in the guinea-pig intestine project primarily into the mucosa[8] whereas those in the rat intestine issue oral and caudal projections within the plexuses as well as to the muscle layers and the mucosa.[5,9]

The other important source of CGRP is extrinsic afferent neurons, most of which originate from cell bodies in the dorsal root ganglia and reach the gut via sympathetic (splanchnic, colonic and hypogastric) and sacral parasympathetic (pelvic) nerves while passing through prevertebral ganglia and

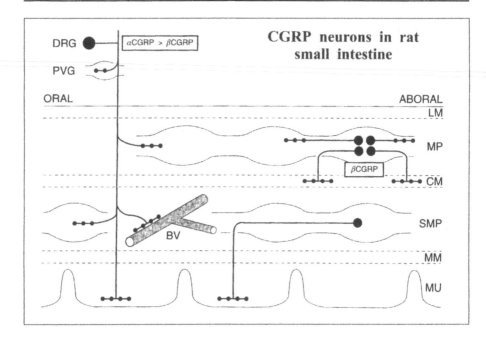

Fig. 12.1. Occurrence of CGRP in extrinsic afferent and intrinsic enteric neurons of the rat intestine. BV, blood vessel; CM, circular muscle; DRG, dorsal root ganglion; LM, longitudinal muscle; MM, muscularis mucosae; MP, myenteric plexus; MU, mucosa; PVG, prevertebral ganglion; SMP, submucosal plexus.

forming collateral synapses with sympathetic ganglion cells.[1,7,10-12] The mostly unmyelinated (C-fiber) axons of spinal CGRP-positive afferents supply the mesenteric arteries and, within the wall of the digestive system, project primarily to submucosal arteries and arterioles where they form a para- and perivascular network.[5-7] Some fibers can also be traced to the lamina propria of the mucosa, to the submucosal and myenteric nerve plexuses and to the circular and longitudinal muscle layers (Fig. 12.1). Vagal afferents that have their cell bodies in the nodose ganglion supply the esophagus and proximal part of the stomach[13-15] but make a relatively small contribution to the content of CGRP in the gastric corpus, antrum and intestine, given that the vast majority (80-90 %) of the CGRP-containing nerve fibers in the rat stomach is derived from dorsal root ganglia.[7] It is a characteristic of many spinal afferents in the rat, guinea-pig and canine gut that CGRP is coexpressed with the tachykinin substance P, whereas CGRP and tachykinins do not coexist in enteric neurons of these species.[1,3,6,8,12,16,17]

Apart from neurons, CGRP can also be found in endocrine cells of the human gastrointestinal mucosa[18] and rat pancreas[19] and in blood-derived or resident immune cells within the lamina propria of the rat gastric mucosa.[20] However, the quantitative and functional significance of these sources is still little known.

As is expected for substances with a vesicular localization, CGRP is released from the gastrointestinal tract if extrinsic afferent or intrinsic enteric neurons are depolarized.[1] The calcium dependency of the release process points to an exocytotic mechanism. Peptide release from extrinsic spinal afferents can specifically be elicited by the excitotoxin capsaicin[1] because receptors for this drug (vanilloid receptors) are exclusively expressed on spinal and trigeminal sensory neurons.[21] In the context of this article it is particularly relevant to note that acidification of the tissue, which frequently occurs in the upper gut, is able to release CGRP from extrinsic afferents in the stomach and duodenum.[22-24]

12.3. CGRP and Gastrointestinal Pathophysiology

12.3.1. Implications of CGRP in the Pathophysiological Regulation of Gastrointestinal Motor Activity

Although CGRP has been found to both stimulate and inhibit gastrointestinal motility, it is muscle relaxation via $CGRP_1$ receptors, which is the most prominent motor action of this peptide in the gut.[1,25,26] The inhibitory action of CGRP, which is seen in all regions of the digestive tract including esophagus, stomach, gallbladder and intestine, arises in most instances from a direct action on the muscle.[1] This site of action is also true for the relaxant action of CGRP on muscle strips taken from the human intestine.[27] In addition, CGRP can inhibit intestinal motor activity by blocking electrically evoked release of acetylcholine from enteric neurons and by activating inhibitory neural pathways.[1,28,29] The effect of CGRP to dampen motility is also seen in vivo and comprises retardation of gastric emptying in the rat.[30] Besides inhibiting motility, CGRP can lead to contraction of the longitudinal and circular muscle of the guinea-pig small intestine. Being insensitive to the $CGRP_1$ receptor antagonist $CGRP_{8-37}$, the contractile response to the peptide is most likely brought about by $CGRP_2$ receptors on enteric neurons.[29-33] Activation of neuronal CGRP receptors leads to excitation of enteric cholinergic motor pathways, which is consistent with the peptide's ability to depolarize after-hyperpolarization/type 2 myenteric neurons and to enhance the release of acetylcholine from the myenteric plexus.[29,31,34,35]

There is still scarce information to attribute to CGRP released from intrinsic enteric neurons a role in the control of gastrointestinal motility but there is ample evidence to infer that CGRP released from extrinsic afferent nerve fibers in the gut disturbs gastrointestinal motility.[25] One study claims that capsaicin-sensitive extrinsic afferents function as stretch detectors and by releasing CGRP activate ascending excitatory and descending inhibitory motor pathways within the enteric nervous system of the rat colon.[36] The physiological significance of this observation, though, is unknown because ablation of capsaicin-sensitive afferents fails to alter propulsive motility in the rat and guinea-pig intestine in vivo.[25,37]

Stimulation of capsaicin-sensitive afferent nerve fibers in the gut exerts both excitatory and inhibitory effects on the motility of the gastrointestinal tract.[25,37] There is good evidence to conclude that the relaxant effect of afferent neuron stimulation involves release of CGRP which, in turn, acts on $CGRP_1$ receptors to attenuate motor activity.[25,26] $CGRP_1$ receptors are also responsible for the inhibition of intestinal peristalsis which ensues after sensory neuron stimulation.[37] It is tempting to hypothesize that CGRP released from extrinsic afferent nerve fibers in the intestine has a bearing on pathological disturbances of intestinal motor activity, particularly when irritant, inflammatory or noxious stimuli excite sensory nerve fibers or sensitize them to other stimuli. Indeed, $CGRP_{8-37}$ and a monoclonal antibody to CGRP attenuate the block of gastric emptying, which in the rat is induced by laparotomy[38] or peritoneal irritation with acetic acid.[39] These observations attribute to CGRP a potential role in the pathological shutdown of gastrointestinal motility as exemplified by postoperative or peritonitis-induced ileus.

12.3.2. Implications of CGRP in the Pathophysiological Regulation of Gastrointestinal Secretory Processes

The actions of CGRP on gastrointestinal ion, enzyme and fluid secretion vary with the region and species under study and in some instances appear to depend on the experimental conditions of the study. For instance, CGRP is able to inhibit secretagogue-evoked secretion of enzyme, bicarbonate and fluid from the pancreas of dog and rat in vivo, an effect that to a large extent is mediated by CGRP-induced release of somatostatin.[1,40-42] In contrast, amylase secretion from isolated acini of the rat and guinea-pig pancreas is enhanced by the peptide.[42,43]

Likewise, ion and fluid secretion in the intestine is mostly stimulated by CGRP, if it is affected at all. Whereas ion and fluid transport in the rat small intestine is not altered by the peptide,[44] the secretion of ions and fluid in the small intestine of the dog[45] and the colon of the guinea-pig and rat[44,46,47] is stimulated by CGRP. The secretory effect of CGRP in the rat colon arises from a direct action of the peptide on enterocytes,[46] as is the case with human epithelial cell lines that respond to CGRP with electrogenic ion secretion.[48] In contrast, CGRP's secretory action in the guinea-pig colon is mediated by enteric neurons.[47] Since the secretory action of CGRP in the rat colon is not antagonized by CGRP$_{8-37}$,[49] it has not yet been possible to examine whether endogenous CGRP participates in the physiological control of intestinal secretion. A role in pathological fluid secretion, though, is indicated by the findings that CGRP in spinal afferent neurons is upregulated after exposure of the rat ileum to *Clostridium difficile* toxin A and that CGRP$_{8-37}$ attenuates the toxin-evoked accumulation of fluid in the intestine.[50]

There is good evidence to implicate CGRP as a transmitter by which extrinsic afferent nerve fibers contribute to the homeostatic regulation of gastric secretory processes.[51] This inference is in keeping with the peptide's high activity in inhibiting basal and secretagogue-evoked output of acid and pepsin in the stomach of man, dog, rabbit and rat.[1,51] The antisecretory action of CGRP is brought about by CGRP$_1$ receptors,[52,53] depends on somatostatin as an essential mediator, and goes along with a depression of the release of acetylcholine, gastrin and histamine (Fig. 12.2).[23,54-57] It should be realized in this context that CGRP regulates not only the release of somatostatin and gastrin but also the transcription of their genes. Thus, CGRP enhances the expression of somatostatin mRNA in the rat gastric antrum whereas the expression of gastrin mRNA is reduced.[58] The peptide's effect on gastrin mRNA is mediated by somatostatin because it is prevented by somatostatin immunoneutralization.[58]

There are two lines of evidence to indicate that CGRP participates in the homeostatic regulation of gastric acidity. One hint at such a role comes from the ability of CGRP$_{8-37}$ to augment basal and stimulated acid output.[52,53] The physiological potential of CGRP in gastric secretory control comes to full light if it is considered that accumulation of acid in the gastric lumen releases CGRP from extrinsic afferent nerve fibers.[22,24] By way of its effects on the release of somatostatin, gastrin, histamine and acetylcholine, CGRP halts further secretion of acid (Fig. 12.2) and thus represents an essential transmitter in the feedback inhibition of gastric acid output.[23]

12.3.3. Implications of CGRP in the Pathophysiological Regulation of the Gastrointestinal Vascular System

The mesenteric arteries and gastrointestinal submucosal arterioles are those structures which receive the densest innervation by extrinsic afferents containing CGRP.[3,5-9] The para- and perivascular localization of the axons points to a vascular regulatory function of the peptide, which is strongly indicated by the expression of CGRP$_1$ receptors on arteries and arterioles,

Role of CGRP in feedback inhibition of gastric acid secretion

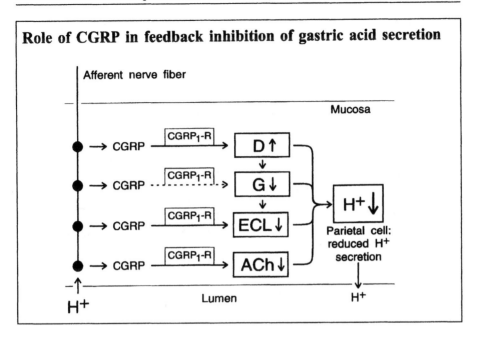

Fig. 12.2. Diagram illustrating the role of CGRP-releasing afferent nerve fibers in the local regulation of gastric acid secretion in the rat stomach. These fibers monitor acidity (H^+) in the gastric lumen and, when the acidity rises, release CGRP which via activation of $CGRP_1$ receptors stimulates the release of somatostatin and inhibits the release of gastrin, histamine and acetylcholine. D, endocrine D cells releasing somatostatin; G, endocrine G cells releasing gastrin; ECL, enterochromaffin-like cells releasing histamine; ACh, neurons releasing acetylcholine.

where they are located on both endothelium and vascular smooth muscle,[2,3] and by the peptide's action on the gastrointestinal vasculature.[1,51]

There is ample evidence to conclude that nonadrenergic noncholinergic dilatation of the rat superior mesenteric artery is mediated by capsaicin-sensitive afferent nerve fibers releasing CGRP.[25,59,60] The peptide, which is released by stimulation of the mesenteric periarterial nerve plexus with electrical impulses or capsaicin application, is very potent in dilating precontracted mesenteric, splenic or hepatic arteries isolated from rats, guinea-pigs, rabbits and humans in an endothelium-independent manner.[25] The mediator role of αCGRP has been corroborated by the findings that the neurogenic dilatation of rat mesenteric arteries is blocked by $CGRP_{8-37}$ and a polyclonal antibody to CGRP.[25,60-62] Further analysis indicates that the CGRP-containing afferent nerve fibers in the rat mesenteric arteries are part of a complex vascular control system. Thus, CGRP inhibits the vasoconstrictor action of noradrenaline at a postjunctional level, whereas the sympathetic transmitters noradrenaline (acting via α_2-adrenoceptors) and neuropeptide Y inhibit the release of CGRP, as do opiates (acting via μ- and possibly δ-opioid receptors), somatostatin and adenosine (acting via A_1 purinoceptors).[25] In addition, the CGRP-containing vasodilator fibers in the rat mesenteric arteries bear CGRP autoreceptors which regulate transmitter release via a negative feedback mechanism.[63]

The functional significance of CGRP in the microcirculation of the small and large intestine is still little known. It is human βCGRP[64] but not αCGRP,[65] which is able to dilate submucosal arterioles in the guinea-pig ileum in which the peptide mediates the vasodilator reaction to afferent neuron stimulation with capsaicin.[66] Mucosal blood flow in the rat small and large intestine, however, does not seem to be altered by rat αCGRP,[25] although αCGRP receptors are present throughout the rat digestive tract.[3] In contrast, αCGRP is very potent in enhancing blood flow in the rat gastric mucosa,[67] an action that is brought about by $CGRP_1$ receptors.[68-70] The vasodilator effect of low doses of CGRP in the rat gastric mucosal microcirculation is mediated by nitric oxide while high doses of the peptide increase blood flow independently of nitric oxide.[71,72] It is in line with this activity that CGRP mediates the nitric oxide-dependent vasodilatation which afferent neuron stimulation by intragastric capsaicin elicits in the rat stomach.[68,71] Like capsaicin, CGRP dilates submucosal arterioles but does not affect the diameter of venules.[69]

It is not known whether CGRP released from perivascular nerve fibers of extrinsic afferents participates in the physiological regulation of gastric blood flow, and it seems as if the peptide comes into play under pathological conditions only.[51] The pathophysiological potential of CGRP in gastric circulatory control is exemplified by the hyperemic response which ensues when the gastric mucosal barrier is disrupted by ethanol or bile salt so that acid can enter the tissue and damage the gastric mucosa. The acid-evoked rise of gastric mucosal blood flow involves CGRP and nitric oxide since it is suppressed by $CGRP_{8-37}$ and a nitric oxide synthase inhibitor.[70,73,74] Further analysis has revealed that the hyperemic response to gastric acid challenge is mediated by a local axon reflex-like circuitry[51] that is triggered by intruding acid itself and by factors that are generated in the injured tissue.[51] Modulated by input from the sympathetic and/or enteric nervous system, the local reflex activity results in release of CGRP and formation of nitric oxide, which

are the major vasodilator messengers (Fig. 12.3).[51] The rise of gastric mucosal blood flow in response to acid backdiffusion serves a homeostatic and protective role in the gastric mucosa as it helps to neutralize and wash away intruding acid and delivers bicarbonate and other factors to defend and repair the mucosa (Fig. 12.3).[51]

Whether CGRP regulates other vascular functions in the digestive tract or serves as an interface between the neural and immune system of the gut has been little studied. It is worth noting in this respect that the inflammation caused by *Clostridium difficile* toxin A in the rat ileum depends on an intact innervation by capsaicin-sensitive extrinsic afferents[75] and involves CGRP because it is inhibited by $CGRP_{8-37}$.[50]

12.3.4. Implications of CGRP in Homeostasis of the Gastrointestinal Mucosa

There are several lines of evidence to indicate that gastrointestinal mucosal integrity and repair are under the control of extrinsic afferent neurons releasing CGRP. The first hint at such a role came from the ability of CGRP to protect the gastrointestinal mucosa in a number of experimental models of gastric injury and colonic inflammation. Both local intra-arterial and systemic administration of the peptide to the rat stomach prevents gastric damage induced by ethanol, acidified aspirin, endothelin-1 or stress.[76-,83] This protective effect extends to the rat colon where the peptide is able to reduce the tissue damage caused by trinitrobenzene sulfonic acid.[84] The action of CGRP to reduce ethanol-induced damage in the gastric mucosa is mediated by $CGRP_1$ receptors and involves both nitric oxide and K_{ATP} channels, whereas prostaglandins do not participate.[79-82,85]

CGRP's activity to strengthen mucosal defense appears to be of pathophysiological significance, given that the peptide mediates the gastroprotective effect of primary afferent neuron excitation with capsaicin and other stimuli. Thus, blockade of $CGRP_1$ receptors with $CGRP_{8-37}$

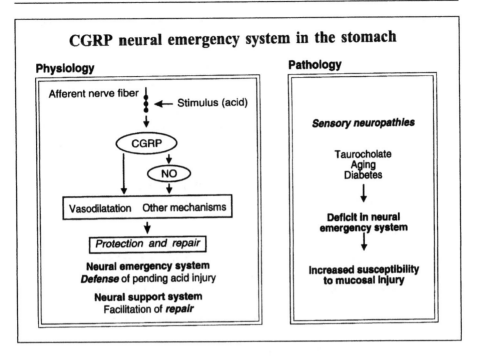

Fig. 12.3. Diagram illustrating the pathophysiological implications of CGRP-releasing afferent nerve fibers in the rat gastric mucosa. The panel headed by "Physiology" depicts the role of these nerve fibers as a neural emergency system in defending the mucosa against pending injury and as a neural support system in aiding the repair of the injured muscosa. The panel headed by "Pathology" lists some conditions which cause malfunction of CGRP-releasing afferent nerve fibers and thereby weaken the resistance of the gastric mucosa to injury. NO, nitric oxide.

prevents the ability of intragastric capsaicin to attenuate experimentally imposed injury[79,82,86] as does immunoneutralization of CGRP with polyclonal[79] and monoclonal[87] antibodies to the peptide. This role of CGRP in gastric mucosal homeostasis is corroborated by the observations that CGRP$_{8-37}$[85] as well as active immunization of rats against CGRP[88] exacerbates experimental injury in the stomach. In addition, CGRP has been found to mediate the gastroprotective effects of a number of other factors that stimulate capsaicin-sensitive afferent nerve fibers, including neurokinin A-related peptides[89] gastrin[90] and amylin.[91] Since the ability of capsaicin to strengthen gastric mucosal defense involves, like that of CGRP, the formation of nitric oxide[79,86,92,93] it would

appear that afferent neuron stimulation results in the local release of CGRP which, via formation of nitric oxide, enforces the resistance of the gastric mucosa against experimental injury (Fig. 12.3).

The parallelism of CGRP and NO as messengers of the hyperemic and protective response to afferent neuron stimulation suggests that vasodilatation is the primary mechanism by which gastric mucosal defense is strengthened.[51] However, there is evidence that protective mechanisms other than hyperemia are also operated by afferent nerve fibers.[51,80,89] In particular, CGRP has been found to counteract the injurious influence of endothelin-1 on the rat gastric mucosa in the absence of hyperemia, which has been taken to suggest that CGRP enhances gastric mucosal resistance by protecting the vascular endothelium from injury.[80] Whether other mechanisms of gastroprotection are operated by CGRP has not yet been elucidated.

Despite these uncertainties it can be hypothesized that the CGRP-mediated rise of gastric mucosal blood flow, which ensues after acid influx into the mucosal tissue, serves an important role in maintaining mucosal homeostasis (Fig. 12.3).[51,73,94] Hyperemia supports a number of gastroprotective mechanisms, including appropriate delivery of bicarbonate to the surface mucus layer, and facilitates the rapid restitution and repair of the wounded mucosa.[95,96] Since they are not tonically active but are stimulated by insults to the mucosa, CGRP-releasing afferent nerve fibers function as a neural emergency system which via vasodilatation and other mechanisms of protection strengthens gastric defense against injury. In addition, these neurons support processes that aid the repair of the injured mucosa (Fig. 12.3).[51]

Pathophysiological evidence for a homeostatic role of CGRP-releasing afferent nerve fibers in the gastric mucosa comes from the observation that sensory neuropathies are liable to weaken the resistance of the tissue to injury (Fig. 12.3). For instance, chronic treatment of rats with oral taurocholate blunts the capsaicin-evoked CGRP release and hyperemia in the gastric mucosa and exacerbates mucosal vulnerability.[97] A similar dysfunction of CGRP-releasing nerve fibers may explain why experimental diabetes induced by streptozotocin impairs gastric mucosal defense.[98] Importantly, the capsaicin- and acid-evoked hyperemia in the rat gastric mucosa declines with age, a change that is associated with a loss of CGRP-containing nerve fibers around submucosal arterioles and a reduction of mucosal resistance to injury.[99,100] The tissue concentrations of CGRP are likewise diminished in the experimentally injured stomach and duodenum,[101] in the experimentally inflamed colon,[102-104] and in inflammatory bowel disease.[105] Although it is not clear whether these peptide changes are primary or secondary to the insult or disease, it is tempting to speculate that CGRP plays a homeostatic role throughout the gut and that disturbance of the CGRP system contributes to the manifestation of the disease.

An exemption to this concept is posed by the pathology which *Clostridium difficile* toxin A induces in the rat ileum. In this model the level of CGRP in afferent neurons increases, and from the effect of CGRP$_{8-37}$ it appears as if CGRP contributes to the diarrhea, inflammation and tissue destruction caused by the toxin.[50,75]

12.3.5. Implications of CGRP in Gastrointestinal Sensitivity and Pain

Since CGRP is a transmitter substance of spinal afferent neurons innervating the gut, it is natural to think of an implication of this peptide in visceral nociception (Fig. 12.4). With the use of CGRP$_{8-37}$ it has indeed been possible to obtain experimental evidence that CGRP participates in visceral pain reactions. For instance, intraperitoneal administration of acetic acid causes a prostaglandin-mediated release of CGRP from afferent nerve fibers and triggers abdominal muscle contractions, a pain reaction that is mimicked by intravenous CGRP.[39,106] The nociceptive response to acetic acid, prostaglandin E_1 and E_2 as well as CGRP is inhibited by capsaicin-induced ablation of extrinsic afferents and pretreatment with CGRP$_{8-37}$.[39,106] Intravenous or intrathecal administration of CGRP$_{8-37}$ is likewise able to prevent acetic acid-evoked colonic inflammation from facilitating the pain response to rectal distention.[107]

While these findings show that CGRP acting via CGRP$_1$ receptors contributes to visceral hypersensitivity and pain, it remains to be examined whether the peptide acts at a peripheral and/or central site within nociceptive afferent pathways. Either possibility is conceivable because, on the one hand, the functional responses to CGRP released within the gut may promote the excitation of nociceptive afferents in the

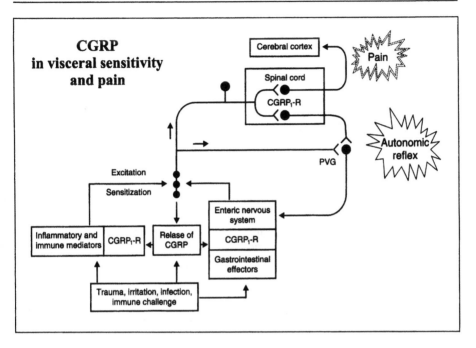

Fig. 12.4. Diagram illustrating the potential implications of CGRP and CGRP$_1$ receptors (CGRP$_1$-R) in visceral sensitivity and pain. The graph shows that, on the one hand, noxious or immunological challenge may release CGRP from afferent nerve endings and the peptide-evoked local reactions could lead to sensitization or excitation of extrinsic afferents. Tissue insult is in addition signalled to the spinal cord or brainstem where CGRP, on the other hand, may participate in the central transmission of nociceptive information to elicit pain and autonomic reflexes. PVG, prevertebral ganglion.

periphery and, on the other hand, CGRP released from central afferent neuron terminals may participate in the transmission of nociceptive information within the spinal cord and brainstem (Fig. 12.4). The former possibility is favored when the pain reaction to peripheral CGRP is considered,[39,106] given that CGRP is unlikely to penetrate the blood-brain barrier to any significant extent. The latter possibility appears likely when the antinociceptive activity of intrathecally administered CGRP$_{8-37}$ is taken into account.[107] It is also worth noting in this context that the spinal afferents which in the gastric mucosa release CGRP in response to acid challenge and thus help maintaining the integrity of the tissue do not seem to participate in the afferent signalling of luminal insults to the spinal cord.[108]

12.4. Conclusion

The experimental findings surveyed in this article indicate that CGRP participates in the pathophysiological regulation of digestive functions, an implication that derives from its role as a transmitter of intrinsic enteric neurons and extrinsic afferent nerve fibers. Enteric neurons expressing CGRP have not yet been fully characterized as to their potential ability to regulate motility and secretory processes within the digestive system. It would appear that, should these roles be verified, CGRP-releasing enteric neurons play a physiological rather than a pathophysiological role in gastrointestinal regulation. In contrast, there is considerable evidence to show that spinal afferent neurons expressing CGRP play an important homeostatic role in gut function. This is related to the fact that CGRP-immunoreactive afferent nerve fibers comprise axons that are sensitive to noxious stimuli. Once stimulated, these nerve fibers release CGRP from their peripheral terminals and in this way can halt motility, increase blood flow, inhibit gastric acid secretion, enforce mucosal resistance to injury and facilitate the repair of the wounded tissue. In addition, they participate in nociception and in pain-activated measures of autonomic homeostasis and thus constitute a neural emergency system with a broad range of actions.

Acknowledgments
The author is grateful to Dr Ulrike Holzer-Petsche for her help with the graphs. Work in the author's laboratory was supported by the Austrian Science Foundation (grant P11834-MED) and the Jubilee Foundation of the Austrian National Bank (grant 6237).

References

1. Holzer P. Calcitonin gene-related peptide. In: Walsh JH, Dockray GJ, eds. Gut Peptides: Biochemistry and Physiology. New York: Raven Press, 1994:493-523.
2. Gates TS, Zimmerman RP, Mantyh CR et al. Calcitonin gene-related peptide-α receptor binding sites in the gastrointestinal tract. Neuroscience 1989; 31:757-770.
3. Sternini C. Enteric and visceral afferent CGRP neurons. Targets of innervation and differential expression patterns. Ann New York Acad Sci 1992; 657:170-186.
4. Mulderry PK, Ghatei MA, Spokes RA et al. Differential expression of α-CGRP and β-CGRP by primary sensory neurons and enteric autonomic neurons of the rat. Neuroscience 1988; 25:195-205.
5. Sternini C, Reeve JR, Brecha N. Distribution and characterization of calcitonin gene-related peptide immunoreactivity in the digestive system of normal and capsaicin-treated rats. Gastroenterology 1987; 93:852-862.
6. Su HC, Bishop AE, Power RF et al. Dual intrinsic and extrinsic origins of CGRP- and NPY-immunoreactive nerves of rat gut and pancreas. J Neurosci 1987; 7:2674-2687.
7. Green T, Dockray GJ. Characterization of the peptidergic afferent innervation of the stomach in the rat, mouse and guinea-pig. Neuroscience 1988; 25:181-193.
8. Costa M, Furness JB, Gibbins IL. Chemical coding of enteric neurons. Prog Brain Res 1986; 68:217-239.
9. Ekblad E, Winther C, Ekman R et al. Projections of peptide-containing neurons in rat small intestine. Neuroscience 1987; 20:169-188.
10. Lee Y, Hayashi N, Hillyard CJ et al. Calcitonin gene-related peptide-like immunoreactive sensory fibers form synaptic contact with sympathetic neurons in the rat celiac ganglion. Brain Res 1987; 407:149-151.
11. Kondo H, Yamamoto M. The ontogeny and fine structure of calcitonin gene-related peptide (CGRP)-immunoreactive nerve fibers in the celiac ganglion of rats. Arch Histol Cytol 1988; 51:91-98.
12. Lindh B, Hökfelt T, Elfvin LG. Distribution and origin of peptide-containing nerve fibers in the celiac superior mesenteric ganglion of the guinea-pig. Neuroscience 1988; 26:1037-1071.
13. Rodrigo J, Polak JM, Fernandez L et al. Calcitonin gene-related peptide immunoreactive sensory and motor nerves of the rat, cat, and monkey esophagus. Gastroenterology 1985; 88:444-451.
14. Bäck N, Ahonen M, Häppölä O et al. Effect of vagotomy on expression of neuropeptides and histamine in rat oxyntic mucosa. Digest Dis Sci 1994; 39:353-361.
15. Suzuki T, Kagoshima M, Shibata M et al. Effects of several denervation procedures on distribution of calcitonin gene-related peptide and substance P immunoreactive fibers in rat stomach. Digest Dis Sci 1997; 42:1242-1254.
16. Gibbins IL, Furness JB, Costa M. Pathway-specific patterns of the coexistence of

substance P, calcitonin gene-related peptide, cholecystokinin and dynorphin in neurons of the dorsal root ganglia of the guinea-pig. Cell Tissue 1987; Res 248:417-437.

17. Kirchgessner AL, Dodd J, Gershon MD. Markers shared between dorsal root and enteric ganglia. J Comp Neurol 1988; 276:607-621.

18. Timmermans J-P, Scheuermann DW, Barbiers M et al. Calcitonin gene-related peptide-like immunoreactivity in the human small intestine. Acta Anat 1992; 143:48-53.

19. Sternini C, De Giorgio R, Anderson K et al. Species differences in the immunoreactive patterns of calcitonin gene-related peptide in the pancreas. Cell Tiss Res 1992; 269:447-458.

20. Jakab G, Webster HF, Salamon I et al. Neural and non-neural origin of calcitonin gene-related peptide (CGRP) in the gastric mucosa. Neuropeptides 1993; 24:117-122.

21. Caterina MJ, Schumacher MA, Tominaga M et al. The capsaicin receptor: A heat-activated ion channel in the pain pathway. Nature 1997; 389:816-824.

22. Geppetti P, Tramontana M, Evangelista S et al. Differential effect on neuropeptide release of different concentrations of hydrogen ions on afferent and intrinsic neurons of the rat stomach. Gastroenterology 1991; 101:1505-1511.

23. Manela FD, Ren J, Gao J et al. Calcitonin gene-related peptide modulates acid-mediated regulation of somatostatin and gastrin release from rat antrum. Gastroenterology 1995; 109:701-706.

24. Ren JY, Wang YP, Liang KX et al. Mechanisms of proton-induced stimulation of CGRP release from rat antrum. Regul Pept 1995; 59:103-109.

25. Holzer P, Barthó L. Sensory neurons in the intestine. In: Geppetti P, Holzer P, eds. Neurogenic Inflammation. Boca Raton: CRC Press, 1996:153-167.

26. Maggi CA, Giuliani S, Zagorodnyuk V. Calcitonin gene-related peptide (CGRP) in the circular muscle of guinea-pig colon: Role as inhibitory transmitter and mechanism of relaxation. Regul Pept 1996; 61:27-36.

27. Maggi CA, Patacchini R, Santicioli P et al. Human isolated ileum: Motor responses of the cirulcar muscle to electrical field stimulation and exogenous neuropeptides. Naunyn-Schmiedeberg's Arch Pharmacol 1990; 341:256-261.

28. Schwörer H, Schmidt WE, Katsoulis S et al. Calcitonin gene-related peptide (CGRP) modulates cholinergic neurotransmission in the small intestine of man, pig and guinea-pig via presynaptic CGRP receptors. Regul Pept 1991; 36:345-358.

29. Sun Y-D, Benishin CG. Effects of calcitonin gene-related peptide on longitudinal muscle and myenteric plexus of guinea pig ileum. J Pharmacol Exp Ther 1991; 259:947-952.

30. Raybould HE, Kolve E, Taché Y. Central nervous system action of calcitonin gene-related peptide to inhibit gastric emptying in the conscious rat. Peptides 1988; 9:735-738.

31. Holzer P, Barthó L, Matusák O et al. Calcitonin gene-related peptide action on intestinal circular muscle. Am J Physiol 1989; 256:G546-G552.

32. Barthó L, Kóczán G, Maggi CA. Studies on the mechanism of the contractile action of rat calcitonin gene-related peptide and of capsaicin on the guinea-pig ileum: effect of hCGRP (8-37) and CGRP tachyphylaxis. Neuropeptides 1993; 25:325-329.

33. Maggi CA, Giuliani S, Santicioli P. CGRP potentiates excitatory transmission to the circular muscle of guinea-pig colon. Regul Pept 1997; 69:127-136.

34. Palmer JM, Schemann M, Tamura K et al. Calcitonin gene-related peptide excites myenteric neurons. Eur J Pharmacol 1986;132:163-170.

35. Mulholland MW, Jaffer S. Stimulation of acetylcholine release in myenteric plexus by calcitonin gene-related peptide. Am J Physiol 1990; 259:G934-G939.

36. Grider JR. CGRP as a transmitter in the sensory pathway mediating peristaltic reflex. Am J Physiol 1994; 266:G1139-G1145.

37. Barthó L, Holzer P. The inhibitory modulation of guinea-pig intestinal peristalsis caused by capsaicin involves calcitonin gene-related peptide and nitric oxide. Naunyn-Schmiedeberg's Arch Pharmacol 1995; 353:102-109.

38. Plourde V, Wong HC, Walsh JH et al. CGRP antagonists and capsaicin on celiac ganglia partly prevent postoperative gastric ileus. Peptides 1993; 14:1225-1229.

39. Julia V, Buéno L. Tachykininergic mediation of viscerosensitive responses to acute inflammation in rats: Role of CGRP. Am J Physiol 1997; 272:G141-G146.

40. Debas HT, Nelson MT, Bunnett NW et al. Selective release of somatostatin by calcitonin gene-related peptide and influence on pancreatic secretion. Ann New York Acad Sci 1992; 657:289-298.

41. Li Y, Kolligs F, Owyang C. Mechanism of action of calcitonin gene-related peptide in inhibiting pancreatic enzyme secretion in rats. Gastroenterology 1993; 105:194-201.

42. Jaworek J, Konturek SJ, Szlachcic A. The role of CGRP and afferent nerves in the modulation of pancreatic enzyme secretion in the rat. Int J Pancreatol 1997; 22:137-146.

43. Zhou Z-C, Villanueva ML, Noguchi M et al. Mechanisms of action of calcitonin gene-related peptide in stimulating pancreatic enzyme secretion. Am J Physiol 1986; 251:G391-G397.

44. Rolston RK, Ghatei MA, Mulderry PK et al. Intravenous calcitonin gene-related peptide stimulates net water secretion in rat colon in vivo. Digest Dis Sci 1989; 34:612-616.

45. Reasbeck PG, Burns SM, Shulkes A. Calcitonin gene-related peptide: Enteric and cardiovascular effects in the dog. Gastroenterology 1988; 95:966-971.

46. Cox HM, Ferrar JA, Cuthbert AW. Effects of α- and β-calcitonin gene-related peptides upon ion transport in rat descending colon. Br J Pharmacol 1989; 97:996-998.

47. McCulloch CR, Cooke HJ. Human α-calcitonin gene-related peptide influences colonic secretion by acting on myenteric neurons. Regul Pept 1989; 24:87-96.

48. Cox HM, Tough IR. Calcitonin gene-related peptide receptors in human gastrointestinal epithelia. Br J Pharmacol 1994; 113:1243-1248.

49. Yarrow S, Ferrar JA, Cox HM. The effects of capsaicin upon electrogenic ion transport in rat descending colon. Naunyn-Schmiedeberg's Arch Pharmacol 1991; 344:557-563.

50. Keates AC, Castagliuolo I, Qiu BS et al. CGRP upregulation in dorsal root ganglia and ileal mucosa during Clostridium difficile toxin A-induced enteritis. Am J Physiol 1998; 274:G196-G202.

51. Holzer P. Neural emergency system in the stomach. Gastroenterology 1998; 114:823-839.

52. Lawson DC, Mantyh CR, Pappas TN. Effect of CGRP antagonist, α-CGRP$_{8-37}$, on acid secretion in the dog. Digest Dis Sci 1994; 39:1405-1408.

53. Kato K, Martinez V, St.-Pierre S et al. CGRP antagonists enhance gastric acid secretion in 2-h pylorus-ligated rats. Peptides 1995; 16:1257-1262.

54. Helton WS, Mulholland MM, Bunnett NW et al. Inhibition of gastric and pancreatic secretion in dogs by CGRP: Role of somatostatin. Am J Physiol 1989; 256:G715-G720.

55. Inui T, Kinoshita Y, Yamaguchi A et al. Linkage between capsaicin-stimulated calcitonin gene-related peptide and somatostatin release in rat stomach. Am J Physiol 1991; 261:G770-G774.

56. Ren JY, Young RL, Lassiter DC et al. Calcitonin gene-related peptide mediates capsaicin-induced neuroendocrine responses in rat antrum. Gastroenterology 1993; 104:485-491.

57. Sandor A, Kidd M, Lawton GP et al. Neurohormonal modulation of rat enterochromaffin-like cell histamine secretion. Gastroenterology 1996; 110:1084-1092.

58. Ren JY, Dunn ST, Tang Y et al. Effects of calcitonin gene-related peptide on somatostatin and gastrin gene expression in rat antrum. Regul Pept 1998; 73:75-82.

59. Kawasaki H, Takasaki K, Saito A et al. Calcitonin gene-related peptide acts as a novel vasodilator neurotransmitter in mesenteric resistance vessels of the rat. Nature 1988; 335:164-167.

60. Han S-P, Naes L, Westfall TC. Calcitonin gene-related peptide is the endogenous mediator of nonadrenergic noncholinergic vasodilation in rat mesentery. J Pharmacol Exp Ther 1990; 255:423-428.

61. Claing A, Télémaque S, Cadieux A et al. Nonadrenergic and noncholinergic arterial dilatation and venoconstriction are mediated by calcitonin gene-related peptide₁ and neurokinin-1 receptors, respectively, in the mesenteric vasculature of the rat after perivascular nerve stimulation. J Pharmacol Exp Ther 1992; 263:1226-1232.

62. Kawasaki H, Nuki C, Saito A et al. NPY modulates neurotransmission of CGRP-containing vasodilator nerves in rat mesenteric arteries. Am J Physiol 1991; 261:H683-H690.

63. Vanner S. Mechanism of action of capsaicin on submucosal arterioles in the guinea pig ileum. Am J Physiol 1993; 265:G51-G55.

64. Nuki C, Kawasaki H, Takasaki K et al. Pharmacological characterization of presynaptic calcitonin gene-related peptide (CGRP) receptors on CGRP-containing vasodilator nerves in rat mesenteric resistance vessels. J Pharmacol Exp Ther 1994; 268:59-64.

65. Galligan JJ, Jiang M-M, Shen K-Z et al. Substance P mediates neurogenic vasodilatation in extrinsically denervated guinea-pig submucosal arterioles. J Physiol 1990; 420:267-280.

66. Vanner S. Corelease of neuropeptides from capsaicin-sensitive afferents dilates submucosal arterioles in guinea pig ileum. Am J Physiol 1994; 267:G650-G655.

67. Holzer P, Guth PH. Neuropeptide control of rat gastric mucosal blood flow. Increase by calcitonin gene-related peptide and vasoactive intestinal polypeptide, but not substance P and neurokinin A. Circ Res 1991; 68:100-105.

68. Li D-S, Raybould HE, Quintero E et al. Role of calcitonin gene-related peptide in gastric hyperemic response to intragastric capsaicin. Am J Physiol 1991; 261:G657-G661.

69. Chen RYZ, Li D-S, Guth PH. Role of calcitonin gene-related peptide in capsaicin-induced gastric submucosal arteriolar dilation. Am J Physiol 1992; 262:H1350-H1355.

70. Holzer P, Wachter C, Jocic M et al. Vascular bed-dependent roles of the peptide CGRP and nitric oxide in acid-evoked hyperaemia of the rat stomach. J Physiol (London) 1994; 480:575-585.

71. Whittle BJR, Lopez-Belmonte J, Moncada S. Nitric oxide mediates rat mucosal vasodilatation induced by intragastric capsaicin. Eur J Pharmacol 1992; 218:339-341.

72. Holzer P, Lippe IT, Jocic M et al. Nitric oxide-dependent and -independent hyperaemia due to calcitonin gene-related peptide in the rat stomach. Br J Pharmacol 1993; 110:404-410.

73. Li D-S, Raybould HE, Quintero E et al. Calcitonin gene-related peptide mediates the gastric hyperemic response to acid back-diffusion. Gastroenterology 1992; 102:1124-1128.

74. Lippe IT, Holzer P. Participation of endothelium-derived nitric oxide but not prostacyclin in the gastric mucosal hyperaemia due to acid back-diffusion. Br J Pharmacol 1992; 105:708-714.

75. Castagliuolo I, LaMont JT, Letourneau R et al. Neuronal involvement in the intestinal effects of Clostridium difficile toxin A and Vibrio cholerae enterotoxin in rat ileum. Gastroenterology 1994; 107:657-665.

76. Maggi CA, Evangelista S, Giuliani S et al. Anti-ulcer activity of calcitonin gene-related peptide in rats. Gen Pharmacol 1987; 18:33-34.

77. Lippe IT, Lorbach M, Holzer P. Close arterial infusion of calcitonin gene-related peptide into the rat stomach inhibits aspirin- and ethanol-induced hemorrhagic damage. Regul Pept 1989; 26:35-46.

78. Evangelista S, Tramontana M, Maggi CA. Pharmacological evidence for the involvement of multiple calcitonin gene-related peptide (CGRP) receptors in the antisecretory and antiulcer effect of CGRP in rat stomach. Life Sci 1991; 50:PL13-PL18.

79. Lambrecht N, Burchert M, Respondek M et al. Role of calcitonin gene-related peptide and nitric oxide in the gastro protective effect of capsaicin in the rat. Gastroenterology 1993; 104:1371-1380.

80. Lopez-Belmonte J, Whittle BJR. The paradoxical vascular interactions between endothelin-1 and calcitonin gene-related peptide in the rat gastric mucosal microcirculation. Br J Pharmacol 1993; 110:496-500.

81. Clementi G, Caruso A, Prato A et al. A role for nitric oxide in the anti-ulcer activity of calcitonin gene-related peptide. Eur J Pharmacol 1994; 256:R7-R8.

82. Kato K, Yang H, Taché Y. Role of peripheral capsaicin-sensitive afferent neurons and CGRP in central vagally mediated gastroprotective effect of TRH. Am J Physiol 1994; 266:R1610-R1614.

83. Evangelista S, Renzi D. A protective role for calcitonin gene-related peptide in water-immersion stress-induced gastric ulcers in rats. Pharmacol Res 1997; 35:347-350.

84. Goso C, Evangelista S, Tramontana M et al. Topical capsaicin administration protects against trinitrobenzene sulfonic acid-induced colitis in the rat. Eur J Pharmacol 1993; 249:185-190.

85. Doi K, Nagao T, Kawakubo K et al. Calcitonin gene-related peptide affords gastric mucosal protection by activating potassium channel in Wistar rat. Gastroenterology 1998; 114:71-76.

86. Merchant NB, Goodman J, Dempsey DT et al. The role of calcitonin gene-related peptide and nitric oxide in gastric mucosal hyperemia and protection. J Surg Res 1995; 58:344-350.

87. Peskar BM, Wong HC, Walsh JH et al. A monoclonal antibody to calcitonin gene-related peptide abolishes capsaicin-induced gastroprotection. Eur J Pharmacol 1993; 250:201-203.

88. Forster ER, Dockray GJ. The role of calcitonin gene-related peptide in gastric mucosal protection in the rat. Exp Physiol 1991; 76:623-626.

89. Stroff T, Plate S, Seyed Ebrahim J et al. Tachykinin-induced increase in gastric mucosal resistance: role of primary afferent neurons, CGRP, and NO. Am J Physiol 1996; 271:G1017-G1027.

90. Stroff T, Plate S, Respondek M et al. Protection by gastrin in the rat stomach involves afferent neurons, calcitonin gene-related peptide and nitric oxide. Gastroenterology 1995; 109:89-97.

91. Clementi G, Caruso A, Cutuli VMC et al. Effect of amylin in various experimental models of gastric ulcer. Eur J Pharmacol 1997; 332:209-213.

92. Peskar BM, Respondek M, Müller KM et al. A role for nitric oxide in capsaicin-induced gastroprotection. Eur J Pharmacol 1991; 198:113-114.

93. Brzozowski T, Drozdowicz D, Szlachcic A et al. Role of nitric oxide and prostaglandins in gastroprotection induced by capsaicin and papaverine. Digestion 1993; 54:24-31.

94. Raybould HE, Sternini C, Eysselein VE et al. Selective ablation of spinal afferent neurons containing CGRP attenuates gastric hyperemic response to acid. Peptides 1992; 13:249-254.

95. Takeuchi K, Ueshima K, Ohuchi T et al. The role of capsaicin-sensitive sensory neurons in healing of HCl-induced gastric mucosal lesions in rats. Gastroenterology 1994; 106:1524-1532.

96. Grønbech JE, Lacy ER. Role of sensory afferent neurons in hypertonic damage and restitution of the rat gastric mucosa. Gastroenterology 1996; 111:1474-1483.

97. Narita M, Takahashi S, Takeuchi K et al. Desensitization of capsaicin-sensitive sensory neurons in rat stomachs on chronic treatment with sodium taurocholate. Jap J Pharmacol 1995; 67:321-328.

98. Tashima K, Konaka A, Kato S, Takeuchi K. Increased susceptibility of the rat gastric mucosa to ulcerogenic stimulation in diabetic conditions: Role of capsaicin-sensitive sensory neurons. Gastroenterology 1997; 112:A310.

99. Grønbech JE, Lacy ER. Role of gastric blood flow in impaired defense and repair of aged rat stomachs. Am J Physiol 1995; 269:G737-G744.

100. Miyake H, Inaba N, Kato S et al. Increased susceptibility of rat gastric mucosa to ulcerogenic stimulation with aging: Role of capsaicin-sensitive sensory neurons. Digest Dis Sci 1996; 41:339-345.

101. Evangelista S, Renzi D, Tramontana M et al. Cysteamine induced-duodenal ulcers are associated with a selective depletion in gastric and duodenal calcitonin gene-related peptide-like immu-

noreactivity in rats. Regul Pept 1992; 39:19-28.

102. Eysselein VE, Reinshagen M, Cominelli F et al. Calcitonin gene-related peptide and substance P decrease in the rabbit colon during colitis. A time study. Gastroenterology 1991; 101:1211-1219.

103. Renzi D, Tramontana M, Panerai C et al. Decrease of calcitonin gene-related peptide, but not vasoactive intestinal polypeptide and substance P, in the TNB-induced experimental colitis in rats. Neuropeptides 1992; 22:56-57.

104. Reinshagen M, Patel A, Sottili M et al. Action of sensory neurons in an experi-mental rat colitis model of injury and repair. Am J Physiol 1996; 270:G79-G86.

105. Kimura M, Masuda T, Hiwatashi N et al. Changes in neuropeptide-containing nerves in human colonic mucosa with inflammatory bowel disease. Pathol Int 1994; 44:624-634.

106. Friese N, Diop L, Chevalier E et al. Involvement of prostaglandins and CGRP-dependent sensory afferents in peritoneal irritation-induced visceral pain. Regul Pept 1997; 70:1-7.

107. Plourde V, St.-Pierre S, Quirion R. Calcitonin gene-related peptide in viscerosensitive response to colorectal distension in rats. Am J Physiol 1997; 273:G191-G196.

108. Schuligoi R, Herzeg G, Wachter C et al. Differential expression of c-fos messenger RNA in the rat spinal cord after mucosal and serosal irritation of the stomach. Neuroscience 1996; 72:535-544.

CHAPTER 13

CGRP and Neurogenic Vasodilatation: Role of Nerve Growth Factor in Diabetes

Gavin S. Bennett, Radhika Kajekar, Neil. E. Garrett, Lara T. Diemel, David R. Tomlinson and Susan D. Brain

13.1. Introduction

N eu'ropeptides are synthesised and released by several nerve types. However, it is now realised that C and Aδ fibres can function as vasodilator neurones as a consequence of their ability to release neuropeptides, of which calcitonin gene-related peptide (CGRP) is of major importance. Sensory C-fibres are unmyelinated and often referred to as capsaicin-sensitive. Some Aδ fibres also show sensitivity to capsaicin. Capsaicin acts via a specific binding site on sensory nerve to excite and then desensitise and, in some cases, destroy sensory nerves.[1] The treatment of neonatal rats with neurotoxic doses of capsaicin leads to a depletion of neuropeptides from capsaicin-sensitive nerves and the appearance of cutaneous lesions.[2,3] This suggests that neuropeptides play a physiological role in the maintenance of a healthy skin. Indeed there is evidence from the study of diseases that affect the peripheral vasculature (e.g. Raynaud's phenomenon and diabetes) that a depletion of neuropeptides is associated with the vascular and cutaneous problems which are observed.

13.2. Raynaud's Disease

In Raynaud's phenomenon a lack of reflex vasodilatation is observed in certain individuals after exposure to specific stresses of which the most common is exposure to the cold. This leads to pain, usually in the digits, as a result of local ischemia. There appears to be a loss of the normal balance between endogenous vasoconstrictors and vasodilators. It has been suggested that there is a deficiency of the release of vasodilator calcitonin gene-related peptide CGRP from sensory nerves and that the peripheral administration of CGRP is of use therapeutically. This is supported by clinical studies where infused CGRP has been compared with prostacyclin and found to be better tolerated and more effective in terms of recovery after cooling of the arm and the subjective scoring by patients.[4,5]

13.3. Diabetes

Neuropathy is often observed in patients with long term diabetes. The neuropathy initially affects longer axons, especially those innervating the digits. There is also a loss of axon-reflex vasodilatation that is associated with loss of nociception and with the vascular complications observed in diabetes.[6] Sensory nerve dysfunction is observed in these patients.[7,8] There is now abundant evidence that sensory nerve dysfunction is related to a deficiency in neuropeptides

The CGRP Family: Calcitonin Gene-Related Peptide (CGRP), Amylin, and Adrenomedullin
edited by David Poyner. ©2000 Landes Bioscience.

including CGRP. However, in the case of diabetes much of this evidence has been obtained from the study of diabetes in animals, in particular the induction in the rat of type-1 insulin-deficient diabetes by streptozotocin. A depletion of substance P and CGRP occurs in this model of diabetes and this is insulin-sensitive.[9]

13.4. Nerve Growth Factor

It is now realised that growth factors, of which nerve growth factor (NGF) has been most often studied, have an important role in the regulation of neuropeptide synthesis and activity in sensory nerves,[10] see figure 13.1. NGF can lead to an upregulation of sensory nerve activity and plasticity in inflammatory conditions such as arthritis.[11,12] By comparison, in states where there is sensory nerve deficiency, endogenous NGF has been shown to be deficient. In diabetic neuropathy there are reduced levels of NGF in skin and muscle, together with a reduced uptake of NGF into sensory nerves and in turn reduced transport of NGF to ganglia.[13] Exogenous NGF acts to increase levels of neuropeptides, including CGRP in diabetic rats.[14,15]

The findings outlined above have been among those which have led to the suggestion that the exogenous administration of NGF may be a useful therapeutic approach in diabetic neuropathy. Many of the studies which have been carried out to date to determine the effectiveness of NGF, in restoring sensory nerve function, have been carried out in vitro. The effect of NGF on sensory nerve function in vivo and in particular on the vasoactive effects of stimulated sensory nerves has not been investigated. We therefore decided to establish an in vivo model of rat diabetes. This has enabled us to investigate the ability of NGF to restore sensory nerve mediated vasoactive function. We examined the vasoactive response observed in the rat made diabetic with streptozotocin and examined the vasoactive responses observed in response to stimulation of the sensory saphenous nerve.

13.5. The Rat Saphenous Model of Neurogenic Inflammation

The saphenous nerve contains sensory nerve fibres. The stimulation of this nerve in the anaesthetised normal rat leads to increased blood flow and oedema formation in the hind paw skin (see ref. 16). The vasodilatation is observed in response to a short stimulation of the sensory nerve (e.g. 10 V, 1 ms, 2 Hz for 30 s) whilst oedema formation, in addition to increased blood flow, is observed with a stronger stimulation (e.g. 10 V, 1 ms, 2 Hz for 5 min). The increased blood flow can be assessed by laser Doppler flowmetry and oedema formation measured by the extravascular accumulation of radiolabelled albumin. The methods are briefly as follows: Surgical anaesthesia is achieved and then the hind limbs are shaved and depilated and both saphenous nerves dissected clear of surrounding tissue. The nerves are then ligated to ablate central input and immersed in mineral oil for the duration of the experiment. A laser Doppler blood flow meter (Moor Instruments, UK) is used to assess cutaneous blood flow. Data is expressed as arbitrary units (flux) that increases in a linear manner with an increase in blood flow. The flux is a measure of Doppler shift in the reflected laser light, caused by the number of red blood cells that pass through the path of the laser probe multiplied by their velocity. The laser Doppler probe is placed on an area of skin innervated by the saphenous nerve.[16] Rats are left until a steady basal blood flow is seen in both paws. One saphenous nerve is then placed on a bipolar platinum electrode and stimulated. The other paw acts as a sham control. The blood flow response in both paws is recorded. Oedema formation can either be measured instead of increased blood flow, or measured after an additional stimulation of the saphenous nerve (10 V, 2 Hz, 1 ms for 5 min). For the measure of oedema formation all rats are given ^{125}I-albumin (1μCi) and Evan's Blue (0.2-0.5 ml of 2.5% w/v in saline) i.v. via the tail vein. The radioactivity allows quantification of the innervated area and the

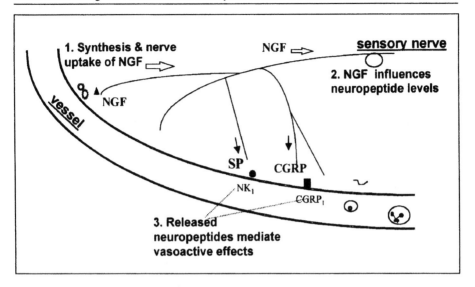

Fig. 13.1. Sensory neuropeptides and microvascular reactivity. NGF is produced in the periphery by a variety of cells and the synthesis is increased at inflammatory sites. The NGF is taken up into sensory nerve terminals by a trk-A-receptor dependent mechanism (shown be the black triangles, 1). The NGF is transported to the dorsal root ganglia (depicted by the open arrows) where it acts to influence neuropeptide levels (2). The neuropeptides can then be released peripherally at sites close to microvascular blood vessels (3). In the case of substance P and neurokinin A, the principle microvascualar response is increased venular permeability, leading to oedema formation, mediated by the NK_1 receptor. By comparison, in the case of CGRP, the principle microvascular response is vasodilatation mediated by the $CGRP_1$ receptor.

Evan's Blue allows a clear visualisation. Then after stimulation, a blood sample is taken and the experiment terminated by killing the rat. The skin of the innervated area in both hind paws is removed and weighed. The ßradioactivity of blood (in the form of plasma) and skin is measured and oedema formation calculated.

Stimulation of the saphenous nerve (10 V, 2 Hz, 1 ms) for 30s in the normal naive untreated rat produces an increase in blood flow in the paw skin, as previously shown. [16,17] Cutaneous blood flow in the innervated area returns to basal at approximately 40 min after stimulation in the naïve rat, see figure 13.2. The vasodilator response is substantially inhibited by the CGRP antagonist, $CGRP_{8-37}$,[16,18] which indicates an involvement of $CGRP_1$ receptors. The increased microvascular permeability response in the normal rat which leads to the oedema formation is considered

to be mediated by a tachykinin (presumed to be substance P), as the response is inhibited by neurokinin (NK) receptor type-1 antagonists.[19]

13.6. Neurogenic Vasoactive Responses in the Diabetic Rat

It has been previously shown that a reduced neurogenic oedema formation is observed in the skin of streptozotocin-induced diabetic rats.[20] We have investigated neurogenic vasodilator responses in rats with streptozotocin-induced diabetes to determine the ability of NGF to restore CGRP vasodilator activity in diabetes.[22]

Male Wistar rats, weighing 200-300 g at the start of the experiment were used. The rats were divided into four groups for the diabetic study and treated as follows: three groups were made diabetic by intraperitoneal (i.p.) injection of streptozotocin (50

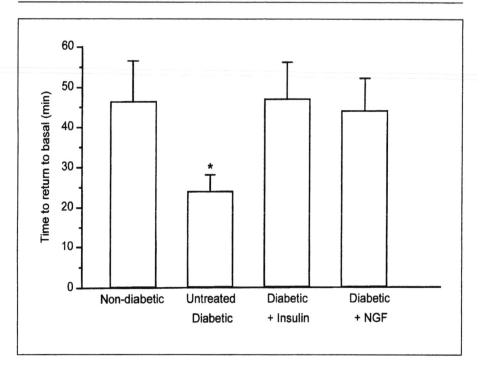

Fig. 13.2. Effect of diabetes on neurogenic vasodilatation. The time taken for blood flow to return to basal for the four treatment groups is shown. All results are mean ± s.e.mean, n=6-10. Results differing significantly from control groups are shown as * P<0.05, as assessed by ANOVA followed by Dunnett's post-test.

mg/kg, Sigma, UK). The fourth group served as a non-diabetic control. After four weeks, one group of diabetic animals received continuous-delivery (4 IU/day) subcutaneous (s.c.) insulin implants (Linplant; Møllegaard, Ejby, Denmark) placed at the back of the neck under halothane anaesthesia.[21] Another group received human recombinant NGF treatment (0.2 mg/kg s.c. at the back of the neck, three times per week, Genentech, San Francisco, USA). The third group of diabetic animals remained untreated. The animals were then maintained for a further four weeks before neurogenic vasodilatation experiments.

Body weights were recorded weekly. Control (non-diabetic) animals grew steadily over the eight week experimental period. The three groups of diabetic animals all failed to grow significantly during the first four weeks of the experiment. Insulin treatment (4 IU/day, s.c., from slow-release implant) during weeks four to eight caused a significant increase in body weight, to a level approaching that seen in the control animals. Treatment with NGF during weeks four to eight had no effect on body weight. A blood sample was taken at four days via the tail vein and blood glucose measured using a glucose oxidase strip-operated reflectance meter (Reflolux II, BCL, Boehringer Mannheim). Any streptozotocin-treated animals with a blood glucose level below 15 mM at four days were excluded from the study. Blood glucose levels were also measured immediately at the start of neurogenic vasodilatation experiments. Blood glucose levels were significantly elevated in all diabetic groups compared with non-diabetic controls, with the insulin-treated diabetic group showing a significantly reduced level compared with uncontrolled diabetics. Glucose levels in NGF-treated diabetics were not significantly different to those seen in untreated diabetics (see ref. 22).

In the diabetic study, the peak increase in blood flow seen following stimulation was similar in all four groups with no significant differences. By comparison the duration of response, was significantly attenuated in the diabetic group as shown in figure 13.2. This indicates that the overall neurogenic vasodilator response, where CGRP is known to be a mediator, is attenuated in diabetic rats. Furthermore, the results shown in figure13.2 demonstrate that both insulin- and NGF-treatment can prevent this attenuation of the neurogenic vasodilatation. Oedema formation was measured during a second stimulation period (10 V, 2 Hz, 1 ms for 5 min) after the blood flow measurements as detailed in.[22] The oedema was significantly reduced in untreated diabetic rats, to an amount that was not different between stimulated and sham paws. Insulin prevented the loss and NGF partially prevented the loss. This is in comparison to the responses to the intradermal injection of exogenous substance P in the presence and absence of CGRP. These responses, to exogenous neuropeptides, were similar between groups.[22]

13.6.1. NGF and the Diabetic Rat

The duration of the vasodilator response was attenuated in the diabetic rat, indicating a reduced activity of this neurogenic vasodilator system. The lack of effect on the peak response in diabetes is of interest. A range of vasodilator mediators, in addition to CGRP, have been studied in diabetes. For example, nitric oxide may be playing an altered vasodilator role in the diabetic rat[23] and influencing these results. Interestingly, whilst sensory nerve-mediated vasodilatation is principally mediated by CGRP,[16] we have evidence that the initial neurogenic vasodilatation may not be[18] and this may be relevant to the present findings. Treatment with insulin or NGF prevented this dysfunction and thus reveals an ability of NGF to restore sensory nerve-mediated microvascular dilatation in diabetes. It is therefore suggested that NGF

is acting, at least in part, to restore the sensory neurogenic component in diabetes. This finding furthers the hypothesis that NGF may have a therapeutic potential in diabetes. The oedema results supports this finding as the loss of neurogenic oedema formation is partially reversed by NGF.

NGF did not markedly affect the failure of the streptozotocin-treated rats to gain weight or suffer from hyperglycaemia whereas insulin significantly attenuated both symptoms. Furthermore the systemic administration of NGF is associated with hyperalgesia. The dosing regime of NGF used in this study has previously been shown to restore levels of substance P and CGRP in the sciatic nerve without the hyperalgesic side effects which are seen with higher doses.[23] The hyperalgesia is related to the effect of NGF on sensory nerves as well as on other cellular components. Thus it is possible that the use of NGF as a therapeutic treatment in humans will be improved if the hyperalgesia is prevented by some additional therapy. However, in the case of diabetes, there is the possibility of coadministering NGF and insulin, and this may be beneficial in chronic diabetes, where polyneuropathy and microvascular complications are most likely to occur. Reduced levels of substance P but not of CGRP have been reported in the skin of patients displaying mild diabetic neuropathy.[24] By comparison, levels of both peptides are reduced in more severe neuropathic patients, indicating that the levels of CGRP may be more tolerant of diabetes-induced NGF depletion. One possibility for this is that substance P is found primarily in small-diameter C-fibres, thought to be a target in diabetic neuropathy, while CGRP is found, in addition, in larger diameter Aδ fibres which may be more resistant to neuropathy in diabetes.

The results provide additional evidence that reduced endogenous NGF levels are central to the neurological and microvascular complications in diabetes.[23,24] Neuropeptides are produced from genes which are regulated by NGF.[25] Experimental evidence suggests that i) expression of NGF is lower in diabetic skin ii) uptake into nerves, retrograde nerve transport and thus delivery of NGF to nerve ganglia is reduced in diabetes[13] and iii) a reduction in the expression of the NGF trkA mRNA and receptors is observed which can be reversed by NGF treatment,[14,15,26] and by treatments which increase NGF production.[27]

13.7. Conclusion

In conclusion there is abundant evidence, as discussed above, to indicate that endogenous NGF is lacking in diabetes. This study provides data to indicate that the restoration of NGF levels is accompanied by a reversal of the neurogenic microvascular vasodilator dysfunction. The consequences of these changes are difficult to analyse at this stage. The neuropeptides have potent vasoactive effects and could have a physiological role in the regulation of microvascular tone and permeability and in wound healing in skin.[28] A lack of NGF in other peripheral vascular diseases is not well characterised. It is possible that if CGRP release is lacking in peripheral vascular conditions such as Raynaud's disease, growth factors are deficient. Thus treatment with NGF, or an associated growth factor may be a therapeutic possibility for a range of vascular diseases.

Acknowlegments

We thank Pfizer (Radhika Kajekar) the British Heart Foundation (Gavin Bennett) the British Diabetic Association (Neil Garrett) and the Medical Research Council (Lara Diemel) for funding.

References

1. Szolcsányi J Neurogenic inflammation: reevaluation of axon reflex theory. In: Geppetti P, Holzer P, eds. Neurogenic Inflammation. Boca Raton: CRC Press, 1996:33-42.
2. Maggi CA, Pattachini R, Rovero P et al. Cutaneous lesions in capsaicin-pretreated rats. A trophic role of capsaicin-sensitive afferents? Naunym-Schmiedeberg's Arch Pharmacol 1987; 336:538-545
3. Thomas DR, Dubner R, Ruda MA. Neonatal capsaicin treatment in rats results in scratching behaviour with skin damage: Potential model of non-painful dyesthesia. Neurosci Lett 1994; 171:101-104.
4. Shawket S, Dickerson C, Hazleman B et al. Prolonged effect of CGRP in Raynaud's patients: A double blind randomised comparison with prostacyclin. Br J Clin Pharmacol 1991; 32:209-213
5. Bunker CB, Reavley C, O'Shaughnessy DJ et al. Calcitonin gene-related peptide in treatment of severe peripheral vascular insufficiency in Raynaud's phenomenon. Lancet 1993; 342:80-83.
6. Johnson PC, Doll SC. Dermal nerves in human diabetic subjects. Diabetes 1984; 33:244-250.
7. Dyck PJ. New understanding and treatment of diabetic neuropathy. N Engl J Med 1992; 326:1287-1288.
8. Levy DM, Terenghi G, Gu X-H et al. Immunohistochemical measurements of nerves and neuropeptides in diabetic skin: Relationship to tests of neurological function. Diabetologia 1992; 35:889-897.
9. Diemel LT, Stevens EJ, Willars GB et al. Depletion of substance P and calcitonin gene-related peptide in sciatic nerve of rats with experimental diabetes; effects of insulin and aldose reductase inhibition. Neurosci Lett 1992; 137:253-256.
10. Lindsay RM, Wiegand SJ, Altar CA et al. Neurotrophic factors: From molecule to man. Trends Neurosci. 1994; 17:182-190.
11. Donnerer J, Schuligoi R, Stein C. Increased content and transport of substance P and calcitonin gene-related peptide in sensory nerves innervating inflamed tissues: Evidence for a regulatory function of nerve growth factor in vivo. Neuroscience 1992; 49: 693-698
12. Woolf CJ, Safieh-Garabedian B, Ma QP et al. Nerve growth factor contributes to the generation of inflammatory sensory hypersensitivity. Neuroscience 1994; 62:327-331.
13. Helleweg R, Raivich G, Hartung H-D et al. Axonal transport of endogenous nerve growth factor (NGF) and NGF receptor in experimental diabetic neuropathy. Exp Neurol 1994; 130:24-30.

14. Apfel SC, Aresso JC, Brownlee M et al. Nerve growth factor administration protects against experimental diabetic sensory neuropathy. Brain Res 1994; 634:7-12.

15. Diemel LT, Brewster WJ, Fernhough P et al. Expression of neuropeptides in experimental diabetes; effects of treatment with nerve growth factor or brain-derived neurotrophic factor. Mol Brain Res 1994; 21:171-175.

16. Escott KJ, Brain SD. Effect of a calcitonin gene-related peptide antagonist (CGRP$_{8-37}$) on skin vasodilatation and oedema induced by stimulation of the rat saphenous nerve. Br J Pharmacol 1993; 110:772-776.

17. Gamse R, Saria A. Antidromic vasodilatation in the rat hindpaw measured by laser Doppler flowmetry: Pharmacological modulation. J Auton Nerv Syst 1987; 19:105-111.

18. Escott KJ, Beattie, DT, Connor HE et al. Trigeminal ganglion stimulation increases facial skin blood flow in the rat: A major role for calcitonin gene-related peptide. Brain Res 1995; 669:93-99.

19. Emonds-Alt X, Doutremepuich J-D, Heaulme M et al. In vitro and in vivo biological activities of SR140333, a novel potent non-peptide tachykinin NK$_1$ receptor antagonist. Eur J Pharmacol, 1993; 250:403-413.

20. Gamse R, Jancsó G. Reduced neurogenic changes in inflammation in streptozotocin-diabetic rats due to microvascular changes but not to substance P depletion. Eur J Pharmacol 1985; 118:175-180.

21. Stephens MJ, Feldman EL, Greene DA. The aetiology of diabetic neuropathy: The combined roles of metabolic and vascular defects. Diab Med 1995; 12:566-579.

22. Bennett GS, Garrett NE, Diemel LT et al. Neurogenic cutaneous vasodilatation and plasma extravasation in diabetic rats: Effect of insulin and nerve growth factor. Br J Pharmacol 1998; 124:1573-1579.

23. Fernyhough P, Diemel LT, Brewster WJ et al. Altered neurotrophin mRNA in peripheral nerve and skeletal muscle of experimentally diabetic rats. J Neurochem 1995; 64:1231-1237.

24. Anand P. Neurotrophins and peripheral neuropathy. Phil Trans R Soc Lond B 1996; 351:449-454.

25. Lindsay RM, Harmar AJ. Nerve growth factor regulates expression of neuropeptide genes in adult sensory neurons. Nature 1989; 337:362-364.

26. Maeda K, Fernhough P, Tomlinson DR. Regenerating sensory neurones of diabetic rats express reduced levels of mRNA for GAP-43, g-preprotachykinin and the nerve growth factor receptors, trkA and p75^{ngfr}. Mol Brain Res 1996; 37:166-174.

27. Garrett NE, Malcangio M, Dewhurst M et al. Alpha-lipoic acid corrects neuropeptide deficits in diabetic rats via induction of trophic support. Neurosci Lett 1997; 222:191-194.

28. Brain SD. Sensory neuropeptides: Their role in inflammation and wound healing. Immunopharmacol 1997; 37:133-152

CHAPTER 14

Neurogenically-Derived Pathological Hyperaemia in Articular Tissues

W.R. Ferrell, J.J. McDougall, J.C. Lockhart and L. McMurdo

14.1. Introduction

Pathology of synovial joints commonly arise through injury or as a consequence of acute or chronic inflammatory diseases. Rupture of the anterior cruciate ligament (ACL) of the knee joint, the most common articular injury in football and other such contact sports, can end a lucrative sporting career whilst the chronic joint inflammation of rheumatoid arthritis can produce crippling deformity of many joints leading to chronic pain, loss of function and disfigurement. The mechanisms responsible for such diseases remain obscure but there is growing evidence to suggest that a component of the pathogenesis in these conditions may be related to changes in vascular function. This article begins with the putative physiological function of calcitonin gene-related peptide (CGRP) in maintaining synovial perfusion in the face of sympathetic vasoconstrictor tone and proceeds to discuss pathophysiological aspects of the regulation of joint perfusion and how CGRP may influence or be influenced by the disease process.

14.2. Physiological Role for CGRP in Regulating Knee Joint Perfusion

Calcitonin gene-related peptide is known to be a potent vasodilator agent in many vascular beds and in many species, including man.[1] Contained within small diameter nerve fibres innervating the ligaments associated with synovial joints in rats[2,3] rabbits[3] and man,[3] CGRP is known to powerfully vasodilate blood vessels in the capsule[4] and ligaments[5] of such joints. Topical application of CGRP to the joint capsule in both the rat and rabbit elicits long-lasting (>10min) dose-dependent vasodilatation (Fig. 14.1A). Blood vessels in the epiligament, which invests the medial collateral ligament (MCL) of the rabbit knee joint and supplies the substance of the ligament, are similarly vasodilated by CGRP (Fig. 14.1B).

More interesting still is the observation that tonic release of CGRP occurs in both the joint capsule and ligament. Local administration of the $CGRP_{8-37}$ fragment, which has been shown to be a selective and potent CGRP receptor antagonist in vivo,[6-8] causes a reversible and dose-dependent fall in joint capsule perfusion in both the rat[4] and rabbit[5] (Fig. 14.2A). In the rabbit knee, $CGRP_{8-37}$ elicits a similar reduction in MCL perfusion (Fig. 14.2B). This provides the first evidence for continuous release of endogenous CGRP in the joint which acts as a vasodilator to oppose vasoconstrictor "tone." Competitive antagonism of endogenously released CGRP would leave sympathetic and any other local vasoconstrictor influences unopposed, resulting in vasoconstriction. This vasoconstrictor effect is substantially attenuated in animals with chronic knee joint

The CGRP Family: Calcitonin Gene-Related Peptide (CGRP), Amylin, and Adrenomedullin, edited by David Poyner. ©2000 EUREKAH.COM.

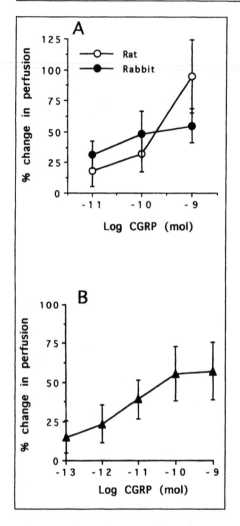

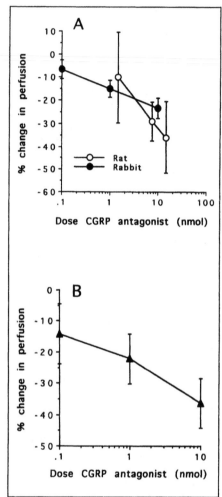

Fig 14.1. Vasodilator responses, measured by laser Doppler perfusion imaging, to topical application of CGRP to the knee joint capsule (A) in rats (n=8) and rabbits (n=8). Topical CGRP application to the rabbit MCL (B) also elicits vasodilatation (n=8). Data shown as means ± SEM.

Fig 14.2. Vascular responses to topical application of CGRP$_{8-37}$ to the knee joint capsule (A) in rats (n=4-8) and rabbits (n=5). Topical CGRP$_{8-37}$ application to the rabbit MCL (B) also elicits vasoconstriction (n=5). Data shown as means ± SEM.

denervation[4,5] and CGRP$_{8-37}$ does not significantly affect responses to other vasodilator agents such as acetylcholine and sodium nitroprusside (Fig. 14.3), ruling out a non-specific effect of the antagonist and providing strong evidence that this phenomenon is dependent on release of CGRP

from sensory nerve terminals. The actual mechanism responsible for such tonic CGRP release and the factors which regulate it have still to be elucidated. It is hypothesised that tonic release of CGRP is likely to play an important physiological role assisting the maintenance of articular

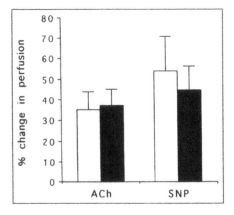

Fig 14.3. Vasodilator responses to topical application of acetylcholine (ACh 10nmol) and sodium nitroprusside (SNP 10nmol) to the rat knee joint capsule did not differ significantly under control conditions (open columns) and in the presence of 15 nmol, $CGRP_{8-37}$ (closed columns). Data shown as means ± SEM; n=4-5.

tissue perfusion by counteracting the known tonic constrictor effect of the sympathetic efferent innervation of the joint and perhaps other locally-derived vasoconstrictor influences. With such a pronounced physiological role, the question of whether CGRP plays a pathological role following joint injury or in inflammatory joint disease is then raised.

14.3. Pathophysiological Role for CGRP in Joint Trauma

Ligaments are essential for maintaining joint stability and preventing abnormal displacements during motion. Injury to a ligament often results in joint laxity and the consequent abnormal changes in joint mechanics culminate in articular cartilage degradation and ultimately osteoarthritis.[9,10] In order to study these effects more fully, a number of different animal models have been developed which aim to mimic the clinical presentation of ligament injury.[11] As mentioned earlier, the most common trauma leading to joint laxity is rupture of the ACL which has been reproduced ex-

perimentally by surgical transection of the ligament[12,13] and ultimately osteoarthritis supervenes.[9,13,14] Loss of joint stability subjects other non-injured ligaments to aberrant loading patterns such that they themselves become injured in a process of microtrauma by proxy.[11] It is known that deterioration of the functional properties of the MCL in injured rabbit knees are closely correlated with substantial increases in ligament perfusion.[15] This is possibly related to the rate of formation of interstitial fluid in the substance of the ligament since change in the water content of a ligament is known to alter its viscoelastic behaviour.[16] Increased perfusion would clearly alter the balance of Starling forces to enhance interstitial fluid formation in the ligament. The mechanism underlying this hyperaemia is obscure but one possibility could be an excessive release of CGRP resulting from repetitive joint injury. Indeed, gap injury to the rabbit MCL results in an increase in CGRP-immunoreactivity in the healing zone of the ligament suggesting that this neuropeptide may be involved in the vascular component of soft tissue repair.[3]

To investigate this possibility, we have examined the effect of knee joint instability, induced by ACL section, on ligament perfusion measured using laser Doppler imaging.[17,18] ACL section of the rabbit knee caused a significant increase in knee joint perfusion at four weeks post-transection (P<0.0024 two-tailed unpaired Student t-test; n=10 and 9 for normal and ACL-sectioned knees respectively) with flow almost doubling compared to control values for the MCL (Fig. 14.4). Denervation ten days prior to ACL section abolished this injury-induced hyperaemia (P=0.013 compared to ACL section alone), the basal perfusion being similar in intact and denervated ACL-sectioned knees (P=0.61). These findings clearly indicate that ligament injury-induced hyperaemia is neurogenically mediated, possibly via neuropeptidergic nerves. Examination of the MCL vascular response to CGRP in the ACL-sectioned

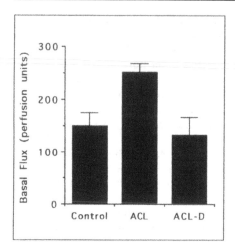

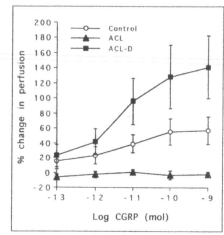

Fig 14.4. Perfusion changes in the rabbit MCL in intact knees (control; n=10), 4 weeks following ACL section (ACL; n=9) and in animals whose knees had been denervated ten days prior to ACL section (ACL-D; n=5). Data shown as means ±SEM.

Fig. 14.5. Dose response relationships to topical application of CGRP to MCL in intact knees (control; n=8), 4 weeks following ACL section (ACL, n=4) and in animals whose knees had been denervated ten days prior to ACL section (ACL-D; n=5). Data shown as means ± SEM.

knee revealed a complete absence of the normal dose-dependent vasodilatation found in this tissue (Fig. 14.5). However, in the denervated ACL-sectioned knee the CGRP response is not only restored but is actually significantly enhanced (Fig. 14.5). Statistical analysis (2-way ANOVA) reveals a significant difference between the CGRP response in intact and ACL-sectioned knees (P<0.0001) and between the intact and the denervated ACL-sectioned knee (P<0.0001). Likewise, the denervated ACL-sectioned and ACL-sectioned joints also differed significantly (P<0.0001). The decreased response to CGRP in ACL-sectioned joints could be explained if there was dysfunction of the sympathetic efferent innervation of the ligament, resulting in loss of constrictor "tone" and hence hyperaemia (Fig. 14.4). Vasodilatation of a blood vessel can only be observed against a background of tonic vasoconstriction, and thus CGRP's lack of effect could be ascribed to such an anomaly. However, electrical stimulation of the saphenous nerve, which supplies

the medial aspect of the rabbit knee joint (including the MCL),[19] shows no difference between intact and ACL-sectioned knees (Fig. 14.6), and so no dysfunction of sympathetic postganglionic fibres innervating the ligament is implicated by these experiments. With sympathetic vasomotor responses preserved in the unstable knee, an alternative explanation is that increased release of CGRP in the injured joint may be responsible for the post-traumatic hyperaemic response to joint instability (Fig. 14.4). Such chronically increased secretion of CGRP could also result in downregulation of CGRP receptors, hence diminishing the response to exogenous CGRP.

Interestingly, in the denervated ACL-sectioned knee the response to CGRP is enhanced, even though the sympathetic fibres responsible for vasoconstrictor tone reaching the MCL via the saphenous nerve have been transected. This suggests that there may be other routes by which sympathetic postganglionic fibres reach the knee joint. One such

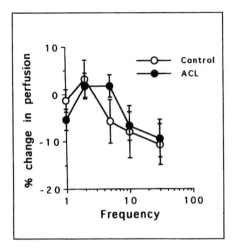

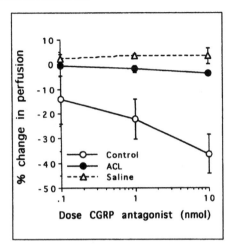

Fig 14.6. Vasoconstrictor responses in the MCL to electrical stimulation (20V, 1msec) of the saphenous nerve in intact knees (control; n=5) and in ACL-sectioned knees (ACL; n=5).

Fig 14.7. Vascular responses to topical application of $CGRP_{8-37}$ to the rabbit MCL in intact knees (control; n=5) and in ACL-sectioned knees)ACL; n=4-5). Administration of saline to ACL-sectioned knees at similar time points is shown for comparison (n=5).

route could be via the sciatic nerve since transection of the saphenous nerve alone leads to only a partial depletion of nerves in the distal medial plantar artery, while the combination of both sciatic and saphenous nerve resection causes a complete loss of perivascular innervation in this region.[20] Alternatively, there could be other vasoconstrictor influences acting on ligament blood vessels. Receptors for angiotensin have been demonstrated in articular tissues,[21] and recent studies in our laboratory have shown that exogenously applied angiotensin II significantly reduces synovial perfusion (unpublished observations). In addition, the endothelins could also contribute to local vasoconstrictor effects. The enhanced response to CGRP in the denervated knee is suggestive of a denervation hypersensitivity response which may be the consequence of CGRP receptor upregulation in the absence of CGRP release in the denervated joint. This further supports the concept of tonic CGRP release from sensory nerve terminals.

If excess production of CGRP does occur in the unstable knee joint, administration of $CGRP_{8-37}$ would be expected to result in dose-dependent vasoconstriction similar to the situation prevailing in the intact knee joint. There appears to be a weak dose-dependent effect of the CGRP antagonist in the ACL-sectioned knee joint (Fig. 14.7), but this is considerably less in magnitude than the response in the intact knee which is also much more significant (P=0.0001; 2-way ANOVA). The effect of $CGRP_{8-37}$ in the ACL-sectioned knee, although small in magnitude, is nevertheless significant (P<0.002, 2-way ANOVA) compared to the effect of application of saline to the knee (Fig. 14.7). This smaller effect of the antagonist in the ACL-sectioned knee could be explained if the production of CGRP in the injured knee is greater than in the intact knee. This would require larger doses of the competitive antagonist $CGRP_{8-37}$ to elicit the same degree of vasoconstriction i.e., the dose-response curve is consistent with a shift to the right.

The functional significance of this excess release of CGRP is obscure. The transitory nature of neuropeptide action on the microvasculature of normal joints may be an important self-regulatory mechanism for the animal, contributing to a short-term but effective healing process following injury. In the unstable joint the potential for repetitive injury is greatly increased and it is therefore possible that with each episode of injury, additional CGRP is released, over and above the physiological basal release, perhaps as a consequence of the "axon reflex".[22,23] Chronically elevated CGRP tissue content would lead to ligament hyperaemia and this would increase interstitial fluid formation, altering the viscoelastic properties of the ligament and decreasing its functional integrity. This leads to further joint laxity and instability, predisposing the joint to further episodes of injury and setting up a positive feedback cycle (Fig. 14.8) based on cumulative episodes of microtrauma. The altered biomechanics of the unstable joint itself results in damage to articular cartilage which ultimately progresses to osteoarthritis. The continued vasodilatation and the presence of raised CGRP and substance P levels also possibly attract immune cells which contribute to the pathogenesis of this arthrosis. Thus, by controlling post-traumatic hyperaemia to permit only acute healing responses, the water content of the recovering ligament could be moderated possibly leading to a functionally more normal tissue and hence degenerative joint disease could potentially be avoided.

14.4. Pathophysiological Role for CGRP in Chronic Joint Inflammation

In a rat adjuvant monoarthritic model of joint inflammation, responsiveness to exogenous CGRP is also diminished. One week after intra-articular injection of Freund's adjuvant, the vasodilator response to CGRP is attenuated (Fig. 14.9). This similarity to the effect observed in the rabbit knee joint injury model suggests that similar mechanisms may be operating i.e., there may also be increased release of CGRP in inflammatory arthritis resulting in receptor downregulation. This concept is supported by a 50% rise in the tissue content of CGRP reported in the ankle joint of a rat adjuvant polyarthritic model.[24] Synovial fluid CGRP content is also elevated in adjuvant monoarthritis[25] providing further evidence that release of CGRP is increased during chronic joint inflammation. This is of relevance to human disease as the CGRP content of synovial fluid has been found to be increased in rheumatoid arthritis.[26,27]

Despite these observations, the contribution of CGRP and the tachykinins to the pathogenesis of inflammatory joint disease remains controversial. Some immunohistochemical investigations have suggested that compared to control tissues there is a reduction in CGRP-immunoreactive fibres in the synovium of adjuvant arthritic rats[28] and the synovium of patients with rheumatoid arthritis.[29,30] However, a more recent investigation found that although the intimal layer of synovium showed a decreased innervation by CGRP-immunoreactive fibres, deeper layers of the joint capsule in fact showed an increase.[24] These conflicting results may be due to the non-quantitative nature of immunohistochemical studies of the synovium which are hindered by the fact that results can be strongly influenced by tissue sampling bias. Actual measurement of CGRP tissue content by radioimmunoassay shows an increase in the joints of adjuvant arthritic rats[24] which argues strongly for increased presence of CGRP in chronic joint inflammation and this is further supported by the finding that CGRP synovial fluid content is increased in rat adjuvant mono - arthritis.[25] Human studies involving examination of synovial fluid from arthritic patients have consistently reported an increased in CGRP levels[26,27,33] but have been more conflicting with regard to substance P, with some able to detect it[31,32] while others could not.[26,27] Further evidence supporting increased CGRP synovial content in arthritis is that CGRP immunoreactivity is

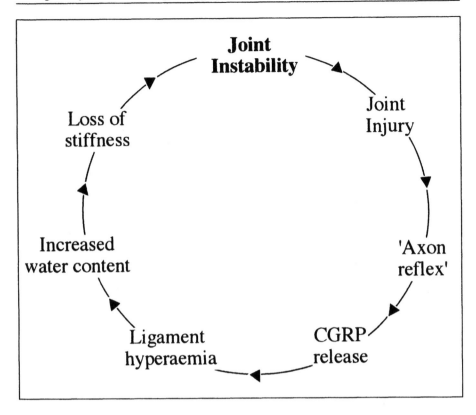

Fig. 14.8. Hypothetical positive feedback cycle in which joint instability leads to injury and triggers the axon reflex, leading to enhanced CGRP release. This in turn leads to ligament hyperaemia, increased interstitial fluid formation which increases the viscoelastic properties of the ligament reducing its stiffness. This increases joint laxity, leading to further joint instability.

increased in rat dorsal root ganglion (DRG) neurones during adjuvant polyarthritis[24,34] and it is well recognized that a larger proportion of sensory neuropeptides formed in the neurones in the DRG are transported peripherally rather than centrally.[35] Collectively, these data indicate that changes both in the neuropeptide content and in the proportion of DRG neurones producing particular neuropeptides are characteristic neuronal changes associated with inflammation and must be reflected by similar changes in the neuropeptide content of peripheral inflamed tissues.

Another controversial issue concerns whether interference with the neuropep-tidergic system can influence the course of arthritis. Acute joint inflammation induced by intra-articular injection of carrageenan is attenuated both by joint denervation and capsaicin pre-treatment.[36-38] Similarly, adjuvant polyarthritis is also significantly reduced by capsaicin pre-treatment.[39] However, chronic surgical denervation fails to inhibit adjuvant arthritis although rather curiously the CGRP content of joint tissues was only halved despite complete sciatic nerve section.[40] This surprising finding could be explained if, as suggested for sympathetic innervation, the joint receives overlapping sensory innervation from several different nerves. In rat medial plantar arteries, for

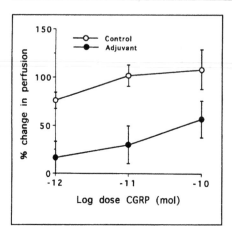

Fig 14.9. Vascular responses to topical application of CGRP to the normal (control) rat knee joint and in adjuvant monoarthritic knees (n=3).

example, CGRP innervation has been found to originate from both the sciatic and saphenous nerves.[20] The continuing presence of sensory neuropeptides despite nerve section, which one would expect to cause complete depletion of sensory neuropeptides in the territory innervated, makes the interpretation of such studies difficult. Using a different approach, it has recently been reported that adjuvant polyarthritis arthritis is diminished in NK_1 receptor "knockout" mice.[41] This, together with our data presented in figure 14.4 revitalise the concept of a neurogenic component to joint inflammation and injury.

14.5. Conclusion

In addition to the joint, a vasoconstrictor effect of $CGRP_{8-37}$ has been demonstrated in rabbit skin,[42] rat mesentery,[43] the kidney and hindquarters[44] as well as the vasa nervosum,[45,46] suggesting that tonic release of CGRP occurs in many vascular beds and may therefore represent a ubiquitous mechanism of physiological control. Because of its potent and long-lasting vasodilator activity, CGRP has the potential to play an important role in inflammatory joint diseases and evidence is now accumulating to implicate its

involvement in the pathogenesis of degenerative joint diseases such as post-traumatic osteoarthritis. Overall, these observations in disparate models argue for important physiological and pathophysiological roles for CGRP in tissues associated with synovial joints and may herald a novel direction for potential drug therapy to help ameliorate the symptoms of various arthritides. Clearly more research is required to elucidate further the function of this sensory neuropeptide and its role in the maintenance of joint integrity.

Acknowledgments

The work included in this articles was supported by the Arthritis & Rheumatism Council, the Canadian Medical Research Council, the Alberta Heritage Foundation for Medical Research, the MacFeat Bequest of the University of Glasgow and the Wellcome Trust. Technical assistance from Craig Sutherland and Heather Collin is gratefully acknowledged.

References

1. Brain SD, Williams TJ, Tippins JR et al. Calcitonin-gene-related peptide is a potent vasodilator. Nature 1985; 313:54-56.
2. Grönblad M, Korkala O, Konttinen YT et al. Immunoreactive neuropeptides in nerves in ligamentous tissue. Clin Orthop Rel Res 1991; 265:291-296.
3. McDougall JJ, Bray RC, Sharkey KA. A morphological and immunohistochemical examination of nerves in normal and injured collateral ligaments of rat, rabbit and human knee joints. Anat Rec 1997; 248:29-39.
4. McMurdo L, Lockhart JC, Ferrell WR. Modulation of synovial blood flow by the calcitonin gene-related peptide (CGRP) receptor antagonist, $CGRP_{(8-37)}$. Br J Pharmacol 1997; 121:1075-1080.
5. Ferrell WR, McDougall JJ, Bray RC. Spatial heterogeneity in the effects of calcitonin gene-related peptide (CGRP) on the microvasculature of ligaments in the rabbit knee joint. Br J Pharmacol 1997; 121:1397-1405.

6. Donoso VS, Fournier A, St-Pierre S et al. Pharmacological characterization of CGRP1 receptor subtype in the vascular system of the rat: Studies with hCGRP fragments and analogs. Peptides 1990; 11:885-889.

7. Gardiner SM, Compton AM, Kemp PA et al. Antagonistic effect of human-CGRP(8-37) on the in vivo regional haemodynamic actions of human-CGRP. Biochem Biophys Res Commun 1990; 171: 938-943.

8. Hughes SR, Brain SD. A calcitonin gene- related antagonist (CGRP8-37) inhibits microvascular responses induced by CGRP and capsaicin in skin. Br J Pharmacol 1991; 104:738-742.

9. Marshall JL, Olsson S. Instability of the knee: A long term experimental study in dogs. J Bone Joint Surg 1971; 53A:1561-1570.

10. Jacobsen K. Osteoarthrosis following insufficiency of the cruciate ligaments in man: A clinical study. Acta Orthop Scand 1977; 48:520-526.

11. McDougall JJ, Bray RC. Animal models of ligament repair. In An YH, Friedman RO. Animal models in orthopaedic research CRC Press LLC, Boca Raton; 1999:461-476.

12. Clayton ML, Weir GJ Jr. Experimental investigation of ligamentous healing. Am J Surg 1959; 98:373-378.

13. Pond MJ, Nuki G. Experimentally-induced osteoarthritis in dog. Ann Rheum Dis 1973; 32:387-388.

14. McDevitt C, Gilbertson E, Muir H. An experimental model of osteoarthritis; early morphological and biochemical changes. J Bone Joint Surg 1977;59-B:24-35.

15. Bray RC, Doschak MR, Gross TS et al. Physiological and mechanical adaptations of rabbit medial collateral ligament after anterior cruciate ligament transection. J Orthop Res 1997; 15:830-836.

16. Chimich D, Shrive N, Frank C et al. Water content alters viscoelastic behaviour of the normal adolescent rabbit medial collateral ligament. J Biomech 1992; 25:831-837.

17. Ferrell WR, McDougall JJ, Bray RC. Altered neuropeptidergic vasoregulation in the ACL deficient rabbit knee. Proc Can Orthop Res Soc 1997; 31:51.

18. McDougall JJ, Ferrell WR, Bray RC. Synthetically-mediated contractor responses in normal and in the ACL-deficient rabbit knee. Proc Can Orthrop Res Soc. 1997; 31:51.

19. Ferrell WR, Khoshbaten A, Angerson WJ. Responses of bone and joint blood vessels in cats and rabbits to electrical stimulation of nerves supplying the knee. J. Physiol. 1990; 431:677-687.

20. Bentzer P, Nielsen N, Arner M et al. Supersensitivity in rat micro-arteries after short-term denervation. Acta Physiol. Scand. 1997; 161:125-133.

21. Walsh D, Suzuki T, Knock GA et al. Receptor characteristics of angiotensin analog binding in human synovium. Brit J Pharmacol 1994; 112:435-442.

22. Lewis T. Experiments relating to cutaneous hyperaldesia and its spread through somatic nerves. Clin Sci 1936, 2:373.

23. Lewis T. The Blood Vessels of Human Skin and Their Responses. London: Shaw, 1927.

24. Ahmed M, Bjurholm A, Schulzberg M et al. Increased levels of substance P and calcitonin gene-related peptide in rat adjuvant arthritis: a combined immunohistochemical and radioimmunoassay analysis. Arth Rheum 1995; 38:699-709.

25. Bileviciute I, Lundberg T, Ekblom A et al. Bilateral changes of substance P-, neurokinin A-, calcitonin gene-related peptide- and neuropeptide Y-like immunoreactivity in rat knee joint synovial fluid during acute monoarthritis, Neurosci Lett 1993: 153:37.

26. Larsson J, Ekblom A, Henrikson K et al. Immunoreactive tachykinins, calcitonin gene-related peptide and neuropeptide Y in human synovial fluid from inflamed knee joints, Neurosci Lett 1989; 100:326.

27. Larsson J, Ekblom A, Henrikson K et al. Concentration of substance P, neurokinin A, calcitonin gene-related peptide, neuropeptide Y and vasoactive intestinal polypeptide in synovial fluid from knee joints in patients suffering from rheumatoid arthritis, Scand J Rheumatol1991; 20:326.

28. Konttinen YT, Rees R, Hukkanen M, et al. Nerves in inflammatory synovium: Immunohistochemical observations on the adjuvant arthritic model. J Rheumatol 1990; 17:1586-1591.

29. Mapp PI, Kidd BL, Gibson SJ et al. Subtance P-, calcitonin gene-related peptide and C-flanking peptide of neuropeptide Y-immunoreactive fibres are present in normal synovium but depleted in patients with rheumatoid arthritis. Neurosci 1990; 37:143-153.

30. Pereira da Silva JA, Carmo-Fonseca M. Peptide containing nerves in human synovium: Immunohistochemical evidence for decreased innervation in rheumatoid arthritis. J Rheumatol 1990; 17:1592-1599.

31. Marshall, KW, Chiu B, Inman RD. Substance P and arthritis: analysis of plasma and synovial fluids. Arthritis Rheum 1989; 33:87.

32. Marabini S, Matucci-Cerinic M, Geppetti P et al. Substance P and somatostatin levels in rheumatoid arthritis, osteroarthritis, and psoriatic arthritis synovial fluid. Ann NY Acad Sci 1991; 632:435.

33. Lygren I, Ostensen M, Burhol PG et al. Gastrointestinal peptides in serum and synovial fluid from patients with inflammatory joint disease. Ann Rheum Dis 1986; 45:637.

34. Hanesch U, Pfrommer U, Grubb BD et al. Acute and chronic phases of unilateral inflammation in rat's ankle joint are associated with an increase in the proportion of calcitonin gene-related peptide-immunoreactive dorsal root ganglion cells. Eur J Neurosci 1993; 5:154.

35. Keen P, Harmar AJ, Spears F et al. Biosynthesis, axonal transport and turnover of neuronal substance P. In: Ciba Foundation Symposium 91. Substance P in the Nervous System London: Pitman; 1982:145-164.

36. Lam FY, Ferrell WR. Inhibition of carrageenan-induced joint inflammation by substance P antagonist. Ann Rheum Dis 1989; 48:928-932.

37. Lam FY, Ferrell WR. Capsaicin suppresses substance P-induced joint inflammation. Neurosci Lett 1989; 105:155-158.

38. Lam FY, Ferrell WR. The neurogenic component of different models of acute inflammation in the rat knee joint. Ann Rheum Dis 1991; 50:747-751.

39. Ahmed M, Bjurholm A, Srinivasan GR et al. Capsaicin effects on substance P and CGRP in rat adjuvant arthritis. Reg Peptides 1995; 55:85-102.

40. Ahmed M, Srinivasan GR, Theodorsson E et al. Effects of surgical denervation on substance P and calcitonin gene-related peptide in adjuvant arthritis. Peptides 1995; 16:569-579.

41. Cruwys SC, de Felipe C, Hunt JP et al. Inhibition of inflammation and hyperalgesia in NK1 receptor knock-out mice. Bri J Rheumatol 1998; 37:92.

42. Hughes SR & Brain SD. A calcitonin gene-related antagonist (CGRP8-37) inhibits microvascular responses induced by CGRP and capsaicin in skin. Br J Pharmacol 1991; 104:738-742.

43. Han SP, Naes L, Westfall TC. Inhibition of periarterial stimulation-induced vasodilatation of the mesenteric arterial bed by $CGRP_{(8-37)}$ and CGRP receptor desensitization. Biochem Biophys Res Commun 1990; 168:786-791.

44. Gardiner SM, Compton AM, Kemp PA et al. Antagonistic effect of human-$CGRP_{(8-37)}$ on the in vivo regional haemodynamic actions of human-CGRP. Biochem Biophys Res Commun 1990; 171:938-943.

45. Zochodne DW, Ho LT. Influence of perivascular peptides on endoneurial blood flow and microvascular resistance in the sciatic nerve of the rat. J Physiol Lond 1991; 444:615-630.

46. Zochodne DW, Ho LT. Vasa nervosum constriction from substance P and calcitonin gene-related peptide antagonists: Sensitivity to phentolamine and nimidopine. Reg Peptides 1993; 47:285-290.

CHAPTER 15

Evidence of the Involvment of CGRP in Migraine and Cluster Headache: A Physiological Perspective

Peter J. Goadsby

15.1. Introduction

Calcitonin gene-related peptide (CGRP) is the most powerful of the vasodilator transmitters and as one of the neuropeptides to be found in trigeminal neurons,[1] may play a role in the physiology and pathophysiology of the migraine and cluster headache.[2] In this chapter the evidence for such a role and the physiological implications that may be gleaned from the putative involvement of CGRP in primary headache is reviewed.

15.2. Why Migraine and Cluster Headache?

By definition head pain involves in some way the trigeminal nerve since it provides the sensory innervation for a substantial portion of the head. Both migraine and cluster headache are primary headache syndromes[3] that have been traditionally known as vascular headaches. Migraine is essentially an episodic often unilateral pain with nausea and associated features of sensory sensitivity, such as sensitivity to light, sound and head movement. Cluster headache is also an episodic, almost invariably unilateral headache with less prominent sensory sensitivity, but with much more prominent symptoms of autonomic dysfunction, such as lacrimation, nasal congestion and ptosis ipsilateral to the pain.

Both are well accepted and recognised in the International Headache Society diagnostic guidelines.[4] Both migraine and cluster headache have, therefore, provided robust clinically important conditions in which to explore the role and effects of CGRP and thus the opportunity for an important interchange of information between basic and clinical neuroscience.

15.3. The Trigeminovascular System

15.3.1. Human Observations

Thomas Willis[5] was probably the first author to draw attention to the sensory innervation of the large vessels and dura mater. It is reasonably clear that it is the large cerebral vessels and the dura mater that receive a nociceptive innervation[6,7] and that their stimulation leads to pain.[8,9] Indeed Penfield[10] proposed that this innervation may offer an opportunity to treat migraine surgically. Stimulation of the dura or large vessels in humans produces pain that is referred substantially to the ophthalmic (first) division of the trigeminal nerve.[11-13] There is a substantial animal experimental literature that provides further information concerning this innervation.

The CGRP Family: Calcitonin Gene-Related Peptide (CGRP), Amylin, and Adrenomedullin, edited by David Poyner. ©2000 EUREKAH.COM.

15.3.2. Observations in Experimental Animals

There are both anatomical and physiological observations that suggest a role for the trigeminal innervation of the cranial circulation and differentiation by both transmitter and anatomical pattern of innervation. The anatomy is covered in this volume (see Edvinsson) and elsewhere,[14] so that a physiological approach will be taken in this section.

Effects of Trigeminal Nerve or Ganglion Lesions

Cerebral blood flow measured with iodoantipyrine and tissue autoradiography is not altered in the cat after trigeminal ganglion section. Indeed, after unilateral section flow is identical to homologous contralateral cortex. Furthermore glucose utilisation is not affected by trigeminal ganglion section and thus the usual close relationship between flow and metabolism is not disturbed.[15] It has been shown that unilateral trigeminal ganglionectomy does not alter cerebrovascular responsiveness to hypercapnia, although it does attenuate the pial arteriolar responses to hypertension.[16] In the cat the cerebral hyperaemia following temporary occlusion of the brachiocephalic and left subclavian vessels is mediated in part (some 60%) via a trigeminal mechanism.[17] Induction of hypertension outside the autoregulatory range or induction of a tonic/clonic seizure with bicuculline in trigeminal ganglionectomised animals leads to a reduced vasodilator response when compared with the non-ganglionectomised side.[18] The effect is seen in the cortical grey matter but not in the diencephalon or white matter and is due to the axon-reflex part of the trigeminovascular system, since trigeminal root section prior to entry of the nerve to the brainstem does not produce the effect whereas ganglionectomy does.[18] Similarly, trigeminal fibres appear to limit noradrenaline-induced constrictor response in the pia and play a role in restoring constricted vessels to their normal calibre.[19] Normal perfusion pressure range autoregulation is unimpaired by trigeminal root section.[16]

These data suggest that the trigeminal system by its sensory role detects changes in perfusion and has in certain settings a role in facilitating blood flow changes. If one considers this system with the other major cerebrovascular vasodilator system that is provided by the parasympathetic nervous system,[20] then perhaps they act to balance the vasoconstrictor actions of the sympathetic nervous system[21] as well as to provide a backup system in times of threat. An imbalance of either mechanisms has implications for pathophysiology in humans in such conditions as stroke, subarachnoid haemorrhage[22] and prolonged seizures.

Effect of Trigeminal Nerve or Ganglion Stimulation

It has been known for some years that stimulation of the trigeminal ganglion in humans by either thermocoagulation or injection of alcohol can cause facial flushing usually in the division or divisions appropriate to the manipulation.[23] In addition it has been shown that this flushing is accompanied by an increase in facial temperature of 1-2°C.[24] Corresponding with this flush there is an increase in the dilator peptides substance P and CGRP in the external jugular but not in the peripheral circulation.[25] Such changes are also seen in the cat. In addition trigeminal ganglion stimulation in the cat leads to a diminution of carotid arterial resistance, with increases in both blood flow and facial temperature through a predominantly reflex mechanism.[26] The afferent limb of this arc is the trigeminal nerve and the efferent the facial/greater superficial petrosal nerve (parasympathetic) dilator pathway.[27] About 20% of the dilatation remains after facial nerve section and is probably mediated by antidromic activation of the trigeminal system directly. The portion running through the parasympathetic outflow traverses the sphenopalatine (pterygopalatine) and otic ganglia[28] and employs vasoactive intestinal polypeptide as its transmitter.[29]

Dura Mater

The dura and vessels of the cerebral circulation are the protection and first portal of entry of any compounds exogenous to the central nervous system. The dura mater has been the subject of detailed study because it and the large the cerebral vessels are the only pain-producing structures within the cranial cavity.[12] The dura mater is innervated by branches of the trigeminal nerve whose cell bodies are to be found in the trigeminal ganglion.[30] The innervation is somatotopic with the first division giving rise to fibres that innervate the anterior fossa and tentorium cerebelli, the second division innervating the orbital roof and the third structures in the middle fossa.[31] The tentorial nerve arises from the ophthalmic nerve proximal to the superior orbital fissure and runs along the edge of the tentorium reflecting over its superior surface to run forward across the dura mater as it covers the forebrain.[6] The supratentorial dural sympathetic innervation arises from the ipsilateral superior cervical ganglion as does that innervating the dura of the posterior fossa.[31] In the posterior fossa the dural innervation has important contributions from both the trigeminal ganglion and the upper two dorsal root ganglia. The dural innervation has been further characterised by examining the distribution of neuropeptide immunoreactivity around the superior sagittal sinus and CGRP is certainly present.[32]

Ultrastructural evidence has suggested that the sympathetic nerves arising from the superior cervical ganglion may make functional connections with the mast cells in the walls of large cerebral vessels.[33] It has been shown in the rat that unilateral trigeminal ganglion stimulation can cause mast cell degranulation in the dura mater.[34] The possible interaction between these highly humorally active cells and the protective covering of the brain is of great interest. The cells are located randomly over the rat dura mater with some concentration at the vessels but in monkey, human and cat they are only located at the branching of vessels and are thus well placed for some regulatory role.

15.4. CGRP in Migraine

Studies of cranial venous neuropeptide levels have proved valuable in elucidating some of the mechanisms involved in primary headaches. Stimulation of the trigeminal ganglion in the cat leads to an increase in cranial levels of both substance P and CGRP. Similarly, stimulation of the trigeminal ganglion in humans undergoing thermocoagulation for trigeminal neuralgia leads to elevation in the cranial venous outflow of both peptides.[25] More specific stimulation of pain-producing intracranial structures, such as the superior sagittal sinus, also results in cranial venous release of CGRP and not substance P.[35] During migraine CGRP is elevated in the external jugular vein blood whereas substance P is not in both adults[36] and in adolescents.[37]

Consistent with these findings substance P (neurokinin-1) receptor antagonists have been studied in migraine and both RPR100893[38] and lanepitant[39] are ineffective. These compounds were orally administered but unpublished parentral administration of highly potent neurokinin-1 antagonists have had the same outcome. This contrasts their blockade of neurogenic plasma protein extravasation,[40] but is consistent with the lack of effect of GR205171, a neurokinin-1 antagonist,[41] on neuronal firing of cells in the trigemino-cervical complex of the cat.[42]

Again in terms of the relative contribution of substance P and CGRP, substance P can induce plasma protein extravasation in the dura mater but CGRP does not.[43] Of further interest is the fact that bosentan, a highly potent endothelin antagonist,[44] is also an effective inhibitor of plasma protein extravasation[45] but is ineffective in migraine.[46] In addition, CP122,288 a conformationally restricted analogue of sumatriptan,[47] which is 1000 times more potent than sumatriptan against neurogenic plasma protein extravasation,[48] is also completely without effect in acute migraine.[49] Consistent with these observations avitriptan, which is a potent anti-migraine compound.[50,51] is relatively impotent in its

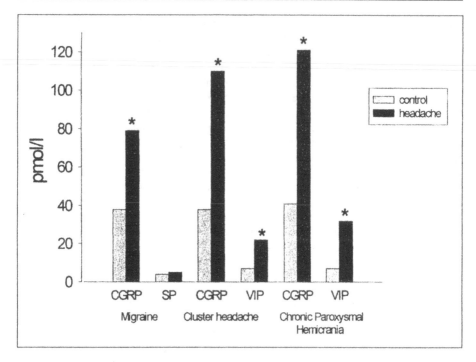

Fig. 15.1. Elevation of CGRP levels in venous blood in patients with migraine, cluster headache and chronic paroxysmal hemicrania. Substance P (SP) is not elevated in migraine while vasoactive intestinal polypeptide (VIP) is elevated in cluster headache and chronic paroxysmal hemicrania in line with the clinical feature of autonomic activation in these syndromes. The ordinate is in pmol/l.

effects on neurogenic plasma protein extravasation,[52] but potently blocks CGRP release in animal studies whereas CP122,288 does not effect CGRP release.[53]

15.5. CGRP in Cluster and Related Headaches

The two key clinical features of cluster headache, in terms of the expression of the attack, are the pain and autonomic features. The link between the two features is seen in a number of clinical syndromes collectively described as trigeminal-autonomic cephalalgias.[54] Given that severe head pain, which is most usually in the distribution of the ophthalmic division of the trigeminal nerve, is a major feature of acute attacks of cluster headache it is not surprising to find that CGRP levels are elevated in acute cluster headache.[55] Similarly in patients

with cluster headache who have their attacks trigger by nitrates, a well recognised feature of the condition,[56] there is also an elevation in venous CGRP.[57] In our studies we also saw elevation of vasoactive intestinal polypeptide, (VIP),[55] a marker of parasympathetic nervous activity,[58] and thus a likely explanation for the autonomic symptoms so characteristic of the disorder.[4] It is interesting in terms of disease activity that the level of CGRP seen in cluster headache is generally greater than that seen in migraine which is consistent with the general experience that the pain is worse.

A form of headache closely related to cluster headache is Chronic Paroxysmal Hemicrania.[4] In this very rare form of headache patients who are more usually female have relatively short and frequent attacks of very severe unilateral headache with again autonomic features. It is remarkable that they respond to indomethacin which can completely stop the attacks.[59,60] We have studied a patient with chronic paroxysmal hemicrania and found elevation of both CGRP and VIP during attacks and normal levels after treatment with indomethacin when the patient's headache was controlled.[61]

15.6 Conclusion

CGRP is an important and potent vasodilator peptide that is localised in the pain-producing innervation of the cranial vessels, the trigeminovascular system. Its release parallels in clinical settings headache in primary neurovascular headaches, such as migraine and cluster headache (Fig. 15.1; see previous page). CGRP is elevated during headache and reduced when headache is controlled. It is possible that CGRP plays a major transmitter role in the expression of these primary headaches and that blockade of its effects may provide relief in such patients.

Acknowledgments

The work of the author reported herein has been supported by the Wellcome Trust and the Migraine Trust. PJG is a Wellcome Senior Research Fellow.

References

1. Edvinsson L, Goadsby PJ. Neuropeptides in headache. Eur J Neurol 1998; 5:329-341.
2. Goadsby PJ. Current concepts of the pathophysiology of migraine. In: Mathew NT, ed. Neurologic Clinics of North America. Philadelphia: W.B. Saunders, 1997; 15:27-41.
3. Goadsby PJ, Olesen J. Migraine: Diagnosis and treatment in the 1990's. BMJ 1996; 312:1279-1282.
4. The Headache Classification Committee of The International Headache Society. Classification and diagnostic criteria for headache disorders, cranial neuralgias and facial pain. Cephalalgia 1988; 8(Suppl 7):1-96.
5. Willis T. The Anatomy of the Brain and Nerves. (Tercentenary Edition- 1964 ed.) Montreal: McGill University Press, 1964.
6. Penfield W. A contribution to the mechanism of intracranial pain. Proceedings of the Association for Research in Nervous and Mental Disease 1934; 15:399-415.
7. Penfield W. Intracerebral vascular nerves. Arch Neurol Psychiatry 1932; 27:30-44.
8. Feindel W, Penfield W, McNaughton F. The tentorial nerves and localisation of intracranial pain in man. Neurology 1960; 10:555-563.
9. McNaughton FL. The innervation of the intracranial blood vessels and the dural sinuses. In: Cobb S, Frantz AM, Penfield W et al, eds. The Circulation of the Brain and Spinal Cord. New York: Hafner Publishing Co. Inc, 1966:178-200.
10. Penfield W. Operative treatment of migraine and observations on the mechanism of vascular pain. Trans Am Acad Ophthal Otol 1932; 3:1-16.
11. Penfield W, McNaughton FL. Dural headache and the innervation of the dura mater. Arch Neurol Psychiatry 1940; 44:43-75.
12. Wolff HG. Headache and other head pain. New York: Oxford University Press, 1963.
13. Ray BS, Wolff HG. Experimental studies on headache. Pain sensitive structures of the head and their significance in headache. Arch Surg 1940; 41:813-856.
14. Edvinsson L, MacKenzie ET, McCulloch J Cerebral Blood Flow Metabol. New York: Raven Press, 1993.
15. Edvinsson L, McCulloch J, Kingman TA et al. On the functional role of the trigemino-cerebrovascular system in the regulation of cerebral circulation. In: Owman C, Hardebo JE, eds. Neural Regulation of the Cerebral Circulation. Stockholm: Elsevier Science Publishers B.V., 1986:407-418.
16. Moskowitz MA, Wei EP, Saito K, Kontos HA. Trigeminalectomy modifies pial a arteriolar responses to hypertension or norepinephrine. Am J Physiol 1988; 255:H1-H6.
17. Macfarlane R, Tasdemiroglu E, Moskowitz MA et al. Chronic trigeminal ganglionectomy or topical capsaicin application to pial vessels attenates postocculsive cortical hyperemia but does not influence postischemia hypoperfusion. J Cereb Blood Flow Metabol 1991; 11:261-271.

18. Sakas DE, Moskowitz MA, Wei EP et al. Trigeminovascular fibers increase blood flow in cortical grey matter by axon-dependent mechanisms during severe hypertension or siezures. Proc Nat Acad Sci USA 1989; 86:1401-1405.

19. McCulloch J, Uddman R, Kingman TA et al. Calcitonin gene-related peptide: Functional role in cerebrovascular regulation. Proc Nat Acad Sci (U.S.A.) 1986; 83:1-5.

20. Goadsby PJ, Edvinsson L. Regulation of cerebral blood flow by the parasympathetic nervous system. In: Welch KMA, Caplan L, Reis DJ, Weir B, Siesjo B, eds. Primer of Cerebrovascular Diseases. San Diego: Academic Press, 1997:63-66.

21. Edvinsson L, Goadsby PJ. Regulation of cerebral blood flow by the sympathetic nervous system. In: Welch KMA, Caplan L, Reis DJ, Weir B, Siesjo B, eds. Primer of Cerebrovascular Diseases. San Diego: Academic Press, 1997:60-62.

22. Edvinsson L, Juul R, Uddman R. Peptidergic innervation of the cerebral circulation. Role in subarachnoid hemorrhage in man. Neurosurgical Reviews 1990; 13:265-272.

23. Onofrio BM. Radiofrequency percutaneous Gasserian ganglion lesions. Results in 140 patients with trigeminal pain. J Neurosurg 1975; 42:132-143.

24. Drummond PD, Gonski A, Lance JW. Facial flushing after thermocoagulation of the gasserian ganglion. J Neurol Neurosurg Psychiatry 1983; 46:611-616.

25. Goadsby PJ, Edvinsson L, Ekman R. Release of vasoactive peptides in the extracerebral circulation of man and the cat during activation of the trigeminovascular system. Ann Neurol 1988; 23:193-196.

26. Lambert GA, Bogduk N, Goadsby PJ et al. Decreased carotid arterial resistance in cats in response to trigeminal stimulation. J Neurosurg 1984; 61:307-315.

27. Goadsby PJ, Duckworth JW. Effect of stimulation of trigeminal ganglion on regional cerebral blood flow in cats. Am J Physiol 1987; 253:R270-R274.

28. Goadsby PJ, Lambert GA, Lance JW. The peripheral pathway for extracranial vasodilatation in the cat. J Auto Nerv Syst 1984; 10:145-155.

29. Goadsby PJ, Macdonald GJ. Extracranial vasodilatation mediated by VIP (Vasoactive Intestinal Polypeptide). Brain Res 1985; 329:285-288.

30. Mayberg MR, Zervas NT, Moskowitz MA. Trigeminal projections to supratentorial pial and dural blood vessels in cats demonstrated by horseradish peroxidase histochemistry. J Comp Neurol 1984; 223:46-56.

31. Keller JT, Marfurt CF, Dimlich RV et al. Sympathetic innervation of the supratentorial dura mater of the rat. J Comp Neurol 1989; 290:310-321.

32. Keller JT, Marfurt CF. Peptidergic and serotoninergic innervation of the rat dura mater. J Comp Neurol 1991; 309:515-534.

33. Dimitriadou V, Aubineau P, Taxi J, Seylaz J. Ultrastructural changes in the cerebral artery wall induced by long-term sympathetic denervation. Blood Vessels 1988; 25:122-143.

34. Dimitriadou V, Buzzi MG, Theoharides TC et al. Ultrastructural evidence for neurogenically mediated changes in blood vessels of the rat dura mater and tongue following antidromic trigeminal stimulation. Neuroscience 1992; 48:187-203.

35. Zagami AS, Goadsby PJ, Edvinsson L. Stimulation of the superior sagittal sinus in the cat causes release of vasoactive peptides. Neuropeptides 1990; 16:69-75.

36. Goadsby PJ, Edvinsson L, Ekman R. Vasoactive peptide release in the extracerebral circulation of humans during migraine headache. Ann Neurol 1990; 28:183-187.

37. Gallai V, Sarchielli P, Floridi A, et al. Vasoactive peptides levels in the plasma of young migraine patients with and without aura assessed both interictally and ictally. Cephalalgia 1995; 15:384-390.

38. Diener HC. Substance-P antagonist RPR100893-201 is not effective in human migraine attacks. In: Olesen J, Tfelt-Hansen P, eds. Proceedings of the VIth International Headache Seminar.

39. Goldstein DJ, Wang O, Saper JR et al. Ineffectiveness of neurokinin-1 antagonist in acute migraine: A crossover study. Cephalalgia 1997; 17:785-790.

40. Moskowitz MA. Neurogenic versus vascular mechanisms of sumatriptan and ergot alkaloids in migraine. Trends Pharmacol Sci 1992; 13:307-311.

41. Gardner CJ, Armour DJ, Beattie DT, et al. GR205171: A novel antagonist with high affinity for the tachykinin NK1 receptor, and potent broad-spectrum anti-emetic activity. Reg Peptides 1996; 65:45-53.

42. Goadsby PJ, Hoskin KL, Knight YE. Substance P blockade with the potent and centrally acting antagonist GR205171 does not effect central trigeminal activity with superior sagittal sinus stimulation. Neuroscience 1998; 86:337-343.

43. Cutrer FM, Garret C, Moussaoui SM et al. The non-peptide neurokinin-1 antagonist, RPR 100893, decreases c-fos expression in trigeminal nucleus caudalis following noxious chemical meningeal stimulation. Neuroscience 1995; 64:741-750.

44. Clozel M, Breu V, Gray AG, et al. Pharmacological characterisation of bosentan, a new potent orally active nonpeptide endothelin receptor antagonist. J Pharmacol Exp Ther 1994; 270:228-235.

45. Brandli P, Loffler B-M, Breu V et al. M. Role of endothelin in mediating neurogenic plasma extravasation in rat dura mater. Pain 1996; 64:315-322.

46. May A, Gijsman HJ, Wallnoefer A et al. Endothelin antagonist bosentan blocks neurogenic inflammation, but is not effective in aborting migraine attacks. Pain 1996; 67:375-378.

47. Gupta P, Brown D, Butler P et al. The in vivo pharmacological profile of a 5-HT1 receptor agonist, CP122,288, a selective inhibitor of neurogenic inflammation. Br J Pharmaco. 1995; 116:2385-2390.

48. Lee WS, Moskowitz MA. Conformationally restricted sumatriptan analogues, CP-122,288 and CP-122,638, exhibit enhanced potency against neurogenic inflammation in dura mater. Brain Res 1993; 626:303-305.

49. Roon K, Diener HC, Ellis P, et al. CP-122, 288 blocks neurogenic inflammation, but is not effective in aborting migraine attacks: Results of two controlled clinical studies. Cephalalgia 1997; 17:245.

50. Couch JR, Saper J, Meloche JP. Treatment of migraine with BMS180048: Response at 2 hours. Headache 1996; 36:523-530.

51. Ryan RE, Elkind A, Goldstein J. Twenty-four hour effectiveness of BMS 180048 in the acute treatment of migraine headaches. Headache 1997; 37:245-248.

52. Yocca FD, Gylys JA, Smith DW, et al. BMS-181885: a clinically effective migraine abortive with peripherovascular and neuronal $5HT_{1D}$ antagonist properties. Cephalalgia 1997; 17:404.

53. Knight YE, Edvinsson L, Goadsby PJ. Blockade of release of CGRP after superior sagittal sinus stimulation in cat: a comparison of avitriptan and CP122,288. Cephalalgia 1997; 17:248.

54. Goadsby PJ, Lipton RB. A review of paroxysmal hemicranias, SUNCT syndrome and other short-lasting headaches with autonomic features, including new cases. Brain 1997; 120:193-209.

55. Goadsby PJ, Edvinsson L. Human in vivo evidence for trigeminovascular activation in cluster headache. Brain 1994; 117:427-434.

56. Ekbom K. Nitroglycerin as a provocative agent in cluster headache. Arch Neurol 1968;19:487-493.

57. Fanciullacci M, Alessandri M, Figini M et al. Increases in plasma calcitonin gene-related peptide from extracerebral circulation during nitroglycerin-induced cluster headache attack. Pain 1995; 60:119-123.

58. Edvinsson L, Fahrenkrug J, Hanko J et al. VIP(vasoactive intestinal polypeptide)-containing nerves of intracranial arteries in mammals. Cell Tissue Research 1980; 208:135-142.

59. Sjaastad O, Dale I. Evidence for a new (?) treatable headache entity. Headache 1974; 14:105-108.

60. Sjaastad O, Apfelbaum R, Caskey W, et al. Chronic paroxysmal hemicrania (CPH). The clinical manifestations. A review. Uppsala J Med Sci 1980; 31(Suppl):27-33.

61. Goadsby PJ, Edvinsson L. Neuropeptide changes in a case of chronic paroxysmal hemicrania- evidence for trigemino-parasympathetic activation. Cephalalgia 1996; 16:448-450.

CHAPTER 16

CGRP, CGRP mRNA and CGRP$_1$ Receptor mRNA and Release from the Human Trigeminovascular System

Lars Edvinsson, Peter J. Goadsby, Inger Jansen-Olesen and Rolf Uddman

16.1. Introduction

In their "classical" studies Ray and Wolff [1] demonstrated that only the large cerebral arteries at the base of the brain, the meningeal (dural) arteries and the veins were sensitive to noxious stimuli. This directed interest to the wall of the intracranial vessels as a putative source of headache. The concept was strengthened following the first series of selective vertebral angiographic studies. The patients complained of acute headache following injection of the contrast media, even visual phenomena, resembling the aura in migraine, were described. [2]

Nerve fibers belonging to the trigeminal system innervate cranial vessels and dura mater via the first ophthalmic division. The cell bodies are bipolar and are located in the trigeminal ganglion. [3] They make functional first-order connections with neurons in the trigeminal nucleus caudalis and in its related extension down to the C2 level. [4] In the trigeminal ganglion numerous nerve cell bodies contain calcitonin gene-related peptide (CGRP), substance P (SP) and neurokinin A. In addition, minor populations of cell bodies contain nitric oxide synthase (NOS)-, pituitary adenylate cyclase activating peptide (PACAP)-, dynorphin- and galanin-immunoreactivity. It is a vasodilator pathway via antidromic release upon activation as well as having a primary involvement in sensory functions.

16.2. Innervation of the Human Intracranial Vessels

The innervation of human intracranial vessels has been studied in some detail. In the cerebral arteries the supply of CGRP- and SP-containing fibers was higher than that of meningeal arteries. [5,6] In human trigeminal ganglia CGRP immunoreactive (-ir) neurons occured in high numbers (in 36-40% of all neuronal cells) and distribute homogeneously throughout the ganglia. A smaller number of the neurons showed SP-(18%), NOS-(15%) and PACAP-(20%)-ir. Double immunostaining revealed that only few CGRP-ir neurons (<5%) were NOS positive. The majority of the CGRP- and SP-ir somata were small and medium sized. Neurons with ir to the C-terminal portion of neuropeptide Y (NPY) were not visible in the ganglia. With in situ hybridization, CGRP mRNA was found in a large portion of the trigeminal neurons. [7]

16.3. Receptor mRNA Analysis

Gel electrophoresis of the PCR products from human trigeminal ganglia[7,8] demonstrated the presence of mRNA for the NPY Y_1, Y_2, $CGRP_1$ and VIP_1 receptors. In another study mRNA for the human NPY Y_1 and $CGRP_1$ receptors were shown in brain vessels.[8,9] (The $CGRP_1$ receptor was CRLR as discussed by Aiyar et al. and Legon, this volume).

16.4. Neurotransmitter Release

The involvement of neuropeptides in primary headache attacks have been studied by analysis of neurotransmitter release into cranial venous outflow.

16.4.1. Trigeminal Ganglion Stimulation

The trigeminal system provides the only known pain-sensitive innervation of the cranial vasculature by virtue of the central convergence of trigeminal and upper cervical pain inputs. This takes place at the level of second-order neurons. These pathways do not seem to have any resting tonic influence on regional cerebral blood flow (rCBF) or regional metabolism. In man, stimulation of the trigeminal ganglion results in an increased cortical blood flow slightly more ipsilateral than contralateral.[10] Trigeminal ganglion stimulation results in release of both CGRP and SP, both acting as markers of trigeminal activation.[11]

16.4.2. Migraine Attacks

The results of trigeminal ganglion stimulation in man led us to examine the levels of various neuropeptides in patients during migraine attacks. Blood samples were drawn from the external jugular vein during the headache. The patients were assessed by using markers for the sympathetic (NPY), the parasympathetic (VIP), and the trigeminovascular systems (CGRP, SP). There were no changes in the peripheral blood levels of the peptides studied or in NPY, VIP or SP in the external jugular venous blood during resting conditions. However, there was a marked increase in

CGRP during migraine headache.[12] Two patients with some symptoms similar to those seen in cluster headache, e.g. nasal congestion and rhinorrhea, had increases in VIP, but this was not significant for the whole group. The changes in VIP suggested that some reflex parasympathetically mediated event was occurring. There was no difference between patients with migraine with aura or without aura: both resulted in substantial increases in CGRP levels as recorded in the jugular venous outflow blood. Neither were there any changes in SP, VIP or NPY in jugular or in peripheral venous blood. The selective release of CGRP might be due to the fact that the cerebral circulation has a preferential CGRP innervation from the trigeminal ganglion. These observations have subsequently been confirmed.[13,14] Following sumatriptan administration, it was seen that CGRP returned to control levels with successful amelioration of the headache.

16.4.3. Cluster Headache

Cluster headache would seem to be an ideal condition to examine in that it is a well-described clear-cut clinical syndrome. Thus, patients with episodic cluster headache, fulfilling the criteria of the International Headache Society (IHS), were examined during acute spontaneous attacks of headache to determine the local cranial release of neuropeptides. During the attacks, the levels of CGRP and VIP were markedly raised, while there were no changes in the levels of NPY or SP. Treatment with oxygen or sumatriptan (subcutaneous) reduced the CGRP levels to normal, while opiate administration did not alter the peptide levels. These data demonstrate activation of the trigeminovascular system and the cranial parasympathetic nervous system in an acute attack of cluster headache. It was particularly noteworthy that all subjects had release of VIP. This is in concert with the facial symptoms well known to the symptomatology of this disorder. Furthermore, it was shown that both oxygen and sumatriptan, while aborting the attacks, terminated the activity in the trigeminovascular system. This agrees well with results demonstrating release of CGRP in nitroglycerine-elicited attacks of cluster headache.[16] CGRP in the external jugular vein homolateral to the pain side in cluster headache patients was found to be augmented during the active period and became elevated further at the peak of the provoked attack. Complete reversal occurred both at the spontaneous and the sumatriptan-induced remission. Interestingly, nitroglycerine neither provoked a cluster headache attack nor altered CGRP levels in patients out of their cluster bout. There were no alterations in the SP levels.

Thus, CGRP, a marker for the trigeminovascular system, and VIP, a marker for parasympathetic activity, are both elevated in the cranial venous blood of patients with an acute spontaneous attack of cluster headache. The termination of the attack with either sumatriptan or oxygen causes normalisation of the CGRP levels, reflecting cessation of activity in the trigeminovascular system, whereas pain relief after administration of an opiate agonist apparently terminates the pain of the attack but does not immediately end the trigeminovascular activity. The finding of both CGRP and VIP in the cranial venous blood suggests that there is activation of a brain stem reflex, the afferent arc of which is the trigeminal nerve and the efferent the cranial parasympathetic outflow from the VIIth nerve.

16.4.4. Trigeminal Neuralgia

In a study of patients with trigeminal neuralgia, there was no difference between the resting levels of neuropeptides during anaesthesia and those of a control population.[11] Stimulation of the trigeminal ganglion during thermocoagulation caused a marked increase in the levels of CGRP and SP in subjects with unilateral facial flushing. After cessation of the stimulation, the levels normalised.

Vasoactive substances may be responsible not only for headache but also in neurally mediated facial flushing. The following was noted in a patient with an unstable trigeminal system. This patient had a 17- year history of intractable left-sided facial pain.[17] The pain occurred daily in 5-second spasms to a maximum of one every 2-3 mins and was restricted to the left upper face including the upper teeth. It was associated with rhinorrhea on the left side and often with ipsilateral facial flushing. Conventional therapy including carbamazepine, baclofen and three posterior lobe explorations had not provided lasting relief. Local facial stimulation by tapping a painful trigger point led to both pain and flushing of the face ipsilaterally. A marked increase (119%) in CGRP levels in the external jugular vein was seen ipsilaterally during the flushing with no change in SP, NPY or VIP. This change was also seen in venous blood from the cubital fossa but to a lesser degree.

16.4.5. Chronic Paroxysmal Hemicrania

Chronic paroxysmal hemicrania (CPH) is a syndrome that is defined by IHS operational diagnostic criteria as frequent short-lasting attacks of unilateral pain usually in the orbital, supra-orbital or temporal region that may last from 2-45 minutes. The attack frequency may vary but would usually be five or more each day. The pain is associated with prominent autonomic symptoms such as conjunctival injection, lacrimation, nasal congestion, rhinorrhoea, ptosis or eyelid oedema and should have at least one of these to fulfil IHS criteria. It is a requirement of the diagnostic criteria that the attacks should rapidly settle on treatment with indomethacin. In one such case, we observed that, during pain, the CGRP level rose to 123 pmol/L compared to a level of 41 pmol/L while on indomethacin.[18] The VIP level rose to 32 pmol/L during an attack, dropping to 7 pmol/L on indomethacin. The case report and accompanying data establish that attacks of CPH are characterised by activation of both the sensory and the parasympathetic cranial innervation. The patient had rather stereotyped attacks of severe pain with autonomic symptoms associated with elevation of both trigeminal (CGRP) and parasympathetic (VIP) neuropeptide markers. These neuropeptides were at normal levels when the headaches were controlled on indomethacin. Thus, studies of neuropeptide markers in CPH may serve two roles. they provide a biological marker and illustrate a basic shared pathophysiology with primary headache syndromes, such as cluster headache. The neuropeptide studies reported here add to the view that there are many similarities between CPH and cluster headache.

16.5. What Causes the Release of CGRP in Headache

The trigeminovascular reflex was demonstrated more than a decade ago.[19,20] Thus, studies of cerebral vessels in situ showed that arteries could sense vasoconstriction and respond to this via the antidromic release of a vasodilator from the sensory nerves. Subsequent studies made it likely that the molecule was CGRP.[21,22] Interestingly, if during the experiments a vasodilator is administrated, this reflex was not activated. Strong support for the involvement of this reflex in cerebrovascular disorders has come from studies of subarachnoid haemorrhage (SAH), both in patients and in laboratory animals.[23,24]

After operation with aneurysm clipping and nimodipine treatment of patients with acute SAH, the degree of vasoconstriction was monitored with Doppler ultrasound recordings bilaterally from the middle cerebral and internal carotid arteries. External jugular vein blood sampling for neuropeptides analysis was analysed every second day. The highest CGRP levels were found in patients with the highest velocity index values (vasospasm). In patients with middle cerebral artery (MCA) aneurysms, a significant correlation was found between the vasospasm index and the CGRP levels. There were no changes observed in the SP and VIP levels. Alterations in cerebrovascular tone and in CBF, induced by changing

arterial CO_2 tension or by lowering of blood pressure (autoregulation test), did not alter the levels of the perivascular peptides (CGRP, SP, VIP or NPY) in the external jugular vein. In individual patients with marked vasoconstriction, due to SAH, increased levels of CGRP and NPY were also observed in the cerebrospinal fluid (CSF), thus, complementing the venous outflow peptide data.

These results demonstrate that, in a strict and clearcut intracranial arterial vasoconstrictory disorder, the trigeminovascular reflex is activated to counterbalance vasospasm by release of the vasodilator CGRP. This is verified both by measurements in the jugular vein outflow and in the CSF. In addition subjects that died due to their vasospasm and cerebral ischemia had no measureable levels of CGRP in their middle cerebral artery while the vessel wall levels of the other peptides were not altered.[25]

Acknowledgments

Supported by grants from the Swedish Medical Research Council.

References

1. Ray BS, Wolff HG. Experimental studies on headaches, pain sensitive structures of the head and their significance in headaches. Arch Surg 1940; 41:813-856
2. Radner S. Intracranial angiography via the vertebral artery. Acta Radiologica 1947; 28:838-843.
3. Edvinsson L, MacKenzie ET, McCulloch J. Cerebral Blood Flow and Metabolism New York: Raven Press, 1993.
4. Goadsby PJ, Zagami AS. Stimulation of the superior sagittal sinus increases metabolic activity and blood flow in certain regions of the brainstem and upper cervical spinal cord of the cat. Brain. 1991; 114:1001-1004.
5. Jansen I, Uddman R, Ekman R et al. Distribution and effects of neuropeptide Y, vasoactive intestinal peptide, substance P and calcitonin gene-related peptide in human middle meningeal arteries: Comparison with cerebral and temporal arteries. Peptides 1992; 13:527-536.

6. Edvinsson L, Gulbenkian S, Barroso CP et al. Innervation of the human middle meningeal artery: Immunohistochemistry, ultrastructure, and role of endothelium for vasomotility. Peptides 1998; 19:1213-1225.

7. Tajti J, Uddman R, Möller S et al. Messenger molecules and receptor mRNA in the human trigeminal ganglion. J Autonomic Nerv Sys 1999; 76:176-183.

8. Edvinsson L, Cantera L, Jansen-Olesen I et al. Expression of calcitonin gene-related peptide-1 receptor mRNA in human trigeminal ganglia and cerebral arteries. Neurosci Lett 1997; 229:209-211.

9. Nilsson T, Cantera L, Edvinsson L. Presence of neuropeptide Y Y1 receptor mediating vasoconstriction in human cerebral arteries. Neurosci Lett 1996; 204:145-148

10. Tran-Dinh YR, Thurei C, Cunin G et al. Cerebral vasodilation after the thermocoagulation of the trigeminal ganglion in humans. Neurosurgery 1992; 31:658-662.

11. Goadsby PJ, Edvinsson L, Ekman R. Release of vasoactive peptides in the extracerebral circulation of humans and the cat during activation of the trigeminovascular system. Annals Neurol 1988; 23:193-196

12. Goadsby PJ, Edvinsson L and Ekman R. Vasoactive peptide release in the extracerebral circulation of human during migraine headache. Annals Neurol 1990; 28:183-197.

13. Gallai V, Sarchielli P, Floridi A et al. Vasoactive peptide levels in the plasma of young migraine patients with and without aura assessed both interictally and ictally. Cephalagia 1995; 15:384-390.

14. Goadsby PJ, Edvinsson L The trigeminovascular system and migraine: Studies characterizing cerebrovascular and neuropeptide changes seen in humans and cats. Annals Neurol 1993; 33:48-56.

15. Goadsby PJ, Edvinsson L. Human in vivo evidence for trigeminovascular activation in cluster headache. Neuropeptide changes and effects of acute attacks during therapies. Brain 1994; 117:427-434.

16. Fanciullacci M, Alessandri M, Figini M et al. Increases in plasma calcitonin gene-related peptide from extracerebral circulation during nitroglycerin-induced cluster headache attack. Pain 1995; 60:119-423.

17. Goadsby PJ, Edvinsson L, Ekman R. Cutaneous sensory stimulation leading to facial flushing and release of calcitonin gene-related peptide. Cephalalgia 1992; 12:53-56.

18. Goadsby PJ, Edvinsson L. Neuropeptide changes in a case of chronic paroxysmal hemicrania-evidence for trigemino-parasympathetic activation. Cephalalgia 1996; 16:448-450.

19. Edvinsson L, McCulloch J, Kingman TA et al. On the functional role of the trigemino-cerebrovascular system in the regulation of cerebral circulation. In: Owman C, Hardebo JE, eds. Neural Regulation of the Cerebral Circulation Amsterdam: Elsevier, 1986:407-418.

20. McCulloch J, Uddman R, Kingman TA et al. Calcitonin gene-related peptide. Functional role in cerebrovascular regulation. Proc Nat Acad Sci USA 1986; 83:5731-5735.

21. Edvinsson L, Jansen I, Kingman T et al. Cerebrovascular responses to capsaicin vitro and in situ. Brit J Pharmacol 1990; 100:312-318.

22. Edvinsson L, Jansen-Olesen I, Kingman T et al. Modification of vasoconstrictor responses in cerebral blood vessels by lesioning of the trigeminal nerve: Possible involvement of CGRP. Cephalalgia 1995; 15:373-383.

23. Juul R, Edvinsson L, Gisvold S et al. Calcitonin gene-related peptide-LI subarachnoid haemorraghe in man. Signs of activation the trigemino-cerebrovascular system? Brit J Neurosurg 1990; 4:171-180.

24. Juul R, Hara H, Gisvold SE et al. Alterations in perivascular dilatory neuropeptides (CGRP, SP, VIP) in the external jugular vein and in the cerebrospinal fluid following subarachnoid haemorrhage in man. Acta Neurochirurgica 1995; 32:32-41.

25. Edvinsson L, Ekman R, Jansen I et al. Reduced levels of calcitonin gene-related peptide-like immunoreactivity in human brain vessels after subarachnoid haemorrhage. Neurosci Lett 1991; 121:151-154.

CHAPTER 17

Investigation of Different Pathways for the Relaxant Responses of Human αCGRP and βCGRP on Guinea Pig Basilar Artery

Anette Sams and Inger Jansen-Olesen

17.1. Introduction

Human (h) αCGRP and βCGRP have previously been shown to induce vasodilation by activation of hαCGRP$_{8-37}$ sensitive CGRP receptors on segments of guinea pig basilar artery in vitro.[1] In this study we have investigated levels of cAMP and cGMP in segments of guinea pig basilar artery following administration of hαCGRP and hβCGRP in presence or absence of hαCGRP$_{8-37}$. We additionally investigated whether the vasodilatory action of hα- and hβCGRP are mediated via the endothelium, nitric oxide synthase (NOS), guanylate cyclase stimulation or by a cyclooxygenase involved mechanism. This was investigated by determination of CGRP induced vasomotor responses following removal of endothelium or treatment by N^G-nitro-L- arginine methyl ester (L-NAME), methylene blue or indomethacin.

17.2 . Materials and Methods

Basilar artery segments of 1-2 mm were mounted on two tissue prongs in a Multimyograph Model 610 M (J. P. Trading, Denmark) and isometric vessel tension (mN/mm) was continuously monitored. Artery segments were normalized and the response to K$^+$ (125 mM) was determined twice.[2] A stable level of tension was reached by 3x10^{-6} M prostaglandin F$_{2\alpha}$ (PGF$_{2\alpha}$) and vessel segments were treated with 1) single doses of 10^{-7} M hα- or hβCGRP for one minute in presence or absence of 10^{-7} or 10^{-6} M hαCGRP$_{8-37}$ or 2) cumulative concentrations of hα or hβCGRP in presence or absence endothelium, 10^{-5} M L-NAME, 3x10^{-5} M methylene blue or 3x10^{-6} M indomethacin. Vessels treated with single doses were used for cyclic nucleotide determinations by RIA kits (Amersham) and vessels treated with cumulative doses were used for vasomotor studies.

17.3. Results

17.3.1. Studies of Cyclic Nucleotide Levels

Single doses of αCGRP or βCGRP induced vasodilation and increased levels of cAMP within one minute of application on PGF$_{2\alpha}$ precontracted segments of guinea pig basilar artery (Fig. 17.1). Both of these effects induced by α as well as βCGRP were inhibited by αCGRP$_{8-37}$. Differences between mean values of cAMP levels representing

The CGRP Family: Calcitonin Gene-Related Peptide (CGRP), Amylin, and Adrenomedullin, edited by David Poyner. ©2000 EUREKAH.COM.

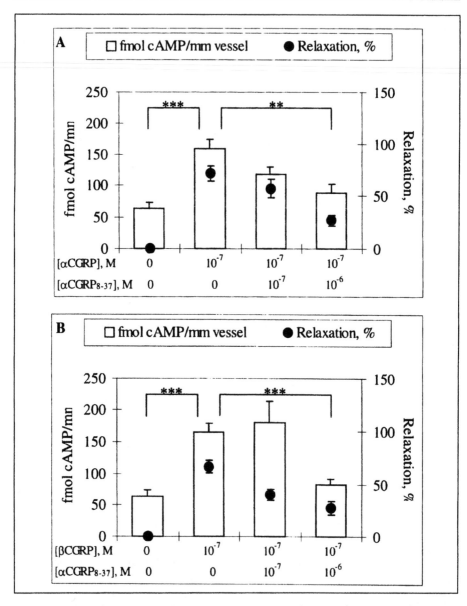

Fig. 17.1. Relaxation and levels of cAMP following one minute administration of 10^{-7} M hαCGRP (A) and 10^{-7} M hβCGRP (B) in presence or absence of hαCGRP$_{8-37}$, 10^{-7} or 10^{-6} M. The cAMP formation shown in empty bars is expressed as fmol cAMP/mm vessel segment and is given as mean values ± s.e.mean. The CGRP induced relaxation shown in dark circles is expressed as % of PGF$_{2\alpha}$ precontraction and is given as mean values ± s.e.mean. Relaxation and cAMP formation are induced by α as well as βCGRP, and the effects of both peptides are specifically inhibited by the CGRP$_1$ antagonist αCGRP$_{8-37}$. Significance of differences between cAMP levels representing different treatments are shown above the bars (**: p < 0,01; ***: p < 0,001). Number of experiments = 12-22. Reprinted from Sams A, Yenidunya A, Engberg J et al. Equipotent in vitro actions of α– and β-CGRP on guinea pig basilar artery are likely to be mediated via CRLR derived CGRP receptors. Regulatory Peptides, 85(1). ©1999, with permission from Elsevier Science.

different treatments were evaluated by non parametric one-way ANOVA and significance was evaluated by a two-tailed t-test. In preliminary studies we found cGMP levels ranging from 2-6 fmol/mm vessel. The level seemed unrelated to administration of α- and βCGRP and we did not investigate cGMP accumulations further.

17.3.2. Studies of Vasomotor Responses

αCGRP and βCGRP induced potent relaxation of $PGF_{2\alpha}$ precontracted guinea pig basilar arteries in vitro (Table 17.1). Neither endothelium removal (15 seconds of Triton X-100 perfusion) nor pretreatment for 15-20 minutes with L-NAME, methylene blue or indomethacin significantly altered E_{max} or pEC_{50} for the relaxant responses to α- or βCGRP as evaluated by a two tailed t-test. A non significant increase in αCGRP sensitivity was seen following endothelium removal and indomethacin pretreatment and a non significant decrease in βCGRP sensitivity was seen following L-NAME pretreatment.

17.4. Conclusion

On segments of guinea pig basilar artery there are no significant differences in potency or mechanisms of action between α- and β-CGRP. An increase in cAMP levels in artery segments treated by 10^{-7} M α- or βCGRP for one minute is evident. The increase is significantly inhibited by 10^{-6} M $h\alpha CGRP_{8-37}$. We conclude that CGRP possibly mediates its effects by cAMP accumulation which agrees well with previous studies showing that the $CGRP_1$ receptor is coupled to a G-protein that stimulates adenylate cyclase. In the studies of vasomotor responses we were not able to relate the CGRP induced relaxation to the cyclo- oxygenase pathways or the NOS/cGMP pathway.

Table 17.1. Relaxant responses of guinea pig basilar artery segments to αCGRP and βCGRP with and without pretreatments

Agonist	hαCGRP			hβCGRP		
Treatment	pEC_{50}	E_{max}	n	pEC_{50}	E_{max}	n
None	8.4 ± 0.2	87 ± 5	8	8.9 ± 0.2	103 ± 10	7
Endothelium removal	9.2 ± 0.2	86 ± 10	6	8.9 ± 0.2	99 ± 7	9
L-NAME, 10^{-5}M	8.5 ± 0.2	95 ± 3	11	7.7 ± 0.3	92 ± 7	5
Methylene blue, $3*10^{-5}$M	8.2 ± 0.1	94 ± 5	7	8.6 ± 0.1	92 ± 5	6
Indomethacin, $3*10^{-6}$M	9.0 ± 0.2	90 ± 5	8	8.6 ± 0.3	99 ± 1	7

E_{max}: maximum relaxant response in % of $PGF_{2\alpha}$ precontraction; pEC_{50}: negative logarithm of the agonist concentration eliciting half maximum relaxation. The data are presented as mean values ± s.e.mean; n: number of vessel segments examined.

References

1. Jansen I. Characterization of calcitonin gene-related peptide (CGRP) receptors in guinea pig basilar artery. Neuropeptides 1992; 21(2):73-79.
2. Mulvany MJ, Halpern W. Contractile properties of small arterial resistance vessels in spontanously hypertensive and normotensive rats. Circulation research 1977; 41(1):19-26.

CHAPTER 18

CGRP Receptors in Rat Intramural Left Coronary Arteries

M. Sheykhzade and N.C.B. Nyborg

18.1. Introduction

Calcitonin gene-related peptide (CGRP) mediates its effect through $CGRP_1$ and $CGRP_2$ receptor subtypes and is a potent coronary vasodilator with positive inotropic action.[1] CGRP induces endothelium-dependent cyclo-oxygenase mediated relaxation in rat epicardial arteries and direct endothelium-independent relaxation in intramural arteries indicating heterogeneity in vasodilator mechanisms in different segments of the coronary circulation.[2]

18.2. Methods

Using an isometric small vessel myograph[2] the vasodilator activity of CGRP was studied in ring segments of rat intramural left coronary arteries. The CGRP-receptor subtype was characterized partly by Schild-plot analysis using human (h) $\alpha CGRP_{8-37}$ and partly by comparison of potencies of rat (r) $\alpha CGRP$ to that of $r\beta CGRP$, [CysACM]CGRP, rat-adrenomedullin and -amylin.

18.3. Results and Discussion

Our results showed that the sensitivity of rat intramural coronary arteries to $r\alpha CGRP$ and its efficacy increased with decreasing vessel calibre. We found that the response to $r\alpha CGRP$ could be restored if the preparations were depolarized twice with 125 mM K^+ between each $r\alpha CGRP$ concentration- response curve.[3] This indicates that the depolarization could prevent ion channel inactivation in the smooth muscle membrane or perhaps prevent internalization of the CGRP receptors.[4] $r\alpha CGRP$ induced endothelium-independent relaxation which was concentration dependently antagonized by the $CGRP_1$-selective antagonist, $h\alpha CGRP_{8-37}$, giving Schild plot slopes equal to unity and pA_2 values of 7 in both female and male rats. The receptor type was constant in coronary arteries with different calibre (Fig. 18.1). The selective $CGRP_2$-receptor agonist, [CysACM]CGRP, failed to induce significant relaxations. Finally, rat-amylin and rat-adrenomedullin which both have relatively high homology with CGRP, were much less potent and efficacious ($\approx$ 16% and $\approx$ 28%, respectively) than CGRP itself.[5] Thus the general potency order was $r\alpha CGRP \approx r\beta CGRP >> $ rat-adrenomedullin > rat-amylin. This strongly indicates that CGRP mediates vasodilatation directly on the vascular smooth muscle cells through the $CGRP_1$-receptor subtype in rat isolated intramural coronary arteries in either sex.[5]

The CGRP Family: Calcitonin Gene-Related Peptide (CGRP), Amylin, and Adrenomedullin, edited by David Poyner. ©2000 EUREKAH.COM.

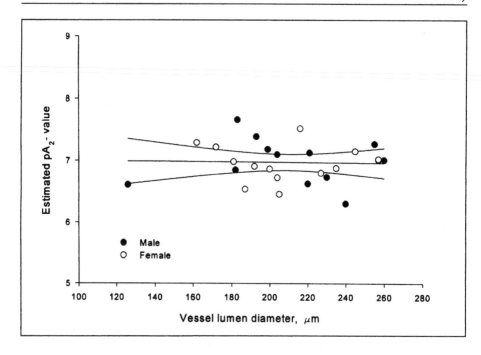

Fig. 18.1. Relationship between estimated pA_2-values for $CGRP_{8-37}$ and vessel calibre in rat intramural coronary arteries from male (●) and female (O) Sprague Dawley rats.

References

1. Bell D, McDermott BJ. Calcitonin gene-related peptide in the cardiovascular system: characterization of receptor populations and their (patho)physiological significance. Pharmacol Rev 1996; 48:253-288.
2. Prieto D, Benedito S, Nyborg NCB. Heterogeneous involvement of endothelium in calcitonin gene-related peptide-induced relaxation in coronary arteries from rat. Br J Pharmacol 1991; 103:1764-1768.
3. Sheykhzade M, Nyborg NCB. Caliber dependent calcitonin gene-related peptide-induced relaxation in rat coronary arteries: effect of K$^+$ on the tachyphylaxis. Eur J Pharmacol, 1998; 351:53-59.
4. Franco-Cereceda A. Effect of colchicine on the development of tachyphylaxis to calcitonin gene-related peptide (CGRP) in the guinea pig heart. Neuroscience Letters 1990; 122: 249-251.
5. Sheykhzade M, Nyborg NCB. Characterization of calcitonin gene-related peptide (CGRP) receptors in intramural coronary arteries from male and female Sprague Dawley rats. Br J Pharmacol 1998; 123:1464-1470.

CHAPTER 19

Discrete Expression of a Putative CGRP Receptor (RDC1) mRNA in the Rat Brain and Peripheral Tissues

Yiai Tong, Yvan Dumont, Denise vanRossum, Adrian J.L. Clark, Shi-Hsiang Shen and Remi Quirion

19.1. Introduction

Calcitonin gene-related peptide (CGRP) is a 37 amino acid peptide that arises from the alternate processing of the RNA transcript of the calcitonin gene.[1,2] In the CNS, CGRP-like immunoreactivity is especially concentrated in hypothalamic areas and certain brainstem nuclei (for detailed reviews see refs. 3,4).

CGRP-like peptides have been shown to exert a wide variety of biological actions. Behavioral effects of CGRP include the modulation of food intake, blood pressure, analgesia, catalepsy and body temperature.[5-8]

Based on in vitro bioassays, structure-activity studies suggested that the various effects of CGRP are mediated by the activation of at least two CGRP receptor subtypes designated as $CGRP_1$ and $CGRP_2$ on the basis of the preferential affinity of the fragment $CGRP_{8-37}$ to antagonize the effect of CGRP ($CGRP_1$) and the agonistic property of the linear analogue of CGRP, [Cys (acetylaminomethyl)^{2,7}] CGRP, to mimic the effect of CGRP ($CGRP_2$) (see Quirion et al, this volume for detailed review).[5,7] Additionally, [^{125}I]hCGRPα binding sites were found to be widely distributed in the rat brain.[4,7,9-12] The existence of [^{125}I]hCGRPα/salmon calcitonin-sensitive

sites has also been reported and appears to be mostly restricted to the nucleus accumbens.[7,13] This unique binding sites may, in fact, represent an amylinreceptor (see Quirion et al, Sexton et al, this volume).[4]

Recently, the cloning of CGRP receptors belonging to this peptide family has been reported, one likely represents the adrenomedullin receptor[14] while the others may be CGRP-like receptors.[15-17] They all belong to the seven transmembrane G-protein coupled receptor superfamily except for the clone reported by Luebke and collaborators which possesses a single transmembrane domain[17] (see Rosenblatt et al, this volume). Interestingly, two of the G-protein type clones, first identified as an orphan receptor the CRLR[18, 19] and the RDC-1[20] while respectively transfected in HEK293[15] and COS-7[16] cell, have a pharmacological profile similar to the $CGRP_1$ receptor but demonstrate very low sequence homologies. Very little is currently known on the respective distribution of the mRNA of these newly cloned putative CGRP receptors. Most recently, one group has studied the expression and distribution of the CRLR (a putative $CGRP_1$-like) receptor mRNA in the brain and peripheral tissues.[21] They observed that the distribution of the CRLR receptor mRNA did not closely resemble the distribution of the CGRP receptors suggesting that this clone may not represent the $CGRP_1$ receptor or only a subtype yet to be fully characterized.

In this study, we investigated the expression and distribution of the rat RDC-1 receptor mRNA (a putative $CGRP_1$- like receptor[16]) in the rat brain and peripheral tissues using an in situ hybridization method and observed its widespread but discrete expression in the rat CNS and some peripheral tissues. Surprisingly, only very limited overlap exists with the distribution of the rat RDC-1 receptor mRNA and that of CGRP binding sites suggesting that this clone may not represent a genuine CGRP receptor or at least not the usual, pharmacologically characterized $CGRP_1$ subtype.

19.2. Materials and Methods

Brain slices from male adult Sprague-Dawley rats were prepared as described previously.[22]

19.2.1. In Situ Hybridization

A 690 base pair fragment derived from the rat RDC-1 cDNA[16] was subcloned into a bluescript KS^{M13-} vector (Invitrogen, San Diego, USA) and used as template to generate [^{35}S]-labeled transcripts in either orientations using T3/T7 transcript kits (Riboprobe Gemini System, Promega, Fisher Scientific, Montréal, QC, Canada). An antisense riboprobe labeled with [^{35}S]UTP (1000 Ci/mmol, Amersham, Mississauga, ON, Canada) was transcribed by T3 RNA polymerase (Promega) using the template linearized by EcoR I (Promega). The sense-strand control riboprobe was generated by T7 RNA polymerase (Promega) using the same template linearized by Hind III (Promega). The in situ hybridization (ISH) protocol was carried our as described in detail elsewhere.[23]

19.2.2. Receptor Autoradiography

[^{125}I]h$CGRP_{8-37}$ was used as radioligand and receptor autoradiography performed as described elsewhere.[22]

19.3. Results

In this study, we have investigated the expression and distribution of the rat RDC-1 (a putative $CGRP_1$-like) receptor mRNA. As shown in figure 19.1, stronger specific ISH signals were obtained with the antisense as compared to sense probe in all brain areas studied here. These data demonstrate further that the rat RDC-1 receptor mRNA is broadly distributed but with levels of expression varying from low to high depending upon the brain region under study. High levels of expression were observed in the dentate gyrus of the hippocampus, ependyma of the ventricles and choroid plexus (Fig. 19.1). Moderate to high ISH signals were detected in the anterior olfactory nuclei, nucleus accumbens, tenia tecta and dorsomedial hypothalamic nuclei (Fig. 19.1). Most other brain areas including the cortex, caudate putamen, thalamus, most hypothalamic and brainstem nuclei, and the cerebellum revealed only very weak specific ISH signals (Fig. 19.1). This pattern is strikingly distinct from that of specific CGRP binding sites which are most enriched in the cerebellum, various brainstem nuclei, nucleus accumbens, amygdaloid body and tail of the caudate putamen (Fig. 19.2)

Sense and antisense incubated rat brain slides were also exposed to liquid emulsion to improve anatomical resolution. At high magnification, diffused specific silver grains representing mRNA were observed in cerebral cortex (Fig. 19.3A). The intensity of the silver grains even revealed a laminar distribution with higher number of grains being found in lamina 5 compared to all other laminae (Fig. 19.1 and 19.3A). Interestingly, most of silver grains are located to pyramidal neurons with no signal being detected in glial cells in these cortical areas (Fig. 19.3A). Similarly, neurons but not glial cells of the nucleus accumbens (Fig. 19.3B) and of the dentate gyrus of the hippocampus (Fig. 19.3C) were found to express the rat RDC-1 receptor mRNA, as well as the ependyma of the ventricles (Fig. 19.3D) and smooth muscle of cerebral vasculature (Fig. 19.3E).

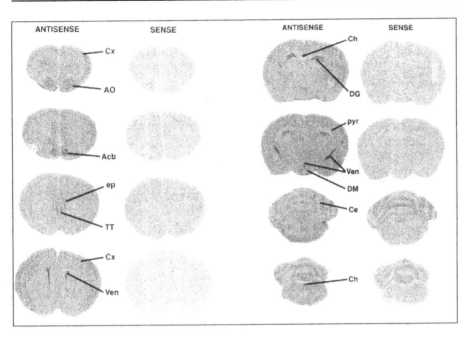

Fig. 19.1. Photomicrographs of coronal sections of the rat brain demonstrating the expression and distribution of the rat RDC-1 (a putative CGRP) receptor mRNA using antisense and sense probes to demonstrate specificity. Abbreviations: anterior olfactory nucleus (AO), cortex (Cx), nucleus accumbens (Acb), ependyma of the ventricles (ep), tenia tecta (TT), ventricle (Ven), dentate gyrus of the hippocampus (DG), choroid plexus (Ch) pyramidal cell layer of the hippocampus (pyr), dorsomedial nucleus of the hypothalamus (DM) and cerebellum (Ce).

In peripheral tissues, sections exposed to dry films failed to reveal strong specific ISH signals (data not shown). However, sections exposed to liquid emulsion revealed light specific signals in the terminal bronchiole (Fig. 19.4A) and blood vessels (Fig. 19.4C) of the rat lung as well as in the kidney with specific ISH signals detected in the pondocytes of the glomeruli (Fig. 19.4E). In all these tissues, the sense probe failed to reveal any signal demonstrating specificity (Fig 19.4B, D and F).

19.4. Discussion

Our results revealed that the highest levels of ISH signals of the rat RDC-1 (a putative CGRP$_1$-like) receptor mRNA are observed in areas of the rat brain that contained only low to very levels of CGRP binding sites. In fact, quantitative receptor autoradiography in the rat brain demonstrated that [^{125}I]hCGRPα and/or [^{125}I]CGRP$_{8-37}$ binding sites are mostly concentrated in the nucleus accumbens (shell), caudate putamen (tail), amygdaloid body, cerebellum, pontine nuclei and inferior olive.[22] These areas are not enriched with the RDC-1 receptor mRNA. In contrast, high ISH signals were observed in the olfactory nuclei, tenia tecta, lining of ventricles, choroid plexus and dentate gyrus of the hippocampus; regions which are not enriched with significant level of binding sites. While technical issues cannot be fully excluded to explain these mismatches, it suggests that the gene (RDC-1 cDNA) identified as a putative CGRP$_1$ receptor may not be related to the

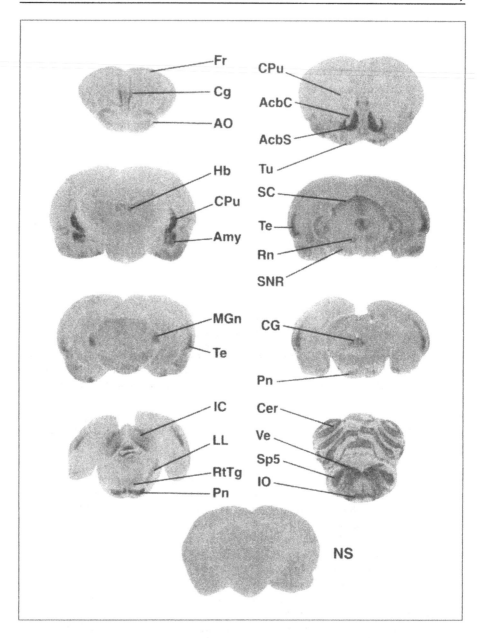

Fig. 19.2. Photomicrographs of the autoradiographic distribution of [^{125}I-Tyr]hCGRP$_{8-37}$ binding sites in coronal sections of the rat brain. Abbreviations used: AcbC, nucleus acccumbens core; AcbS, nucleus accumbens shell; Amy, amygdaloid body; AO, anterior olfactory nucleus; Cg, cingulate cortex; CPu, caudate putamen; CG, central gray; Cer, cerebellum; Fr, frontal cortex; Hb, habenular nucleus; IC, inferior colliculus; IO, inferior olive; LL, lateral lemniscus; MGn, medial geniculate; Pn, pontine nuclei; Rn, red nucleus; RtTg, reticulotegmental nucleus of the pons; SC, superior colliculus; SNR, substantia nigra; Sp5, spinal trigeminal nucleus; Te, temporal cortex; Tu olfactory tubercle; Ve vestibular nuclei; NS, non-specific binding.

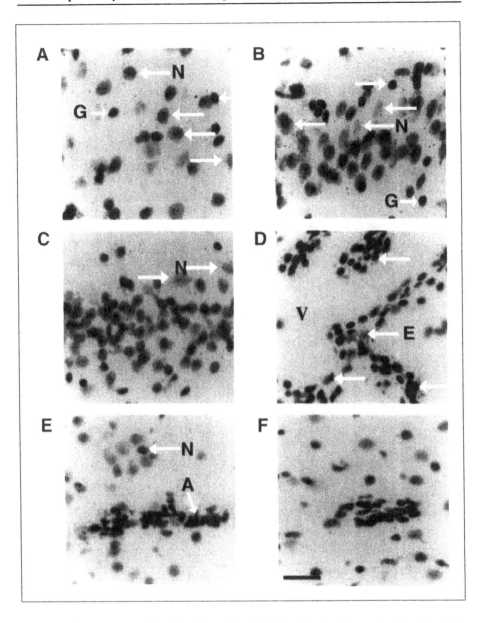

Fig. 19.3. Photomicrographs of liquid emulsion ISH showing the discrete localization of the rat RDC-1 (a putative CGRP) receptor mRNA signal using antisense probes in layer 5 of the cortex (A), nucleus accumbens (B), dentate gyrus of the hippocampus (C), choroid plexus (D) and small arteries of the cerebral cortex (E). F: Sense signal to demonstrate specificity. Scale bar: 50 mm. Symbols represent pyramidal cells (N), glial cells (G), epithelial cells of the ventricle (E), smooth muscle of small arteries (A) and ventricle (V).

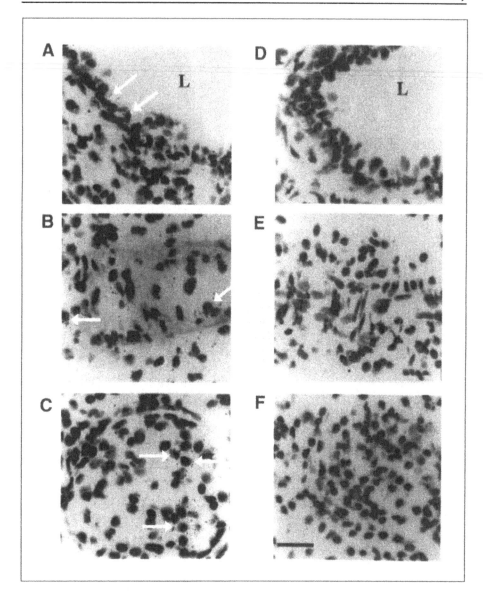

Fig. 19.4. Photomicrographs revealing the expression and distribution of the rat RDC-1 (a putative CGRP) receptor mRNA (see arrows) in the epithelial cells of terminal bronchioles (A), the endothelial cells of small arteries of the lung (C) and the podocytes of glomeruli of the kidney (E) (see arrows). B, D and F represent sense signal to demonstrate specificity. Scale bar: 50 mm.

pharmacologically characterized CGRP$_1$ receptor.

The cloned used in this study was originally identified as an orphan receptor from canine thyroid cDNA.[20] Northern blot analysis revealed major sites of expression in the heart and kidney with the brain and spleen showing much lower levels of expression and only low to very low amounts being present in the stomach, liver, lung and salivary gland. Subsequently, Kapas and Clark[16] suggested that this clone, upon transfection in COS-7 cells, demonstrated a pharmacological profile similar to the CGRP$_1$ receptor subtype. However, these authors recently indicated that they failed to replicate their original findings (personal communication). Together with the present data, it would appear that this clone does not represent the CGRP$_1$ receptor and hence it returns to the status of "orphan" receptor. As to the other putative CGRP$_1$ receptor clone, CRLR, while early studies failed to identify it,[21] the recent characterization of a novel family of a single transmembrane proteins, called receptor-activity-modifying-proteins or RAMPs seemingly shows that this clone with RAMP1 can form a biologically relevant CGRP$_1$ receptor[24] (see also Foord et al this volume).

Acknowledgments

This research is supported by a joint Industry/University grant from the Medical Research Council of Canada and BioMega/ Karl Thomae GmbH. Rémi Quirion is "Chercheur Boursier" of the "Fonds de la Recherche en Santé du Québec."

References

1. Amara SG, Jonas V, Rosenfeld MG et al. Alternative RNA processing in calcitonin gene expression generates mRNAs encoding different polypeptide products. Nature 1982, 296:240-244.
2. Rosenfeld MG, Mermod JJ, Amara SG et al. Production of a novel neuropeptide encoded by the calcitonin gene via tissue-specific RNA processing. Nature 1983, 304:129-135.
3. Ishida-Yamamoto A, Tohyama M. Calcitonin gene-related peptide in the nervous tissue. Prog Neurobiol 1989; 33:335-386.
4. van Rossum D, Hanisch UK, Quirion R. Neuroanatomical localization, pharmacological characterization and functions of CGRP and related peptide receptors. Neurosci Biobehav Rev 1997; 21:649-678.
5. Dennis T, Fournier A, Cadieux A et al. hCGRP8-37 a calcitonin gene-related peptide antagonist revealing CGRP receptor heterogeneity in brain and periphery. J Pharmacol Expt Ther 1990; 254:123-128.
6. Jolicoeur FB, Menard D, Fournier A et al. Structure activity analysis of CGRP's neurobehavioral effects. Ann NY Acad Sci 1992; 657:155-163.
7. Quirion R, van Rossum D, Dumont Y et al. Characterization of CGRP$_1$ and CGRP$_2$ receptor subtypes. Ann NY Acad Sci 1992; 657:88-105.
8. Krahn DD, Gosnell BA, Levine AS et al. Effects of calcitonin gene-related peptide on food intake. Peptides 1984; 5:861-864.
9. Sexton PM, McKenzie JS, Mason RT et al. Localization of binding sites for calcitonin gene-related peptide in rat brain by in vitro autoradiography. Neuroscience 1986; 19:1235-1245.
10. Sexton PM, McKenzie JS, Mendelsohn FAO. Evidence for a new subclass of calcitonin/calcitonin gene-related peptide binding sites in rat brain. Neurochem Int 1988; 12:323-335.
11. Henke H, Sigrist S, Lang W et al. Comparison of binding sites for the calcitonin gene-related peptides I and II in man. Brain Res 1987; 410:404-408.
12. Inagaki S, Kite S, Kubota Y et al. Autoradiographic localization of calcitonin gene related peptide binding sites in human and rat brains. Brain Res 1986; 374:287-298.
13. Dennis T, Fournier, A, Guard S et al. Calcitonin gene related peptide (hCGRPα) binding sites in nucleus accumbens. Atypical structural requirements and marked phylogenic differences. Brain Res 1991; 539:59-66.
14. Kapas S, Catt KJ, Clark AJL. Cloning and expression of cDNA encoding a rat adrenomedullin receptor. J Biol Chem 1995; 270:25344-25347.
15. Aiyar NJ, Rand IC, Elshourbagy NA et al. A cDNA encoding the calcitonin gene-related peptide type 1 receptor. J Biol Chem 1996; 271:11325-11329.
16. Kapas S., Clark AJL. Identification of an orphan receptor gene as type 1 calcitonin gene-related peptide receptor. Biochem Biophys Res Commun 1995; 217:832-838.

17. Luebke AE, Dahl GP, Ross BA et al. Identification of a protein that confers calcitonin gene-related peptide responsiveness to oocytes by using a cystic fibrosis transmembrane conductance regulator assay. Proc Natl Acad Sci USA 1996; 93:3455-3460.

18. Chang CP, Pearse RV, O'Connel S et al. Identification of a seven transmembrane helix receptor for corticotropin-releasing factor and sauvagine in mammalian brain. Neuron 1993; 11:1187-1195.

19. Njuki F, Nicholl CG, Howard A et al. A new calcitonin-receptor-like sequence in rat pulmonary blood vessels. Clin Sci 1993; 85:385-388.

20. Libert F, Parmentier M, Lefort A et al. Selective amplification and cloning of four new members of the G protein-coupled receptor family. Science 1989; 244:569-572.

21. Fluhmann B, Lauber M, Lichtensteiger W et al. Tissue-specific mRNA expression of a calcitonin receptor-like receptor during fetal and postnatal development. Brain Res 1997; 774:184-192.

22. van Rossum D, Ménard DP, Fournier A et al. Binding profile of a selective calcitonin gene-related peptide (CGRP) receptor antagonist ligand, $[^{125}I]hCGRP_{8-37}$, in rat brain and peripheral tissues. J Pharmacol Expt Ther 1994; 269:846-853.

23. Tong Y, Rheaume E, Simard J et al. Localization of peripheral benzodiazepine binding sites and diazepan-binding inhibitor (DBI) mRNA in mammary tumors in the rat. Regul Pept 1991; 33:263-273.

24. McLatchie LM, Fraser NJ, Main MJ et al. RAMP's regulate the transport and ligand specificity of the calcitonin-receptor-like receptor. Nature 1998; 393:333-339.

CHAPTER 20

Amylin-Immunoreactivity in Monkey Hypothalamus

L. D'Este, R. Vaccaro, S. Wimalawansa, T. Maeda and T. G. Renda

20.1. Introduction

Amylin (or Islet Amyloid Polypeptide, AMY) is a 37 amino acid peptide first isolated from pancreatic islet amyloid of Type II (non-insulin dependent) diabetic patients. It shares about 50% homology with calcitonin gene-related peptide (CGRP) and is involved in several regulatory functions including glucose homeostasis, food intake and memory processing (for a review see refs.1,2). Because several studies had reported little or no amylin mRNA in the brain, amylin has usually been considered as being an exclusively peripheral peptide that acts centrally because of its ability to cross the blood brain barrier. Recent reports, however, have described amylin expression in chicken brain,[3] rat sensory neurons[4] and in the rat brain, where it has a widespread immunohistochemical distribution.[5] In this study the amylin-like immunoreactivity of the monkey hypothalamus was investigated.

20.2. Methods

Brain specimens of monkey *(Macaca fuscata japonica)* were immunohistochemically investigated using a polyclonal antiserum raised in rabbits against the carboxy amidated terminal tridecapeptide of human amylin (AMY$_{25-37}$). Previous trials had excluded cross-reactivity of this antiserum to CGRP, adrenomedullin, calcitonin and other known biologically active peptides.

Animals were deeply anesthetized and perfused via the ascending aorta with PAF mixture (4% paraformaldehyde, 0.2% picric acid and 0.35% glutaraldehyde in 0.1 M phosphate buffer, pH 7.4). Each brain was coronally sliced into several specimens that were postfixed in the same fixative without glutaraldehyde for a further 48 hours. The slices corresponding to the hypothalamic areas were frozen and cut into 20 μm thick sections that were treated as free-floating sections and submitted to the immunofluorescent or ABC-peroxidase indirect immunohistochemical methods. Controls included the pre-absorption test: preincubation of primary antiserum at a working dilution (1: 30,000) with synthetic amylin (20 μg/ml) overnight at +4°C .

20.3. Results and Conclusions

Amylin-immunoreactive nerve cell bodies were found in discrete areas of monkey hypothalamus, mainly in the magnocellular portion of the supraoptic nucleus, in perifornical nuclei and in medium size neurons of the paraventricular nucleus. Positive small neurons were also detected in the periventricular, ventromedial and arcuate nuclei. Positive nerve fibers were shown along the hypothalamo-hypophysial tracts, the median eminence and the tubero-mammillary region.

The CGRP Family: Calcitonin Gene-Related Peptide (CGRP), Amylin, and Adrenomedullin,
edited by David Poyner. ©2000 EUREKAH.COM.

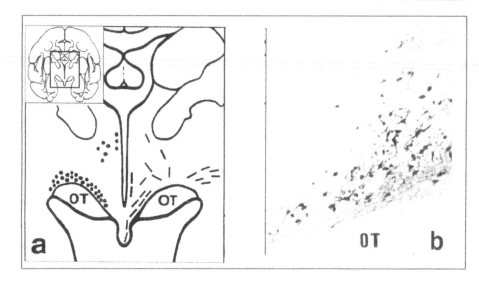

Fig. 20.1. a: Drawing of the hypothalamic region from a coronal section of monkey brain (inset) showing the distribution patterns of the amylin-immunoreactive nerve cell bodies (dots on the left side) and fibers (lines on the right side). b: Photomicrograph showing a cluster of amylin-immunoreactive neurons in the supraoptic nucleus. ABC-peroxidase method. Original magnification: 80x. OT= optic tract.

Acknowledgments

This study received financial support from the Italian Ministry of the University (MURST 40 and 60%).

References

1. Van Rossum D, Hanisch U-K, Quirion R. Neuroanatomical localization, pharmacological characterization and functions of CGRP, related peptides and their receptors. Neurosci Biobehav Rev 1997; 21:649-678.
2. Wimalawansa SJ. Amylin, calcitonin gene-related peptide, calcitonin and adrenomedullin: a peptide superfamily. Crit Rev Neurobiol 1997; 11:167-239.
3. Fan L, Westermark G, Chan SJ et al. Altered gene structure and tissue expression of islet amyloid polypeptide in the chicken. Mol Endocrinol 1994; 8:713-721.
4. Mulder H, Leckstrsm A, Uddman R et al. Islet amyloid polypeptide (Amylin) is expressed in sensory neurons. J Neurosci 1995; 15:7625-7632.
5. Skofisch G, Wimalawansa SJ, Gubish W. Comparative immunohistochemical distribution of amylin-like and calcitonin gene-related peptide-like immunoreactivity in rat central nervous system. Can J Physiol Pharmacol 1995; 73:945-956.

CHAPTER 21

Analysis of Responses to hAmylin and hCGRP in Isolated Resistance Arteries from the Mesenteric Vascular Bed of the Rat

Hunter C. Champion, Robert L. Pierce, Trinity J. Bivalacqua, William A. Murphy, David H. Coy and Philip J. Kadowitz

21.1. Introduction

Amylin is a 37 amino acid peptide that is a normal constituent of pancreatic βcells. It is generally considered to be a glycoregulatory peptide that inhibits the actions of insulin in in vivo and in vitro assays. It belongs to the calcitonin gene-related peptide (CGRP) family of peptides that also includes adrenomedullin (ADM) (see Sexton et al and Young et al for reviews, this volume).

It is well known that CGRP and ADM are potent vasodilators in human and animal models. (see Kadowitz et al, this volume). However, the vascular actions of amylin are less well understood. It has been reported that amylin induces vasodilator responses in the rabbit, the perfused rat mesenteric vascular bed and kidney of the rat, and it appears that vasodilator responses to amylin are dependent upon species and vascular bed studied.[1-6] In the cat amylin has significant vasodilator activity in the pulmonary vascular bed,[2] but not in the mesenteric and hindlimb vasculature.[3,4]

Although amylin, CGRP and ADM are structurally related peptides, little is known about the receptors mediating responses to these agonists. (see Smith et al, Kadowitz et al, this volume) It has been reported that amylin and CGRP produce vasodilator activity that is mediated via the activation of the CGRP$_1$ receptor, since responses were blocked by CGRP$_{8-37}$ in the isolated perfused mesenteric vascular bed and kidney of the rat.[2,5] Although it has been shown that amylin produces vasodilator responses via CGRP$_{8-37}$-sensitive receptors in the perfused mesenteric vascular bed and kidney of the rat, little if anything is known about the effects of amylin$_{8-37}$ on vasodilator responses to CGRP and amylin in the mesenteric vasculature of the rat.[2,5] Moreover, the mechanism mediating this vasodilator response is not known. Therefore, the purpose of the present investigation was to study the effects of amylin on isolated, precontracted resistance vessels from the mesenteric vascular bed of the rat and to evaluate the role of CGRP$_{8-37}$- and amylin$_{8-37}$-sensitive receptors in mediating the responses to human (h) CGRP and amylin.

21.2. Methods

Methods were essentially as described previously (See Kadowitz et al, Champion et al, this volume)

The CGRP Family: Calcitonin Gene-Related Peptide (CGRP), Amylin, and Adrenomedullin, edited by David Poyner. ©2000 EUREKAH.COM.

21.3. Results

21.3.1. Responses to hCGRP and hAmylin

Administration of hCGRP and hAmylin to U46619 precontracted arteries caused dose-related increases in arterial diameter (Fig 21.1). The dose-response curve for hADM was very similar to hCGRP (data not shown). Administration of hCGRP and hamylin did not alter resting tone of the vessels (data not shown).

21.3.2. Mechanisms of Vasodilation

The role of nitric oxide synthesis/release and soluble guanylate cyclase in mediating vasodilator responses to hCGRP and hamylin was investigated in isolated resistance arteries from the rat mesenteric vascular bed (Fig. 21.2). Administration of the nitric oxide synthase inhibitor L-NAME or the soluble guanylate cyclase inhibitor ODQ did not alter the vasodilator responses to hCGRP and hamylin at a time when vasodilator responses to acetylcholine were significantly reduced (Fig. 21.2). Vasodilator responses to hADM were reduced significantly after administration of L-NAME or ODQ (data not shown). A similar pattern of responsiveness was seen after removal of vascular endothelium or addition of the K_{ATP} channel antagonist U37883A (Fig. 21.3).

21.3.3. Role of the hCGRP₈₋₃₇ - and hAmylin₈₋₃₇ -Sensitive Receptors

Vasodilator responses to hCGRP and hamylin were significantly reduced after administration of the $CGRP_1$ receptor antagonist hCGRP$_{8-37}$ and also amylin$_{8-37}$ (Fig. 21.4), whereas the vasodilator response to hADM was not altered (data not shown).

21.4. Discussion

The results of the present investigation show hamylin and hCGRP induce dose-related vasodilator responses in isolated resistance arteries from the mesenteric vascular bed of the rat. In terms of relative vasodilator activity, hCGRP was similar in potency to hADM and both were slightly

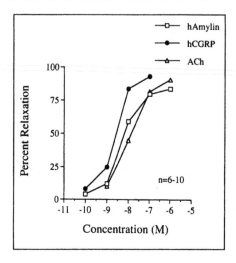

Fig. 21.1. Dose-response curves comparing increases in mesenteric artery diameter (percent relaxation) in response to superfusion of human (h) amylin, human calcitonin gene-related peptide (hCGRP) and acetylcholine (ACh). The arteries were precontracted with the thromboxane A_2 mimic U46619. n indicates number of experiments.

more potent than hamylin. The mechanism by which hamylin and hCGRP dilates rat mesenteric resistance arteries does not involve the NO/cGMP pathway or membrane hyperpolarization by way of activation of K^+_{ATP} channels. Responses to hamylin and hCGRP were reduced by hCGRP$_{8-37}$ and hamylin$_{8-37}$.

The results of the present study are consistent with previous studies in the perfused rat mesenteric vascular bed in that hamylin has significant vasodilator activity in the isolated resistance artery from the rat mesenteric vascular bed.[5] Human amylin has vasodilator activity that is slightly less potent than hCGRP in this preparation, and both peptides are comparable in potency to acetylcholine. It is interesting to note that hamylin does not have vasodilator activity in the hindquarters vascular bed of the rat (unpublished observations) in contrast to hCGRP.[7] The differences in responses to hCGRP and hamylin in the mesenteric and hindquarters vascular beds of the rat are uncertain but may reflect

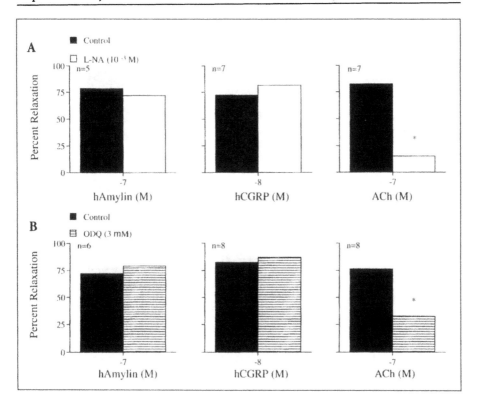

Fig. 21.2. A) Increases in arterial diameter (percent relaxation) in response to administration of human (h) amylin (hamylin), human calcitonin gene-related peptide (hCGRP) and acetylcholine (ACh) before and after administration of the nitric oxide synthase inhibitor L-NA. B) Increases in arterial diameter in response to administration of hamylin, hCGRP and ACh before and after administration of the soluble guanylate cyclase inhibitor ODQ. The arteries were precontracted with the thromboxane A_2 mimic U46619. n indicates number of experiments. *P <0.05.

differences in receptor subtype population. It may be hypothesized that multiple receptors are present in the vasculature that bind hCGRP and hamylin to varying degrees. It has also been shown that hamylin does not have vasodilator activity in the mesenteric vascular bed of the cat providing further evidence to suggest that there exists species differences in responses to these peptides.[3]

It has previously been reported that vasodilator responses to CGRP are mediated by the release of nitric oxide in the rat.[8] However, other studies report that responses to CGRP are not altered by the nitric oxide synthase inhibitors in the pulmonary, mesen-teric and hindquarters vascular beds of the cat and rat and suggest that responses to these peptides are mediated by the activation of adenylate cyclase and the subsequent increase in cAMP.[2,7] It has also been reported that amylin has activity that is related to nitric oxide synthase. The results of the present study are consistent with previous studies showing that NO is not involved in the vasodilation,[2,7,9] in contrast to hADM. It has previously been shown that hamylin has endocrine and paracrine activity that is mediated by cAMP (Sexton et al, this volume) and that hADM and hCGRP have vasodilator activity that is mediated by a

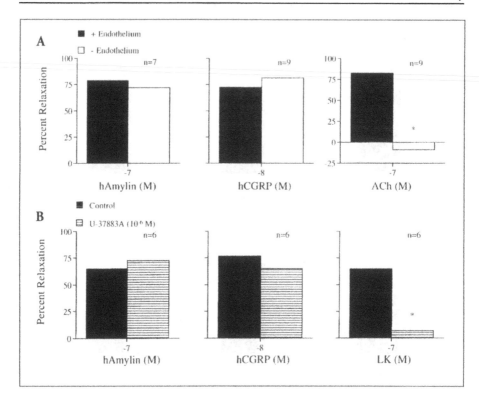

Fig. 21.3. A) Increases in arterial diameter (percent relaxation) in response to administration of human (h) amylin (hamylin), human calcitonin gene-related peptide (hCGRP) and acetylcholine (ACh) in vessels with intact (+) vascular endothelium and after the vascular endothelium was denuded. B) Increases in arterial diameter in response to administration of hamylin, hCGRP and levcromakalim (LK) before and after administration of the vascular selective K^+_{ATP} channel antagonist U-37883A. The arteries were precontracted with the thromboxane A_2 mimic U46619. n indicates number of experiments. *$P <0.05$.

cAMP-dependent mechanism (Kadowitz et al, this volume). Based on the present data, it is not certain whether the vasodilator activity of hCGRP and hamylin is mediated by the activation of adenylate cyclase. Further experiments are needed to further evaluate this mechanism in the isolated mesenteric arteries of the rat.

Previous studies have reported that CGRP mediates vasodilation in rabbit vascular smooth muscle by opening K^+_{ATP} channels.[10] The present study suggests that the K^+_{ATP} channels do not play a significant role in mediating vasodilator responses to these peptides in the isolated resistance

arteries from the rat mesenteric vascular bed and this is consistent with studies in the hindlimb and mesenteric vascular beds of the cat. (Kadowitz et al, Champion et al, this volume).

While it has been reported that vasodilator responses to hADM are blocked by the $CGRP_1$ receptor antagonist $hCGRP_{8-37}$, other studies demonstrate that responses to hADM were not inhibited by this antagonist in a dose that inhibited vasodilator responses to CGRP (Kadowitz et al, this volume). Moreover, it has been reported that responses to hamylin are mediated by a $CGRP_{8-37}$-sensitive receptor in the perfused

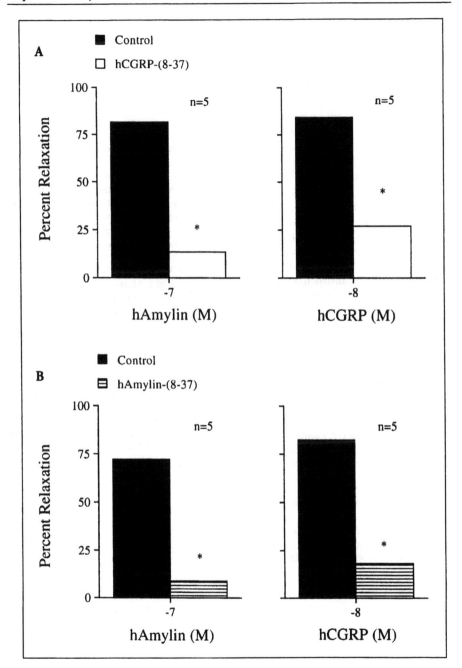

Fig. 21.4. A) Increases in arterial diameter (percent relaxation) in response to administration of human (h) amylin (hamylin) and human calcitonin gene-related peptide (hCGRP) before and after administration of $hCGRP_{8-37}$. B) Increases in arterial diameter in response to administration of hamylin and hCGRP before and after administration of $hamylin_{8-37}$. The arteries were precontracted with the thromboxane A_2 mimic U46619. n indicates number of experiments. $*P < 0.05$.

mesenteric vascular bed and kidney of the rat.[10,20] The results of the present study suggest that responses to hamylin and hCGRP, but not hADM, are mediated at least in part by the activation of a hCGRP$_{8-37}$-sensitive receptor in the isolated mesenteric artery from the rat. Responses to hamylin and hCGRP, but not hADM, are mediated by an hamylin$_{8-37}$-sensitive receptor. It is not known, however, if hamylin$_{8-37}$ is blocking the CGRP$_1$ receptor or a CGRP/amylin receptor subtype. It is possible that a number of receptors exist with varying affinity for hADM, hCGRP and hamylin.

In summary, these data suggest that hCGRP and hamylin- have direct vasodilator activity in the isolated mesenteric resistance artery that is mediated by receptors that are sensitive to hamylin-$_{8-37}$ and hCGRP$_{8-37}$. The responses are not mediated by NO or K$^+_{ATP}$ channels.

References

1. Chin SY, Hall JM, Brain SD et al. Vasodilator responses to CGRP and amylin in the rat isolated perfused kidney are mediated via CGRP$_1$ receptors. J Pharmacol Exp Ther 1994; 2690:8989-992.

2. DeWitt BJ, Cheng DY, Caminiti GN et al. Comparison of responses to adrenomedullin and calcitonin gene-related peptide in the pulmonary vascular bed of the cat. Eur J Pharmacol 1994; 257:303-306.

3. Santiago JA, Garrison EA, Purnell WL et al. Comparison of responses to adrenomedullin and adrenomedullin analogs in the mesenteric vascular bed of the cat. Eur J Pharmacol 1995; 272:115-118.

4. Santiago JA, Garrison EA, Ventura VL et al. Synthetic human adrenomedullin and adrenomedullin-(15-52) have potent short-lived vasodilator activity in the hindlimb vascular bed of the cat. Life Sci 1994; 55:PL-85-PL-90.

5. Westfall TC, Curfman-Falvey M. Amylin-induced relaxation of the perfused mesenteric arterial bed: mediation by CGRP receptors. J Cardiovasc Pharmacol 1995; 26:932-936.

6. Brain SD, Wimalawansa S, McIntyre I et al. The demonstration of vasodilator activity of pancreatic amylin amide in the rabbit. Am J Pathol 1990; 136:487-490.

7. Feng CJ, Kang B, Kaye AD et al. L-NAME modulates responses to adrenomedullin in the hindquarters vascular bed of the rat. Life Sci 1994; 55:PL-433-PL-438.

8. Marshall I. Mechanism of vascular relaxation by the calcitonin gene-related peptide. Ann NY Acad Sci 1992; 657:204-215.

9. Gumusel B, Hao Q, Hyman AL et al. Analysis of responses to adrenomedullin-(13-52) in the pulmonary vascular bed of the rat. Am J Physiol 1998; 274:1255-1263.

10. Quayle JM, Bonev AD, Brayden JE et al. Calcitonin gene-related peptide activated ATP-sensitive K$^+$ currents in rabbit arterial smooth muscle via protein kinase. Am J Physiol 1994; 475:9-13.

CHAPTER 22

Adrenomedullin in the Rat Placenta and Uterus

G.M. Taylor, J.D. Hurley, M.A. Ghatei, S.R. Bloom and D.M. Smith

22.1. Introduction

We have shown that adrenomedullin (ADM) peptide, mRNA and receptors are all present in the rat uterus, indicating that ADM may have a physiological role in reproduction.[1] Other vasoactive peptides present in the uterus, such as endothelin (ET), are known to show variations in their immunoreactivity and mRNA during the menstrual cycle.[2,3] To establish whether ADM has a role in the uterus during the oestrous cycle, rat uteri from each stage of the oestrous cycle were assessed for changes of ADM peptide and mRNA.

Placenta has previously been shown to have high levels of ADM mRNA.[1] However, no ADM binding sites or immunoreactivity were detected in placenta. It is well known that ADM is released from other tissues and cells, for example endothelial cells, which contain high levels of ADM mRNA, but little immunoreactivity.[4] Therefore isolated rat placentae were investigated for release of ADM immunoreactivity.

22.2. Methods

Female Wistar rats were vaginally smeared to determine the stage of the oestrous cycle (n=4/5 per group). The uterus was removed and divided into horns. One horn from each rat was extracted in boiling acetic acid (0.5M) and aliquots of the extract assayed

for ADM peptide by radioimmunoassay. The remaining uterus horn was used for total RNA extraction by the acid guanidinium thiocyanate/phenol/chloroform method.[5] RNA was fractionated on a 1% agarose gel and transferred to nylon membranes. A 150 base-pair cDNA probe corresponding to the entire sequence of rat ADM was used to probe the membrane.[6] The probe was labelled with $[\alpha\text{-}^{32}P]$-dCTP to a specific activity of $> 7 \times 10^8$ dpm/μg. Signals were normalised for RNA loading by reprobing with labelled oligo (dT).

Placentae from 15 day pregnant rats were removed and incubated in 95% O_2: 5% CO_2 Krebs buffer at 37°C for 3 x 10 minutes. Each placenta was then placed in a clean flask containing 3 mls of fresh buffer and incubated for 30-60 minutes either alone or with a test substance. Buffer was removed from the flask and assayed for ADM peptide by radioimmunoassay.[1]

22.3. Results and Discussion

There was no difference in uterine ADM peptide content between the stages of the oestrous cycle (Fig. 22.1). However, there was a significant decrease in ADM mRNA in the dioestrous stage compared to the mid/late prooestrous stage (Fig. 22.2). ADM may have a role in proliferation of the

The CGRP Family: Calcitonin Gene-Related Peptide (CGRP), Amylin, and Adrenomedullin, edited by David Poyner. ©2000 EUREKAH.COM.

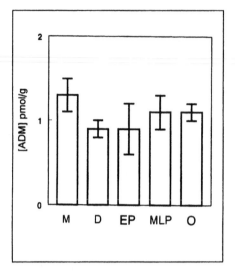

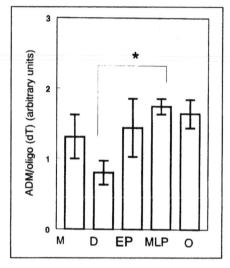

Fig. 22.1. ADM peptide content in rat uterus at each stage of the oestrous cycle. n=4/5 per group. M=metoestous, D=dioestrous, EP=early prooestrous, MLP=mid/late prooestrous, O=oestrous.

Fig. 22.2. ADM mRNA in rat uterus at each stage of the oestrous cycle. The filter was probed with ADM followed by oligo (dT). Data was quantified by capture on a Phosphorimager screen (Phosphorimager SR, Molecular Dynamics) and data analysed using Imagequant software. Data is expressed in arbitrary units with data representing mean ± s.e. mean and analysed statistically by ANOVA, * $p < 0.05$, n=4/5 per group. M=metoestous, D=dioestrous, EP=early prooestrous, MLP=mid/late prooestrous, O=oestrous.

endometrium immediately prior to ovulation. Recent studies have shown that ADM is an angiogenic factor in human endometrium.[7]

ADM was released from isolated rat placentae in a time dependent manner at a mean rate of 6.62 ± 0.37 fmol/g/minute (n=22) (Fig. 22.3). Calculations show that this is roughly equivalent to the rate of ADM release from vascular smooth muscle cells.[8] Neuropeptide Y (NPY) significantly increased the rate of ADM release (155% of control, p<0.05, n = 8) from placentae (Fig. 22.4). The thromboxane mimic, U46619, also significantly increased ADM release from isolated placenta (200% of control p<0.01 n=6) (Fig. 22.5). Thromboxane and NPY are vasoconstrictors in the uteroplacental vasculature.[9,10] Interestingly angiotensin II, a powerful peripheral vasoconstrictor but ineffective in the placental vasculature, was without effect on ADM release (Fig. 22.5). Thus ADM may be involved in regulation of vascular tone in the placental unit and therefore be a

possible factor in the pathogenesis of preeclampsia and other pregnancy related complications.

References

1. Upton PD, Austin C, Taylor G.M et al. Expression of adrenomedullin (ADM) and its binding sites in the rat uterus: Increased number of binding sites and ADM messenger ribonucleic acid in 20-day pregnant rats compared with non pregnant rats. Endocrinology 1997; 138 (6):2508-14.
2. Cameron IT, Plumpton C, Champeney R et al. Identification of endothelin-1, endothelin-2 and endothelin-3 in human endometrium. J Repro Fertil 1993; 98 (1):251-255.

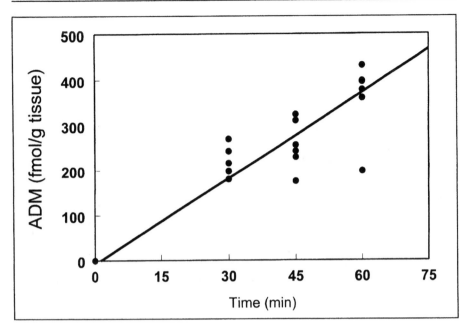

Fig. 22.3. Time course of ADM release from isolated rat placenta. n=5/6 per time point

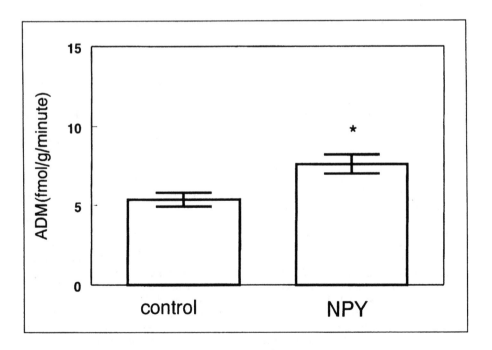

Fig. 22.4. Effect of NPY (1mM) on ADM release from isolated rat placenta. * p < 0.05. n=8.

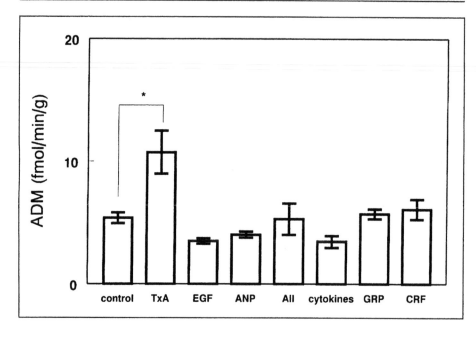

Fig 22.5. Effect of possible regulators on ADM release from isolated rat placenta. * p < 0.05. TxA=thromboxane mimic, U46619 (n=6), EGF=epidermal growth factor (n=4), ANP=atrial natriuretic peptide (n=3), AII=angiotensin II (n=8), cytokines= lipopolysaccaride, tumour necrosis factor-α and interleukin-1b (n=4), GRP=gastrin releasing peptide (n=3) and CRF=corticotropin releasing factor (n=4). All substances were used at 1 mM.

3. O'Reilly G, Charnock-Jones DS, Davenport AP et al. Presence of messenger ribonucleic acid for endothelin-1, endothelin-2 and endothelin-3 in human endometrium and a change in the ratio of ETA and ETB receptor subtype across the menstrual cycle. J Clin Endocrin Metab 1992; 75 (6):1545-1549

4. Sugo S, Minamino N, Kangawa K et al. Endothelial cells actively synthesize and secrete adrenomedullin. Biochem Biophys Res Commun 1994; 201(3):1160-1166.

5. Chomczynski P, Sacchi N. Single-step method of RNA isolation by acid guanidinium thiocyanate-phenol-chloroform extraction. Anal Biochem 1987; 162:156-159.

6. Sakata J, Shimokubo T, Kitamura K et al. Molecular cloning and biological activities of rat adrenomedullin, a hypotensive peptide. Biochem Biophys Res Commun 1993; 195:921-927.

7. Zhao Y, Hague S, Manek S et al. PCR display identifies tamoxifen induction of the novel angiogenic factor adrenomedullin by a non estrogenic mechanism in the human endometrium. Oncogene 1998; 16:409-415.

8. Sugo S, Minamino N, Shoji H et al. Production and secretion of adrenomedullin from vascular smooth muscle cells: Augmented production by tumor necrosis factor α. Biochem Biophys Res Commun 1994; 203(1):719-726.

9. Morris JL, Murphy R. Evidence that neuropeptide Y released from noradrenergic axons causes prolonged contraction of the guinea pig uterine artery. J Auton Nerv Syst 1988; 24(3):241-249.

10. Bjoro K, Stray-Pedersen S. Effects of vasoactive autocoids on different segments of human umbilicoplacental vessels. Gynecol Obstet Invest 1986; 22(1):1-6.

CHAPTER 23

Effects of ADM and CGRP on Erectile Activity in the Cat

Hunter C. Champion, Trinity J. Bivalacqua, Run Wang,
Wayne J.G. Hellstrom, William A. Murphy, David H. Coy
and Philip J. Kadowitz

23.1. Introduction

Penile erection is caused by relaxation of the blood vessels and trabecular mesh work of smooth muscle that constitutes the corpora cavernosa.[1,2] Although parasympathetic innervation is involved in erectile function, a nonadrenergic, noncholinergic (NANC) mechanism may also play a role in penile erection.[1,2] Having been found localized in nerve fibers around cavernous smooth muscle and blood vessels, vasoactive intestinal peptide has been postulated to be a neurotransmitter and/or neuromodulator for erectile function.[1] Similarly, calcitonin gene-related peptide (CGRP), a potent vasodilator and smooth muscle relaxant, has been shown to be localized in high concentrations in the genitofemoral nerve and penile tissue of the rat.[3] CGRP has recently been reported to induce penile erections in primates and may serve to modulate penile erection.[4]

Our previous investigations have demonstrated that the adult cat is a useful and reliable model for the study of penile erectile responses to intracavernous injections of vasoactive agents.[5-10] The present study was undertaken to compare the effects of adrenomedullin (ADM) and CGRP on the erectile response in the cat.

23.2. Materials and Methods

Experimental methods were essentially as described previously.[5-10] The data were expressed as mean the standard error of the mean and analyzed by Student's t-test for single group comparison and by one-way analysis of variance for multiple-group comparison.[11] The value of $P < 0.05$ was established as the criteria for statistical significance.

23.3. Results

23.3.1. Responses to ADM and CGRP

Intracavernosal injections of ADM (0.1 to 1 nmol) and CGRP (0.01-0.3 nmol) caused a dose-dependent increase in cavernosal pressure and penile length. With the 1 nmol dose of ADM there was approximately an 8-fold increase in cavernosal pressure and a 43% increase in penile length when compared to preinjection baseline values for cavernosal pressure and penile length (Fig. 23.1). The 1 nmol dose of ADM elicited a similar increase in penile length when compared to the response obtained by the administration of triple drug therapy.

The CGRP Family: Calcitonin Gene-Related Peptide (CGRP), Amylin, and Adrenomedullin,
edited by David Poyner. ©2000 EUREKAH.COM.

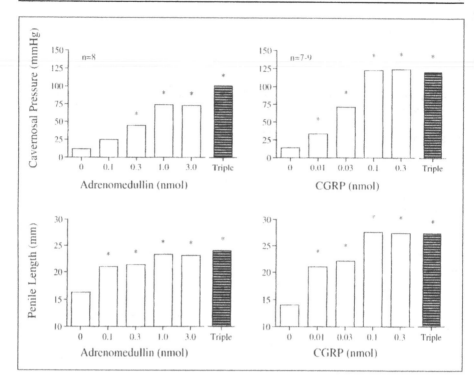

Fig. 23.1. Dose-dependent increases in penile pressure and penile length in response to intracavernosal injection of adrenomedullin (ADM) and calcitonin gene-related peptide (CGRP). Triple denotes response to control triple-drug combination (papaverine, phentolamine and PGE$_1$) administered at the end of the experiment. n indicates number of animals. Asterisk indicates that response is significantly different from baseline. *P <0.05.

Increases in cavernosal pressure to ADM, ADM$_{15-52}$, ADM$_{22-52}$, ADM$_{40-52}$ and CGRP were compared to vasoactive intestinal peptide (VIP) and sodium nitroprusside, and these data are shown in Figure 23.2. The potency order was hCGRP>VIP=hADM$_{15-52}$ =hADM>nitroprusside. The 22-52 and 40-52 fragments of ADM were inactive.

23.3.2. Effect of ADM and CGRP on Systemic Arterial Pressure

Intracavernous injections of ADM (0.1 to 1.0 nmol) did not significantly decrease systemic arterial pressure. When ADM was injected in a dose of 3 nmol, systemic arterial pressure was decreased significantly (Fig. 23.3). Decreases in systemic arterial pressure in response to intracavernosal

injection were similar in magnitude and duration to ADM (data not shown). A similar pattern of responsiveness was seen with CGRP (0.01 to 0.3 nM).

23.3.3. Role of the CGRP Receptor

Intracavernosal injection of the CGRP receptor antagonist CGRP$_{8-37}$ in a dose of 30 nmol significantly reduced increases in cavernosal pressure and penile length in response to CGRP but not ADM or ADM$_{15-52}$ (Fig. 23.4).

23.3.4. Role of Nitric Oxide Synthesis/Release and K$^+$$_{ATP}$ Channels

Intracavernosal injection of the nitric oxide synthase inhibitor L-NAME or the

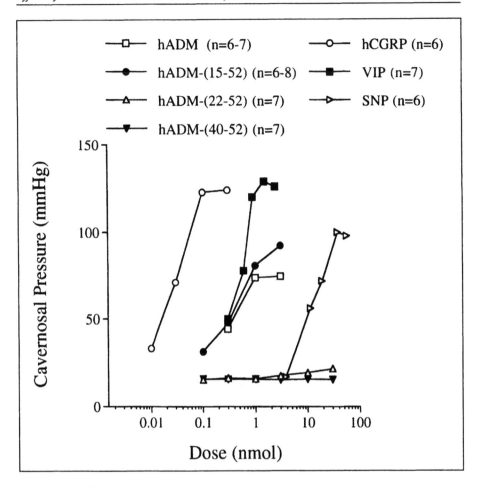

Fig. 23.2. Dose-response curves comparing increase sin intracavernosal pressure in response to adrenomedullin (ADM), ADM$_{15\text{-}52}$, ADM$_{22\text{-}52}$, ADM$_{40\text{-}52}$, calcitonin gene-related peptide (CGRP), vasoactive intestinal peptide (VIP) and sodium nitroprusside (SNP) when doses are compared on a molar basis. n indicates number of animals.

K$^+$$_{ATP}$ channel blocker glybenclamide did not significantly alter the change in cavernosal pressure and penile length in response to ADM and CGRP in the doses given (Figs. 23.5 and 23.6), although they were effective against acetylcholine and laevocromakalin respectively.

23.4. Discussion

The feline model has been successfully employed in our investigations of the physiology and pharmacology of penile erection.[5-10] The present study, however, is the first to investigate erectile responses to ADM and CGRP in the feline penis in vivo. Results of the present investigation show that ADM, ADM$_{15\text{-}52}$ and CGRP caused significant dose-related increases in intracavernous pressure and penile length when the peptides were injected directly into the corpus cavernosum in doses from 0.1-3.0 nmol and 0.01-0.3 nmol, respectively. However, ADM$_{22\text{-}52}$ and ADM$_{40\text{-}52}$ did not induce erectile responses when injected in

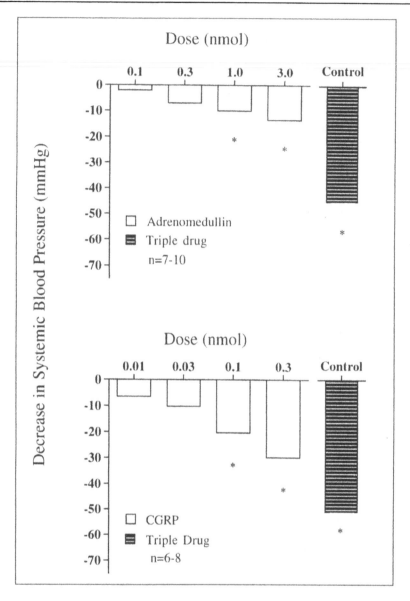

Fig. 23.3 Decrease in systemic arterial pressure in response to intracavernosal injection of adrenomedullin (ADM) and calcitonin gene-related peptide (CGRP). Triple denotes response to control triple-drug combination (papaverine, phentolamine and PGE₁) administered at the end of the experiment. n indicates number of animals. Asterisk indicates that response is significantly different from baseline. *P <0.05

similar doses. These data suggest that the six-membered ring structure in the ADM sequence is required for erectile activity, and these data are similar to those obtained in the systemic, pulmonary, hindlimb and mesenteric vascular beds of the rat and cat

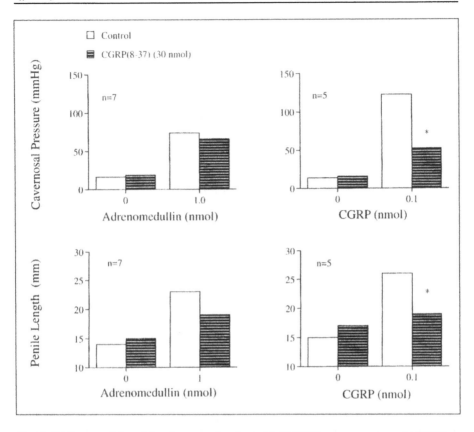

Fig. 23.4 Influence of the calcitonin gene-related peptide (CGRP) receptor antagonist CGRP$_{8-37}$ on erectile responses to intracavernosal injection of adrenomedullin (ADM) and CGRP. n indicates number of animals. Asterisk indicates that response is significantly different from control. *P <0.05

(See Kadowitz et al, this volume for more detail).

Penile erection can be induced by stimulation of pelvic parasympathetic nerves.[1] The discovery that nerve-stimulated erection can be reduced, but not blocked, by atropine suggests that acetylcholine is not the only transmitter mediating the erectile response.[2] Earlier studies suggested that VIP was the additional neurotransmitter responsible for the erectile response.[1,6] However, further studies with VIP did not prove significant.[1,6] CGRP has recently been postulated to be a new candidate for NANC responses in mediating penile erection.[4] It has been shown that CGRP is localized in the genitofemoral nerve and in rat penile tissue and that the peptide induces penile erection in the monkey.[3,4] Based on the finding that ADM and CGRP are found in a number of organ systems and human plasma, these peptides may serve as circulating hormones that are involved in the regulation of systemic arterial pressure.[12,13] These peptides may have a natural role in mediating and augmenting the penile erectile response.

Based on the finding that CGRP and ADM can induce penile erection that is comparable to that seen with the clinically used triple combination (papaverine, PGE$_1$ and phentolamine), these data suggest that

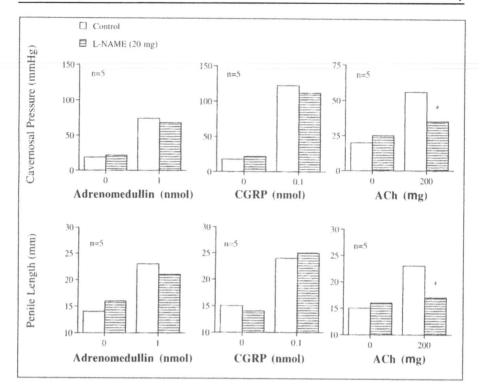

Fig. 23.5 Influence of the nitric oxide synthase inhibitor L-NAME on erectile responses to intracavernosal injection of adrenomedullin (ADM), calcitonin gene-related peptide (CGRP) and acetylcholine (ACh). n indicates number of animals. Asterisk indicates that response is significantly different from control. *P <0.05

the peptides may be useful in the clinical treatment of physiological impotence. Furthermore, the systemic effects on blood pressure and the duration of action are not as great as those seen with the triple drug combination, thus suggesting a possible advantage to the alternative treatment of impotence with these peptides.

Nitric oxide and K^+_{ATP} channels have been previously shown to mediate penile erection in the cat.[5,7] The present study suggests that erectile responses to ADM and CGRP are not mediated by these mechanisms in the cat.[9,10] Further studies will be required to further delineate the intracellular mechanism involved in erectile responses to ADM and CGRP.

The receptors with which ADM and CGRP interact remain uncertain. Several studies in different systems report that responses to ADM are reduced by the CGRP receptor antagonist CGRP$_{8-37}$. However, other studies (e.g., in the hindlimb vascular bed of the cat)[14,15] have shown that vasodilator responses to ADM are not altered by CGRP$_{8-37}$. The present study shows that erectile responses to CGRP are significantly reduced by CGRP$_{8-37}$ at a time when erectile responses to ADM are unaltered, supporting the hypothesis that ADM may interact with its own receptor.

In summary, the present study shows that intracavernosal injections of ADM, ADM$_{15-52}$ and CGRP, but not ADM$_{22-52}$ or

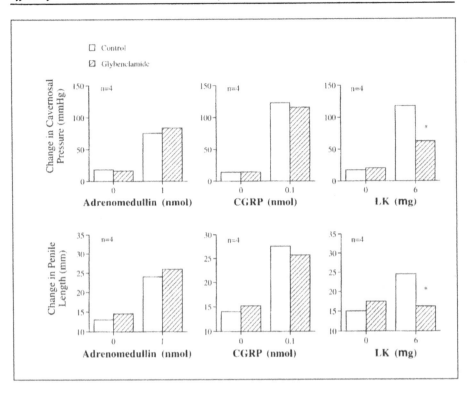

Fig. 23.6. Influence of the K^+_{ATP} channel antagonist glibenclamide on erectile responses to intracavernosal injection of adrenomedullin (ADM), calcitonin gene-related peptide (CGRP) and levcromakalim (LK). n indicates number of animals. Asterisk indicates that response is significantly different from control. *P <0.05

ADM$_{40-52}$, produce dose-related increases in cavernosal pressure and penile length. These data show that CGRP is more potent than ADM, VIP and nitroprusside in increasing cavernosal pressure. Intracavernosal injection of ADM and CGRP decreased systemic arterial pressure in the higher doses administered. These data further suggest that erectile responses to ADM and CGRP are not dependent on the synthesis/release of nitric oxide or the opening of K^+_{ATP} channels in the corpora and that erectile responses induced by ADM are not mediated by the CGRP receptor.

References

1. Anderson KE. Pharmacology of lower urinary tract smooth muscles and penile erectile tissues. Pharmacol Rev 1993; 45:253-308.
2. Lugg JA, Gonzales-Cadavid NF, Rajfer J. The role of nitric oxide in erectile function. J Androl 1995; 16:2-4.
3. Terada M, Hutson JM, Farmer PJ et al. The role of the genitofemoral nerve and calcitonin gene-related peptide in congenitally cryptorchid mutant ts rats. J Urol 1995; 154:734-737.

4. Steif CG, Benard F, Bosch R et al. Calcitonin gene-related peptide: Possible neurotransmitter contributes to penile erection in monkeys. Urology 1993; 41:397- 401.

5. Hellstrom WJG, Wang R, Kadowitz PJ et al. Potassium channel agonists cause penile erection in rats. Int J Impotence Res 1992; 4:35-43.

6. Wang R, Higuera TR, Sikka SC et al. Penile erection induced by vasoactive intestinal peptide and sodium nitroprusside. Urol Res 1993; 21:75-78.

7. Wang R, Domer FR, Sikka PJ et al. Nitric oxide mediates penile erection in cats. J Urol 1994; 151:234-237.

8. Champion HC, Wang R, Hellstrom WJG et al. Nociceptin, a novel endogenous ligand for the ORL_1 receptor, has potent erectile activity in the cat. Am J Physiol 1997; 273:E214-E219.

9. Champion HC, Santiago JA, Wang R, et al. Adrenomedullin induces penile erection in the cat. Eur J Pharmacol 1997; 319:71-75.

10. Champion HC, Wang R, Santiago JA et al. Comparison of responses to adrenomedullin and calcitonin gene-related peptide in the feline erection model. J Androl 1997; 18:513-521.

11. Snedecor GW, Cochran WG. Statistical Methods. 6th ed. Ames: Iowa State University Press, 1967.

12. Kitamura K, Kangawa K, Kawamoto M et al. Adrenomedullin: a novel hypotensive peptide isolated from human pheochromocytoma. Biochem Biophys Res Commun 1993; 192:553-560.

13. Ichiki Y, Kitamura K, Kangawa K et al. Distribution and characterization of imunoreactive adrenomedullin in human tissue and plasma. FEBS Lett 1994; 338:6- 10.

14. Champion HC, Murphy WA, Coy DH et al. Proadrenomedullin NH_2-terminal 20 peptide has direct vasodilator activity in the cat. Am J Physiol 1997; 272:R1047-R1054.

15. Champion HC, Santiago JA, Murphy WA et al. Adrenomedullin-(22-52) antagonizes vasodilator responses to calcitonin gene-related peptide but not adrenomedullin in the cat. Am J Physiol 1997; 272:R234-R242.

CHAPTER 24

Evidence for a Specific Endothelial Cell Adrenomedullin Receptor Regulating Endothelin-1 Gene Expression and Synthesis

Stewart Barker, Delphine Lees, Elizabeth G.Wood and Roger Corder

24.1. Introduction

Adrenomedullin (ADM) has been classified as a member of the calcitonin gene-related peptide (CGRP) family of peptides on the basis of structural homology and because in some tissues ADM and CGRP appear to share a common receptor. ADM exerts potent vasodilator effects in a number of vascular beds,[1] and reverses the vasoconstrictor action of endothelin-1 (ET-1).[2-3] The aim of these studies was to determine whether ADM affects the level of preproET-1 messenger RNA (mRNA) or ET-1 release from bovine aortic endothelial cells (BAEC). In addition, the pharmacological nature of the receptor through which these effects are mediated was also addressed.

24.2. Methods

Rat ADM and CGRP were obtained from Peptide Institute (Osaka, Japan). Cyclic AMP was measured by specific enzyme immunoassay (Amersham International plc, Little Chalfont, UK) in cell-free supernatants after 30 min incubation.[4] ET-1 release was measured by direct RIA.[5] Reverse transcriptase polymerase chain reaction (RT-PCR) was carried out as in ref.

6, using putative intron-spanning primers based on the published bovine preproET-1 cDNA sequence.

24.3. Results

Using BAE-1 cells (ECACC, Porton Down, UK) we found a concentration-dependent increase in cyclic AMP in response to ADM with a threshold value of 0.1 nM. In addition, over the same concentration range, ADM resulted in a reduction in ET-1 release into the culture medium over a 6h incubation period. Using 100 nM ADM, which caused a maximal increase in cyclic AMP, ET-1 accumulation was reduced by 30% compared to basal levels (Figure 24.1). RT-PCR using total RNA extracted from these cells after removal of medium for ET-1 RIA showed that steady state preproET-1 mRNA levels were reduced to a similar extent (Figure 24.2). CGRP produced a significant reduction in ET-1 release at 100 nM. However, only the response to CGRP was blocked by the CGRP receptor antagonist $CGRP_{8-37}$ (1 μM) (Figure 24.1), suggesting that these cells express a specific ADM receptor.

The CGRP Family: Calcitonin Gene-Related Peptide (CGRP), Amylin, and Adrenomedullin,
edited by David Poyner. ©2000 EUREKAH.COM.

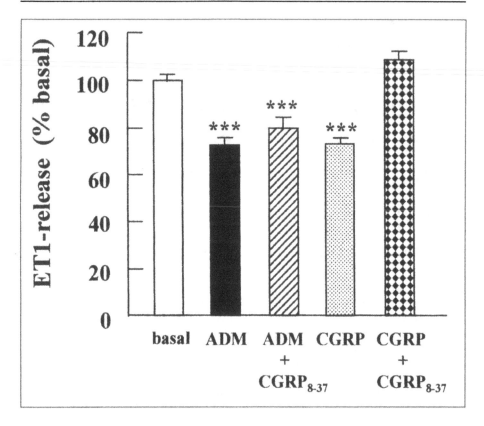

Fig. 24.1 ET-1 release from BAE-1 cells following 6h treatment with 100 nM ADM or 100 nM CGRP in the presence or absence of 1 μM CGRP$_{8-37}$ (n=4-16). *** P<0.001, one-way ANOVA and Tukey-Kramer post-test.

24.4. Discussion

These results show that ADM directly inhibits ET-1 synthesis in endothelial cells at the level of preproET-1 mRNA expression, resulting in diminished ET-1 secretion. ADM therefore joins a number of other vascular mediators which attenuate ET-1 vasoconstrictor activity. As ADM is secreted from vascular smooth muscle cells (VSMC),[7] one role for ADM might be to modulate the impact of ET-1 on these cells. Indeed, ET-1 itself has been reported to increase ADM synthesis in VSMC.[7] We, therefore, propose a protective role for adrenomedullin as a physiological antagonist to ET-1 as part of a homeostatic mechanism providing local regulation of vascular tone.

There is considerable controversy in the literature regarding which receptors mediate the actions of ADM. The pharmacological profile of the inhibitory effect of ADM on ET-1 synthesis in BAEC, described here, matches that of the endothelium-dependent ADM-induced vasodilation of the pulmonary vascular bed in rats[8] with respect to its resistance to blockade by CGRP$_{8-37}$. However, none of the cloned receptors which bind ADM display the dual criteria of vascular cell expression and ADM specificity, see refs.9-10. Thus the cloning of an endothelial cell ADM-specific receptor is likely to provide further insights into the role of ADM in vascular homeostasis.

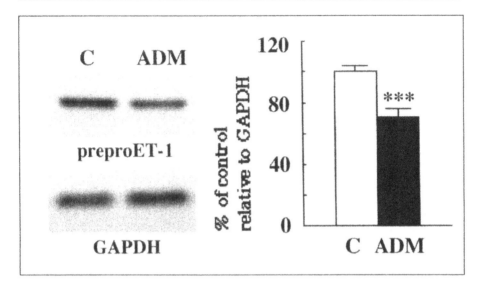

Fig. 24.2 RT-PCR results for preproET-1 mRNA in BAE-1 cells incubated with or without 100 nM ADM for 6h (*** *P*<0.001; Student *t*-test, n=4). Representative bands of expected size (left), and optical density expressed relative to GAPDH (right).

References

1. Kangawa K, Kitamura K, Minamino N et al. Adrenomedullin: a new hypotensive peptide. J Hypertension 1996; 14(Suppl.5): S105-S110.
2. Nishimura J, Seguchi H, Sakihara C et al. The relaxant effect of adrenomedullin on particular smooth muscles despite a general expression of its mRNA in smooth muscle, endothelial and epithelial cells. Br J Pharmacol 1997; 120:193-200.
3. Wang X, Yue T-L, Barone FC et al. Discovery of adrenomedullin in rat ischemic cortex and evidence for its role in exacerbating focal brain ischemic damage. Proc Natl Acad Sci USA 1995; 92:11480-11484.
4. Barker S, Corder R. Adrenomedullin acts as a local mediator of vascular homeostasis through interactions which lead to reduced endothelin-1 synthesis and secretion. J Hum Hypertension 1997; 11:605-606.
5. Corder R, Harrison VJ, Khan N et al. Effects of phosphoramidon in endothelial cell cultures on the endogenous synthesis of endothelin-1 and on conversion of endogenous synthesis of endothelin-1 and on conversion of exogenous big endothelin-1 to endothelin-1. J Cardiovasc Pharmacol 1993; 22(Suppl.8):S73-S76.
6. Barker S, Wood EG, Clark AJL, Corder R. Cloning of bovine preproadrenomedullin and inhibition of its basal expression in vascular endothelial cells by staurosporine. Life Sci. 1998; 62:1407-1415.
7. Sugo S, Minamino N, Shoji H et al. Effects of vasoactive substances and cAMP related compounds on adrenomedullin production in cultured vascular smooth muscle cells. FEBS Letters 1995; 369:311-314.
8. Gumusel B, Hao Q, Hyman AL et al. Analysis of responses to adrenomedullin-(13-52) in the pulmonary vascular bed of rats. Am J Physiol 1998; 274:H1255-H1263.
9. Kapas S, Catt KJ, Clark AJL. Cloning and expression of cDNA encoding a rat adrenomedullin receptor. J Biol Chem 1995; 270:25344-25347.
10. Han Z-Q, Coppock HA, Smith DM et al. The interaction of CGRP and adrenomedullin with a receptor expressed in the rat pulmonary vascular endothelium. J Endocrinol 1997; 18:267-272.

CHAPTER 25

Antimicrobial Effects of Adrenomedullin and CGRP

C. Zihni, S. Kapas and R.P. Allaker

25.1. Introdcution

The epithelium acts as a first barrier of defence against potentially pathogenic microorganisms, essentially by creating an interface between external and internal environments. Additionally, epithelial cells are involved in the expression of antimicrobial peptides, cytokine production and have an active immunological role including antigen processing and presentation. The production of antimicrobial peptides may be considered as an active ingredient of the protective 'epithelial wall.' For example the defensin family of antimicrobial peptides have recently been identified in tracheal and gastrointestinal mucosa.[1,2] Adrenomedullin (ADM) is similar to calcitonin gene-related peptide (CGRP) in terms of effects in vitro and in vivo, and modest structural similarity. Plasma levels of CGRP are difficult to measure (except in the case of neuroendocrine tumours). Plasma concentrations of ADM, on the other hand, are known to rise in a variety of pathological conditions including sepsis.[3] The aim of this study was to investigate the antimicrobial effects of ADM and CGRP based on these observations. Supporting evidence for an antimicrobial role by ADM, in particular, is seen by the accumulation of this peptide at the apical regions of normal human bronchial epithelium and human skin.[4,5]

25.2. Methods and Materials

Unless stated organisms used were laboratory strains: *Propionibacterium acnes*, *Staphylococcus aureus*, *Micrococcus luteus*, *Porphyromonas gingivalis* (ATCC 33277 and W 50), *Actinomyces naeslundii*, *Streptococcus mutans* (NCTC 10440), *Candida albicans* (ATCC 24433), *Eikenella corrodens* (NCTC 10596), *Streptococcus pneumoniae*, *Haemophilus influenzae*, *Streptococcus pyogenes*, *Bacteriodes fragilis* (NCTC 9560), *Escherichia coli* (NCTC 9001) and *Helicobacter pylori* (NCTC 11637).

CGRP/ADM was used to soak Whatman (Grade AA) 6mm discs to give a concentration of 3.5 µg for each disc. Bacterial suspensions were made to a density of 2×10^7 colony forming units (CFU)/ml. Blood agar base with defibrinated horse blood (5% v/v) was seeded using a swab moistened in the bacterial suspension and the discs placed onto the surface of the agar. The plates were incubated at 37°C in the air with 5% CO_2 for 24h. In the case of *B. fragilis*, *P. gingivalis* and *P. acnes* incubation was under anaerobic conditions (80% N_2; 10% CO_2; 10% H v/v) for 48 h at 37°C. In the case of *H. pylori* incubation conditions were microaerophilic (85% N_2; 10% CO_2; 5%0_2 v/v) for 48 h at 37°C. All determinations were carried out in triplicate.

The CGRP Family: Calcitonin Gene-Related Peptide (CGRP), Amylin, and Adrenomedullin, edited by David Poyner. ©2000 EUREKAH.COM.

Table 25.1. Effect of ADM and CGRP on skin, oral, respiratory tract and gut microorganisms measured by agar diffusion assay (3.5 μg/disc)

Origin of isolate	Isolate	ADM (Zone size - mm)	CGRP (Zone size - mm)
Skin	P. acnes	20	15
	S. aureus	4	8
	M. luteus	25	10
Oral cavity	P. gingivalis (ATCC)	12	resistant
	P. gingivalis (W50)	15	resistant
	E. corrodens	11	11
	A. naeslundii	15	11
	S. mutans	14	8
	C. albicans	0	0
Respiratory tract	S. pneumoniae	14	6
	S. pyogenes	14	7
	H. influenza	15	11
	B. fragilis	<1	1
	E. coli	10	8
	H. pylori	25	20

25.3. Results and Discussion

Table 25.1 demonstrates that both ADM and CGRP possess antimicrobial activity against most strains of Gram positive and negative bacteria tested (zone sizes of inhibition ranged from 1 to 25mm). *P. gingivalis* (W50 and ATCC 33277) displayed a high degree of sensitivity to ADM, but was resistant to CGRP. No activity by ADM or CGRP against the yeast *C. albicans* was demonstrated.

The synthesis and secretion of numerous antimicrobial peptides by epithelial cells is well documented. Expression of ADM in bronchial epithelium and macrophages of the respiratory tract suggests a possible protective action against microbial pathogens, similar to that observed for other peptides, such as magainins present in the airway epithelium or the tracheal antimicrobial peptide. ADM has 30% homology at the genetic level to the cecropin group of antimicrobial peptides and therefore may have a similar mechanism of action; via channel formation in membranes and subsequent lysis. It is suggested that ADM along with other related peptides including CGRP represent a new peptide family that contributes to mucosal host defence systems.

Novel antimicrobial agents based upon these molecules, that are not susceptible to existing mechanisms of resistance, may provide useful alternatives to conventional antibiotics.

References

1. Schonwetter BS, Stolzenberg ED, Zasloff MA. Epithelial antibiotics induced at sites of inflammation. Science 1995; 269:1645-1648.

2. Diamond G, Zasloff M, Eck H et al. Tracheal antimicrobial peptide, a cysteine-rich peptide from mammalian tracheal mucosa: Peptide isolation and cloning of a cDNA. Proc Natl Acad Sci USA 1991; 88:3952-3956.

3. Hirata Y, Mitaka C, Sato K et al. Increased circulating adrenomedullin, a novel vasodilatory peptide, in sepsis. J Clin Endocrin Metab 1996; 81:1449-1453.

4. Martínez A, Miller MJ, Unsworth EJ et al. Expression of adrenomedullin in normal lung and in pulmonary tumors. Endocrinol 1997; 136:4099-4105.

5. Martínez A, Elsasser TH, MuroCacho C et al. Expression of adrenomedullin and its receptor in normal and malignant human skin: A potential pluripotent role in the integument. Endocrinol 1997; 138:5597-5604.

CHAPTER 26

The Comparative Activity of Adrenomedullin and CGRP to Potentiate Inflammatory Oedema Formation

D.Q. Chu, T. Cao and S.D. Brain

26.1. Introduction

The intradermal injection of calcitonin gene-related peptide (CGRP) and adrenomedullin (ADM) leads to increased microvascular blood flow in rat skin.[1] ADM was suggested to be 3-30 times less potent than CGRP in these studies which used the ^{133}Xe-clearance technique as a measure of cutaneous blood flow.

Vasodilator concentrations of CGRP potentiate oedema formation induced by mediators of increased microvascular permeability.[2] In this study we have compared the ability of vasodilator doses of CGRP and ADM to i) induce oedema formation when injected alone and ii) to potentiate inflammatory oedema formation induced by substance P.

26.2. Results

Male Wistar rats (200-300g) were anaesthetised with 60 mg/kg sodium pentobarbitone (intraperitoneal) and the dorsal fur shaved. In initial experiments the comparative ability of CGRP and ADM to increase skin blood flow was measured by laser Doppler flowmetry. Results are determined by arbitrary units (flux), and indicate a similar significant increased blood flow over 30 min with 30 pmol CGRP and 100 pmol ADM suggesting a three-fold difference in potency between the two peptides.

In the second set of experiments, oedema formation was measured by the extravascular accumulation of intravenously injected [^{125}I]albumin. Evans blue dye was also used, as a visual marker. Results show that CGRP (30 pmol/site) and ADM (100 pmol/site) did not induce significant oedema formation when injected alone. However CGRP potentiated oedema induced by substance P (SP), which is in keeping with previous findings.[2] Results are as follows: Tyrode (vehicle control) alone $3.6 \pm 0.8 \, \mu l/site$; CGRP (30 pmol) alone $4.6 \pm 1.1 \, \mu l/site$; ADM (100 pmol) alone $4.3 \pm 1.2 \, \mu l/site$; SP (10 pmol) alone $20.9 \pm 3.3 \, \mu l/site$; SP + CGRP $47 \pm 7.6 \mu l/site$; SP + ADM $25.8 \pm 3.9 \, \mu l/site$. Results are expressed as mean $\pm$ s.e.mean, n=7. CGRP significantly potentiated SP-induced oedema (P<0.05, Bonferroni's Modified t-test) when compared with SP-induced oedema alone. By comparison the results show that ADM (100 pmol) did not exhibit a similar ability to potentiate SP-induced oedema formation,

The CGRP Family: Calcitonin Gene-Related Peptide (CGRP), Amylin, and Adrenomedullin, edited by David Poyner. ©2000 EUREKAH.COM.

although in further experiments potentiation was seen with the higher dose of ADM (300 pmol).

26.3. Conclusion

It can be deduced that in this blood flow model in the rat skin where laser Doppler flowmetry was used, CGRP is approximately 3 times more potent than ADM. Interestingly, this 3-fold difference was not observed in terms of potentiating SP-induced oedema formation using the same concentrations of ADM (100 pmol) and CGRP (30 pmol), although a higher dose of ADM did cause potentiation. Possible mechanisms for this difference in potency merits further investigation which is currently being carried out.

Acknowledgments

This study was supported by the British Heart Foundation.

References

1. Hall JM, Siney L, Lippton H et al. Interaction of human adrenomedullin13-52 with calcitonin gene related-peptide receptors in the microvasculature of the rat and hamster. Br J Pharmacol 1995; 114:592-597.
2. Brain SD, Williams TJ. Inflammatory oedema induced by synergism between calcitonin gene-related peptide (CGRP) and mediators of increased vascular permeability. Br J Pharmacol 1985; 86:855-860.

CHAPTER 27

The Effects of a Selective CGRP$_1$ Receptor Antagonist, and a Selective Neuronal Nitric Oxide Synthase Inhibitor on Neurogenic Vasodilatation in the Rat

P.K. Towler and S.D Brain

27.1. Introduction

Calcitonin gene-related peptide (CGRP) is now considered to play a major role in neurogenic vasodilatation, as a result of its release from primary afferent nerve terminals.[1] Vasodilatation due to CGRP is thought to be independent of nitric oxide (NO),[2] but NO has been suggested to trigger the release of CGRP from sensory nerve terminals.[3] We have studied the effects of CGRP$_{8-37}$, and a selective neuronal nitric oxide (nNOS) inhibitor, 1-(2-trifluoromethylphenyl)imidazole (TRIM), on vasodilatation induced by stimulation of the rat saphenous nerve.

27.2. Methods

Male Wistar rats were anaesthetised with intraperitoneal sodium pentobarbitone (60 mg/kg), and their hind legs were shaved and depilated. The saphenous nerves of both legs were carefully exposed, ligated centrally, and immersed in paraffin mineral oil. The animals were allowed to rest for 20-30 min. Laser Doppler blood flow probes were then placed over the hind paws, and baseline flux was measured for 5-10 min. The exposed nerve of one leg was then draped over a bipolar platinum electrode, and electrically stimulated at 10 V, 1 ms, 2 Hz for 30 s (stimulation 1). Changes in flux were monitored on a Laser Doppler blood flow monitor. The flux was then allowed to return to baseline, and vehicle or test drug (intrapertioneal TRIM, 50 mg/kg or intravenous CGRP$_{8-37}$, 400 nmol/kg) was administered. The nerve was then stimulated for a second time (stimulation 2). The unstimulated paw served as a sham control. Results were expressed as peak blood flux (flux units) and area under the curve (AUC, mm^2). Data is expressed as mean ± s.e. mean. Statistics are by paired Student t-test, *P<0.05 and **P<0.01, n=6-8.

27.3. Results

In agreement with published data[1] the CGRP antagonist, CGRP$_{8-37}$ (400 nmol/kg) caused a significant inhibition of blood flow on stimulation 2 (peak size: 48.0 ± 12.8 flux units*; AUC: 66.7 ± 21.4 mm^2**), when compared with stimulation 1 (peak size: 87.3 ± 15.8 units, AUC: 531.1 ± 111.9 mm^2). TRIM (50 mg/kg) caused a similar inhibition of

The CGRP Family: Calcitonin Gene-Related Peptide (CGRP), Amylin, and Adrenomedullin, edited by David Poyner. ©2000 EUREKAH.COM.

neurogenic vasodilatation. Regarding peak size, before TRIM (stimulation 1) a response of 91.6 ± 13.8 units was noted, and 66.5 ± 16.4 units* was seen on stimulation 2. TRIM also reduced the AUC, with 374.6 ± 56.6 mm^2 noted before drug, and 166.6 ± 23.4 mm^{2**} after treatment. The saline treated animals showed no significant differences between first and second stimulations, with peak flux being 91.8 ± 10.8 units and 111.0 ± 17.6 units, and AUC being 399.7 ± 38.5 mm^2 for the first stimulation, and 439.6 ± 77.0 mm^2 for the second. Sham control paws maintained a steady flux level throughout the experiments.

27.4. Conclusion

TRIM significantly inhibits neurogenic vasodilatation induced by stimulation of the saphenous nerve. We suggest that this may be due to inhibition of the release of CGRP from nerve terminals, as NO is not thought to contribute to CGRP-induced vasodilatation.[2]

The above data has been published in Neuroreport, 1998, 9, (7), 1513-1518.

Acknowledgments
P.K. Towler is the recipient of an MRC/Pfizer studentship.

References

1. Escott KJ, and Brain SD. Effect of a calcitonin gene-related peptide antagonist (CGRP$_{8-37}$) on skin vasodilatation and oedema induced by stimulation of the rat saphenous nerve. Br J Pharmacol 1993; 110:772-776.
2. Ralevic V, Khalil Z, Dusting GJ et al. Nitric oxide and sensory nerves are involved in the vasodilator response to acetylcholine, but not calcitonin gene-related peptide in rat skin microvasculature. Br J Pharmacol 1992; 106:650-655.
3. Hughes SR, Brain SD. Nitric oxide-dependent release of vasodilator quantities of calcitonin gene-related peptide from capsaicin-sensitive nerves in rabbit skin. Br J Pharmacol 1994; 111:425-430.

CHAPTER 28

CGRP and NO in the Regulation of Blood Flow in Rabbit Knee Joints

F.Y. Lam and A.L.M. Yip

28.1. Introduction

Calcitonin gene-related peptide (CGRP) is a potent vasodilator that has been shown to play a physiological role in the regulation of blood flow in the rat knee joint.[1] Endothelium-derived nitric oxide (NO) is well known to participate in vascular control but its role in mediating CGRP-induced vasodilatation in joints is uncertain. This study investigates the role of CGRP and NO in the regulation of blood flow in the rabbit knee joint.

28.2. Methods

New Zealand white rabbits of either sex (2.5-3.5 kg) were deeply anesthetized by intravenous bolus injection (40 mg/kg) followed by continuous infusion (0.6 mg/min) of sodium pentobarbitone. Relative changes in knee joint blood flow were monitored by a laser Doppler perfusion imager as previously described.[2] Drugs were administered as a bolus of 0.1 ml applied to the surface of the joint capsule. Tissue perfusions were expressed in term of the mean voltage ($\pm$ s.e.mean.) of a chosen rectangular area on the image. Mean values were compared by paired Student's t-test, and differences between curves were analyzed by repeated measures two-factor analysis of variance (ANOVA). P values < 0.05 were considered statistically significant.

28.3. Results

Topical administrations of the $CGRP_1$ receptor antagonist $CGRP_{8-37}$ caused dose-dependent fall in synovial blood flow (Fig. 28.1). The highest dose (1 nmol) of the antagonist used produced a marked 1.01 $\pm$ 0.18V (representing 34%; P<0.01) fall in blood flow. Administration of 100 nmol N^G-nitro-L-arginine methyl ester (L-NAME), a nitric oxide synthase inhibitor, also caused a substantial 1.67 $\pm$ 0.13V (57%; P<0.001) reduction in blood flow, whereas, administration of 100 nmol D-NAME produced a much smaller 0.41 $\pm$ 0.14V (16%; P<0.01) drop (Fig. 28.2). Cumulative applications of CGRP produced dose-dependent increases in synovial blood flow with maximum rise of 3.87 $\pm$ 0.27V (187%; P<0.001) attained at 3 nmol. Co-administration of CGRP with 1 nmol or 10 nmol $CGRP_{8-37}$ caused equivalent (P>0.05) parallel rightward-shifts of the dose response curves to CGRP (P<0.05 and P<0.01, respectively; Fig. 28.3). In the presence of 100 nmol L-NAME or 100 nmol D-NAME, the dose response curves to CGRP were not affected (P>0.05 for both; Fig. 28.4).

The CGRP Family: Calcitonin Gene-Related Peptide (CGRP), Amylin, and Adrenomedullin, edited by David Poyner. ©2000 EUREKAH.COM.

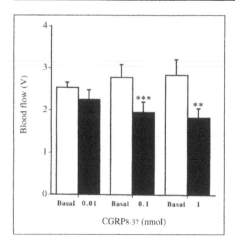

Fig. 28.1. (left) Effect of the CGRP$_1$ receptor antagonist, CGRP$_{8-37}$, on basal blood flow of the rabbit knee joint. The drug was administered as a bolus of 0.1 ml applied to the surface of the joint capsule. Data are shown as means ± s.e.mean. (shown by vertical bars) expressed as voltage that corresponds to tissue perfusion. CGRP$_{8-37}$ had no effect at 0.01 nmol (n=6) but produced significant vasoconstriction on the articular blood vessels at 0.1 nmol (n=7) and 1 nmol (n=5). **= $P<0.01$; ***= $P<0.001$.

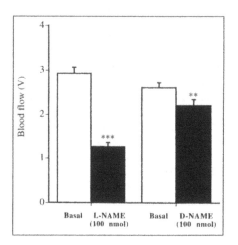

Fig. 28.2. (left) Effect of the nitric oxide synthase inhibitor, N^G-nitro-L-arginine methyl ester (L-NAME), and its enantiomer, D-NAME, on basal blood flow of the rabbit knee joint. Drugs were administered as a bolus of 0.1 ml applied to the surface of the joint capsule. Data are shown as means ± s.e.mean. (shown by vertical bars) expressed as voltage that corresponds to tissue perfusion. L-NAME (n=23) had a more potent vasoconstrictor action than D-NAME (n=18) on the articular blood vessels. **= $P<0.01$; *** = $P<0.001$.

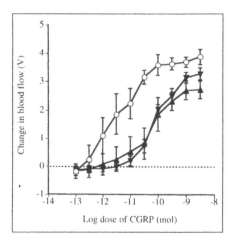

Fig. 28.3. (left) Dose-response curves of CGRP alone (° ;n=5) and in the presence of 1 nmol (▲;n=6) and 10 nmol (▼;n=6) of the CGRP$_1$ receptor antagonist, CGRP$_{8-37}$. Drugs were administered cumulatively as a bolus of 0.1 ml applied to the surface of the joint capsule. Data are shown as means ± s.e.mean. (shown by vertical bars) expressed as voltage difference of test image minus control image. The dose-response curves for CGRP in the presence of 1 nmol and 10 nmol CGRP$_{8-37}$ are not significantly different ($P>0.05$). However, they are significantly different to the dose-response curve of CGRP alone ($P<0.05$ and $P<0.01$, respectively).

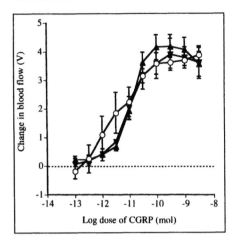

Fig. 28.4 (above) Dose-response curves of CGRP alone (° ;n=5) and in the presence of 100 nmol of the nitric oxide synthase inhibitor, L-NAME (▲;n=5) and its enantiomer, D-NAME (▼;n=4). Drugs were administered cumulatively as a bolus of 0.1 ml applied to the surface of the joint capsule. Data are shown as means± s.e.mean. (shown by vertical bars) expressed as voltage difference of test image minus control image. The dose-response curves for CGRP in the absence and in the presence of L-NAME or D-NAME are not significantly different (P>0.05 for both).

28.4. Discussion

The present study demonstrated that CGRP$_{8-37}$ and L-NAME possess potent vasoconstrictor action in the rabbit knee joint. These results imply that there are continuous, endogenous release of CGRP and NO that serve to regulate blood flow in the rabbit knee joint. CGRP-induced vasodilatation was not modified by L-NAME but it was partially inhibited by CGRP$_{8-37}$. These findings indicate that CGRP-induced vasodilatation is independent of NO and it probably involves activation of both CGRP$_1$ and CGRP$_2$ receptors.

Acknowledgments

This research was supported by the Research Grants Council of Hong Kong.

References

1. McMurdo L, Lockhart JC, Ferrell WR. Modulation of synovial blood flow by the calcitonin gene-related peptide (CGRP) receptor antagonist, CGRP$_{8-37}$. Br J Pharmacol 1997; 121:1075-1080.
2. Lam FY, Ferrell WR. Acute inflammation in the rat knee joint attenuates sympathetic vasoconstriction but enhances neuropeptide-mediated vasodilatation assessed by laser Doppler perfusion imaging. Neuroscience 1993; 52:443-449.

CHAPTER 29

CGRP Modulates LPS- and Cytokine-Induced iNOS/NO Formation

Warwick A. Arden, Ruth Oremus and Gloria Gellin

29.1. Introduction

Septic shock is characterized by a progressive loss of microvascular, and specifically resistance arteriolar tone and control. Numerous mediation systems have been implicated in the pathogenesis of such dysfunction. Recent evidence points to a central role of lipopolysaccharide (LPS)-induced leukocyte cytokine elaboration, with subsequent upregulation of the inducible form of nitric oxide synthase (iNOS) within vascular smooth muscle cells (VSMC) of the arteriolar wall.[1] Over the last 10 years, we and other groups have demonstrated a marked release of vasoactive neuropeptides during the acute phases of endotoxemia and sepsis. Of these, calcitonin gene-related peptide (CGRP) seems to be the best documented[2-4] and potentially the most potent inhibitor of arteriolar contractile function.[5] There is little doubt that CGRP, together with NO generated from iNOS upregulation, is capable of profound direct actions on VSMC and arteriolar contractile function during sepsis. Recently, however, we became interested in the possibility that, in addition to its acute, direct actions on VSMC contractile function, CGRP may be responsible for modulating the cytokine-iNOS-NO axis itself. The central hypothesis of this investigation was therefore that CGRP is capable of modulating LPS- and cytokine-stimulated NO formation. Further, it was hypothesized that such modulation occurs via cAMP/protein kinase A (PKA)-dependent influences on VSMC iNOS gene transcription.

29.2. Methods

This hypothesis was examined in both animal and tissue culture models. In the former, 48 male Sprague-Dawley rats were randomly allocated to four groups: Control (n=12), Capsaicin (n=12), LPS (n=12) and Capsaicin/LPS (n=12). Capsaicin-treated animals received 50 mg/kg subcutaneous capsaicin (under anesthesia) 7-10 days prior to LPS challenge to selectively eliminate unmyelinated sensory (C-fiber) neurons. LPS-treated animals received 10 mg/kg intravenous E.coli LPS. Plasma samples were collected for nitrite/nitrate (NO) assay prior to and 6 or 16 hours following LPS or vehicle treatment. In the second study, cultured rat aortic vascular smooth muscle cells (VSMC) in 7-10 passage were exposed to LPS (1ng/ml or 1mg/ml), interleukin-1b (IL-1: 3 ng/ml), interferon-γ (IFN: 100U/ml), tumor necrosis factor-α (TNF: 10 ng/ml), and combinations of the above, in the presence and absence of 10^{-8} or 10^{-10}M CGRP. After 24 hours incubation, media were assayed for nitrite/nitrate accumulation. These incubations were also repeated in the presence of isoproterenol (3 μM), 8-bromo-cAMP

The CGRP Family: Calcitonin Gene-Related Peptide (CGRP), Amylin, and Adrenomedullin,
edited by David Poyner. ©2000 EUREKAH.COM.

(3 μM), Rp-cAMP (25 μM) and CGRP$_{8-37}$ (1μM). Select combinations were also assessed for VSMC iNOS mRNA expression.

29.3. Results

In the first investigation, LPS infusion caused a marked accumulation of NO in plasma over both 6 and 16 hours (p<0.01). Pretreatment with capsaicin significantly attenuated this accumulation (p<0.01). In the second investigation, neither CGRP, isoproterenol, nor 8-bromo-cAMP alone had an effect on VSMC NO production. Coincubation with CGRP (10^{-8}, 10^{-10} M) enhanced IL-1 and TNF/IFN stimulated VSMC NO production (Fig. 29.1). Isoproterenol (3μM), a receptor coupled cAMP-dependent agonist, enhanced VSMC NO production in response to TNF/IFN (Fig. 29.2). Eight-bromo-cAMP, a cell permeant cAMP analogue, markedly enhanced VSMC NO production in response to LPS, TNF/IFN, IL-1 and combinations of these (Fig. 29.3). Rp-cAMP, a competitive inhibitor

of cAMP, partially inhibited CGRP enhancement of TNF/IFN and IL-1 stimulated responses (data not shown). CGRP$_{8-37}$, a competitive inhibitor of CGRP action at CGRP$_1$ receptors, inhibited CGRP enhancement of both TNF/IFN and IL-1 stimulated responses (data not shown). Coincubation with CGRP (10^{-8} M) enhanced VSMC iNOS mRNA expression in response to TNF/IFN and IL-1 (Fig. 29.4).

29.4. Discussion and Conclusion

The in vivo data, while not specific for CGRP, suggests a potent role for sensory neuropeptides in modulating LPS-induced NO formation in intact animals. Future availability of a specific non-peptide CGRP receptor antagonist should allow clarification of the specific role of CGRP in this model. The in vitro data strongly support the hypothesis that CGRP is capable of modulating LPS- and cytokine-induced NO formation in VSMC. Such modulation would allow for a potent and extended role

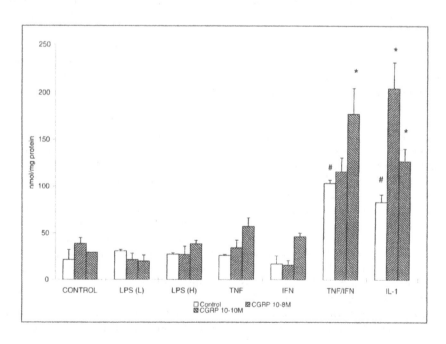

Fig. 29.1. Effect of coincubation with CGRP 10^{-8}M on VSMC NO production (ordinate) in response to stimulation with LPS, TNF, IFN and IL-1 (# significantly different from control, * significantly different from cytokine alone; p<0.05).

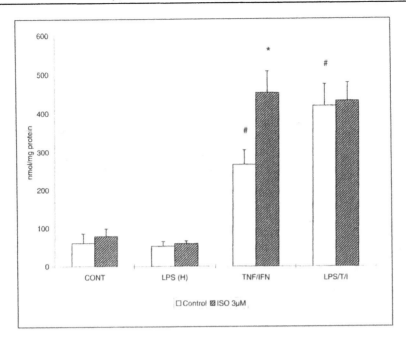

Fig. 29.2. Effect of coincubation with isoproterenol (3μM) on VSMC NO production (ordinate) in response to stimulation with LPS, TNF and IFN (# significantly different from control, * significantly different from cytokine alone; p<0.05).

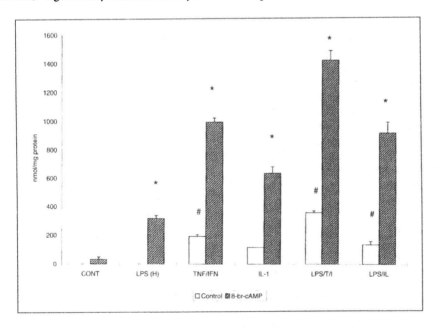

Fig. 29. 3. Effect of coincubation with 8-br-cAMP(3mM) on VSMC NO-production (ordinate) in response to stimulation with LPS, TNF and IFN (# significantly different from control, * significantly different from cytokine alone; p<0.05).

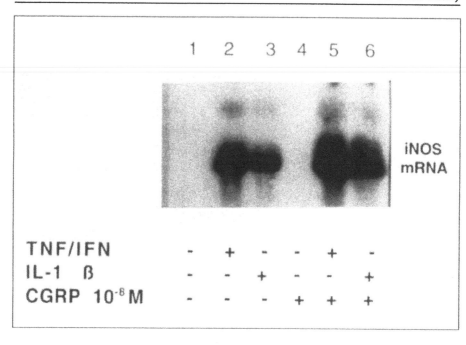

Fig. 29.4. Effect of coincubation with CGRP (10^{-8}M) on VSMC iNOS mRNA expression in response to stimulation with TNF/IFN or IL-1.

for CGRP in the progression of microvascular dysfunction during sepsis. Further, our data suggest the effect of CGRP is most likely by cAMP/PKA modulation of VSMC iNOS gene expression or message stability. This data is consistent with that reported by other investigators for IL-1β, but does not appear to be limited to an effect on the action of this cytokine. Future studies will focus on the molecular mechanisms by which CGRP modulates LPS- and cytokine-induced acute phase gene expression.

References

1. Wright CE, Rees DD, Moncada S. Protective and pathologic roles of nitric oxide in endotoxic shock. Cardiovasc Res 1992; 26:49-57.
2. Wang X, Jones SB, Zhou ZZ et al. Calcitonin gene-related peptide (CGRP) and neuropeptide Y (NPY) levels are elevated in plasma and decreased in vena cava during endotoxic shock in the rat. Circ Shock 1992; 36:21-30.
3. Arden WA, Fiscus RR, Wang X et al. Elevation in circulating calcitonin gene-related peptide correlate with hemodynamic deterioration during endotoxin shock in pigs. Circ Shock 1994; 42:147-153.
4. Arden WA, Fasces RR, Beihn L et al. Skeletal muscle microcirculatory response to rat calcitonin-gene related peptide. Neuropeptides 1994; 27:39-51.
5. Schini-Kerth V, Fisslthaler B and Busse R. CGRP enhances induction of NO synthase in vascular smooth muscle cells via a cAMP-dependent mechanism. Am J Physiol 1994; 267:H2483-2490.

CHAPTER 30

Identification and Regulation of CGRP in T Lymphocytes of the Rat

Xian Wang, Liyu Xing and Yutong Xing

The nervous and immune systems share important informational molecules such as cytokines, hormones, neuropeptides and their receptors. These molecules serve as bi-directional signals of communication. They provide an important link between the nervous and immune systems. Evidence is accumulating that the immune system can produce some neuropeptides. In the light of these facts, there are suggestions that immunocytes might contain calcitonin gene-related peptide (CGRP). However, there is no direct evidence to show whether CGRP exists in immunocytes and if so, which type of immunocytes contain CGRP. In the present study, we obtained direct evidence to prove that T lymphocytes synthesize and release CGRP, a neuropeptide localized within primary sensory nerves.

By using CGRP specific radioimmunoassay (RIA), CGRP-like immunoreactivity (LI) was found in the extracts of rat lymphocytes from thymus and mesenteric lymph nodes. The intracellular concentration of lymphocyte-derived CGRP-LI in rat thymus and mesenteric lymph node was 75.5 ± 8.2 pg/10^8 cells and 63.5 ± 5.3 pg/10^8 cells, respectively. Increasing amounts of lymphocyte extract displaced $[^{125}I]$-CGRP from a CGRP antibody in a manner parallel to that of synthetic rat CGRP, suggesting immunological similarity between the rat lymphocyte CGRP-LI- and the synthetic rat

CGRP, and implying a complete cross reactivity between lymphocyte CGRP-LI and CGRP antibody. In addition, CGRP-LI in lymphocytes was shown to co-elute with synthetic rat CGRP and CGRP derived from dorsal root ganglia (DRG) when analyzed by reverse-phase HPLC. The retention times were all 62 minutes, indicating that these molecules have identical molecular polarities and correspond to full length CGRP.

The CGRP mRNA detected by CGRP-specific reverse transcriptase polymerase chain reaction (RT-PCR) was also present in these lymphocytes and was similar to that in sensory neurons. The level of CGRP mRNA in lymphocytes was lower than that in the DRG determined by the semi-quantitative RT-PCR technique. The results suggest that rat lymphocytes express CGRP mRNA and synthesize CGRP at a lower rate than the DRG.

Furthermore, the presence of CGRP-LI in T lymphocytes was also demonstrated by immunohistochemical and immunocytochemical methods using both light and electron microscopy. Some thymus epithelial cells and epithelial-reticular cells of thymic corpuscles in cortical-medullary and medullary zones were CGRP-LI positive. In addition, some thymocytes within cortical-medullary and cortical zones were CGRP-LI positive. The CGRP-LI positive lymphocytes were also found within the cortical zone of mesenteric lymph nodes. By using

The Calcitonin Gene-Related Peptide Family: CGRP, Amylin, and Adrenomedullin,
edited by David Poyner. ©2000 EUREKAH.COM.

immunocytochemical staining, CGRP-LI was found predominantly in lymphocytes, particularly in those cells not adhering to nylon fiber. The percentage of CGRP positive cells was $1.5 \pm 0.14\%$ in these thymocytes and $1.0 \pm 0.09\%$ in lymph node lymphocytes. The CGRP-LI was located in cytoplasm of lymphocyte from these cells shown by pre-embedding immunoelectronmicroscopy. The results suggest that CGRP is found predominantly in T lymphocytes and in thymus, primarily in pre-mature T lymphocytes. These lymphocytes were capable of secreting CGRP spontaneously. However, CGRP was secreted spontaneously from DRG cells at about 1000 times the rate of that from lymphocytes. Furthermore, Concanavalin-A (Con-A), a T lymphocyte mitogen, and recombinant human interleukin-2 (rhIL-2) caused a time- and a concentration-dependent release of CGRP-LI from these lymphocytes. After 3 days of stimulation, the levels of CGRP synthesis and release in these cells were increased significantly, and further increased after 5 days of stimulation. However, Con-A and rhIL-2 could not cause CGRP release from DRG. LPS and high concentration of potassium rapidly triggered CGRP release from DRG, but had no effect on the lymphocytes. The results suggest that the stimulants which induce CGRP synthesis and release from sensory neurons and lymphocytes are different. CGRP synthesis and release in lymphocytes requires a long period of stimulation and is induced by some stimuli not recognized by sensory nerves.

Finally, rat CGRP (1 nM-1 µM) inhibited Con-A (1 µg/ml) stimulated thymocyte proliferation in a concentration-dependent manner in the rat. The $CGRP_1$ receptor antagonist, $hCGRP_{8-37}$ at 100 nM and 2 µM, significantly enhanced Con-A (1 µg/ml) stimulated lymphocyte proliferation by 27.3% and 41.3%, respectively. The data suggest that endogenous CGRP may inhibit thymocyte proliferation induced by Con-A partially.

In conclusion, the present study provides direct evidence, for the first time, that the neuropeptide CGRP is synthesized and secreted in T lymphocytes of both thymus and lymph nodes in the rat, and is identical to CGRP in neuronal tissue. CGRP from rat lymphocytes is biologically active; it may decrease IL-2 production and inhibit the proliferation of T lymphocytes. Lymphocyte-derived CGRP may act in an autocrine/paracrine mode and play an important role in certain physiological and pathophysiological conditions.

CHAPTER 31

CGRP and Pregnancy

Chandrasekhar Yallampalli and Sunil J. Wimalawansa

31.1. Introduction

The objective of this communication is to analyze the evidence that calcitonin gene-related peptide (CGRP), an endogenous relaxant peptide of smooth muscle, is involved in two crucial adaptations that occur during mammalian pregnancy:

1. vasodilation of maternal systemic circulation, and

2. uterine quiescence before parturition.

Potential compensatory roles of CGRP in vascular adaptations that could occur in response to hypertension during pregnancy, will also be evaluated.

During pregnancy, circulating immunoreactive CGRP (i-CGRP) levels increase up to the time of delivery, followed by a sharp reduction during labor and in the postpartum period.[1,2] In addition, during pregnancy, sensitivity of uterine arteries and systemic vasculature to CGRP seems to be higher when compared to the non-pregnant state.[2,3] Taken together, it is likely that CGRP is involved in regulating not only utero-placental blood supply (promoting fetal survival and growth) during pregnancy, but also in regulating vascular adaptations that occur during normal pregnancy and perhaps in the pathophysiology of pre-eclampsia.[4] In addition, CGRP may be involved in the maintenance of uterine quiescence during pregnancy.

31.2. CGRP and Vascular Adaptations During Pregnancy

31.2.1. Introduction

Vasodilation of maternal systemic circulation occurs during normal pregnancy in humans and other mammalian species. Pressor responsiveness or vascular reactivity to various vasoconstrictors is attenuated. Hypertension with its various complications constitutes bad prognostic indices during pregnancy. The treatment of hypertension in pregnancy has been and is still a matter of obstetrical enigma in the prevention of pre-eclampsia, eclampsia and neonatal morbidity. Pre-eclampsia occurs in approximately 4-8% of human pregnancies. The only effective treatment, to date, is to induce delivery coupled with placental removal.

Reduced perfusion in multiple organ systems accompanied by profound vasoconstriction and activation of a coagulation cascade are some of the pathophysiological changes occurring during pre-eclampsia.[5] In addition, sensitivity to pressor effects of angiotensin II may antedate clinical pre-eclampsia by several weeks.[6] Epidemiological data and animal experiments, so far, point to poor placental perfusion as the initiating factor. According to Roberts and Redman[5] a poorly perfused feto-placental unit is the source of metabolic products that

The CGRP Family: Calcitonin Gene-Related Peptide (CGRP), Amylin, and Adrenomedullin, edited by David Poyner. ©2000 EUREKAH.COM.

adversely affect vascular endothelial cell and maternal systemic functions. The loss of normal endothelial depressor function leads to increased sensitivity to circulating pressor agents resulting in profound vasospasm. Although this scheme is supported by available literature, it is not consistent in all cases. Thus, there could be other mechanisms responsible for the hypertensive disorders of pregnancy. We propose therefore, that hypertension during pregnancy could result from failure of compensatory vasodilator functions mediated through CGRP. This proposed failure of CGRP-induced vascular compensatory functions may be a common occurrence irrespective of etiological factors that could lead to hypertensive disorders during pregnancy.

We have recently developed an animal model of hypertension during pregnancy (i.e. pre-eclampsia) to examine the role of CGRP in vascular adaptations occurring during pregnancy.[7,8] In rats, inhibition of NO synthesis during pregnancy causes hypertension, proteinuria, and fetal growth retardation without affecting gestational length (see Refs. 7-9 for review). Using this animal model, we have investigated the role of CGRP in pregnancy induced hypertension and on fetal growth.

31.2.2. CGRP And Fetal Growth

Continuous infusion of L-NAME, an inhibitor of NO synthesis, to pregnant rats from day 17 of gestation produced signs similar to that of pre-eclampsia, i.e. hypertension, proteinuria and fetal growth retardation .[6] Co-infusion of CGRP (10 μg/rat. day) in L-NAME-treated rats abolished hypertension (Fig. 31.1) and reduced both growth retardation and fetal mortality (Fig. 31.2).[10,11] These studies indicated that in pregnant rats CGRP could not only prevent hypertension, but could also decrease adverse gestational fetal effects associated with pre-eclampsia-like conditions. In these studies, infusion of CGRP alone had no effect on blood pressure or fetal characteristics.

31.2.3. CGRP and Compensatory Vasodilation

A biphasic change occurs to systolic blood pressure when L-NAME is continuously infused into pregnant rats.(Fig. 31.1)[10,11] An initial robust increase in blood pressure within 24-48 h after infusion is followed by a substantial decline in blood pressure through days 19-21 of gestation and a return to a hypertensive state after birth. The elevated blood pressure is substantial throughout the period of L-NAME infusion during postpartum period. The animals are normotensive on days 20 and 21 of gestation with blood pressure levels similar to those of controls (without L-NAME) and this occurs despite continuous infusion of L-NAME. These observations suggests that there appears to be a compensatory vasodilatory mechanism that comes into play in L-NAME-treated pregnant rats. These compensatory vasodilator effects are reversed soon after birth. This mechanism does not involve the NO system since it is still apparent while L-NAME is continuously infused.

If endogenous CGRP is involved in this compensatory vasodilator mechanism, then one can inhibit these vascular changes with an antagonist of CGRP. CGRP$_{8-37}$ induced a further increase in the already elevated mean arterial pressure (MAP) in the L-NAME-treated rats on days 19 to 21 of gestation (Table 31.1), indicating that in this setting CGRP is perhaps acting as a compensatory vasodilator which buffers elevated blood pressure.[12] This effect was absent during postpartum. Thus endogenous CGRP may play an important role in compensatory vasodilation that may be responsible for reducing hypertensive insults during pregnancy.

31.2.4. Progesterone and Vasodilator Effects of CGRP

As CGRP prevents L-NAME-infused hypertension in pregnant rats only until birth, we postulated that the vasodilator effects of CGRP are progesterone dependent.

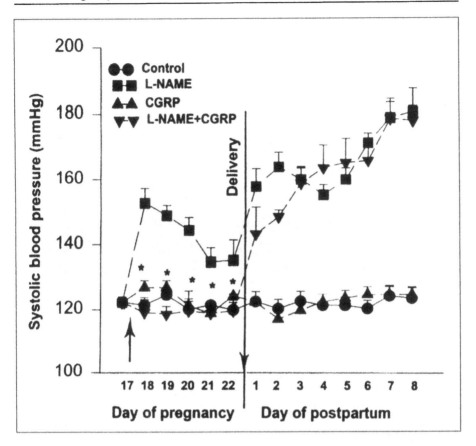

Fig. 31.1. Effect of CGRP (10 µg/day per rat) either alone or in combination with L-NAME (50 mg/day per rat) on systolic blood pressure during pregnancy and post partum in rats. Compounds were dissolved in sterile saline and infused (with osmotic mini pumps) subcutaneously from day 17 of gestation. Blood pressure was measured daily until day 8 postpartum. Values are expressed as mean ± SEM for five animals in each treatment group. * p value <0.05 was considered significant. (Adapted from Gangula et al 1997a).

This was tested by progesterone administration in both postpartum and ovariectomized rats. Figure 31.3 shows blood pressure readings in animals receiving L-NAME, CGRP, and progesterone during the postpartum period. CGRP administration was effective in reducing blood pressure only in the presence of progesterone. Thus, CGRP was effective in reducing blood pressure induced by L-NAME during pregnancy, only when progesterone levels are elevated, but not during the postpartum period, when the circulating progesterone levels are minimal. We also demonstrated that the vasodila-tory effects of CGRP in the postpartum period were restored with progesterone administration.

Hypertension during pregnancy is a life threatening obstetrical problem, and the pathophysiology of pre-eclampsia is rather unclear. From studies using animal model for hypertension during pregnancy, we provide evidence that CGRP may play a role in vascular adaptations that occur during pregnancy. Therefore, we propose that, one of the possible mechanisms in persistent

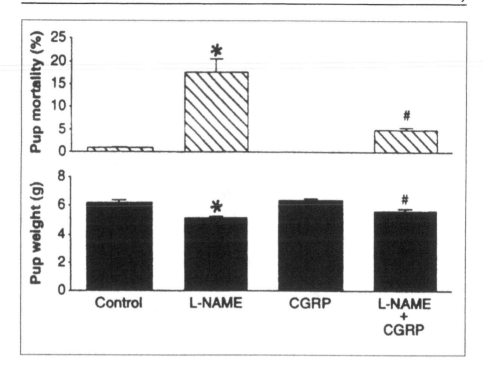

Fig. 31.2. Effect of calcitonin gene-related peptide (CGRP), infusion, alone or in combination with 50 mg N^G-nitro-L-arginine methyl ester (L-NAME) per day, on A) pup mortality rate, B) pup weights, delivered spontaneously at term. Drugs were administered s.c. in osmotic minipumps dissolved in a saline solution, starting on day 17 of gestation. Each bar represents the mean ± SEM value of the percentage of pups dead at birth for five rats in each group. (*) $P < 0.01$, L-NAME versus control and CGRP; (#) $P < 0.01$, L-NAME + CGRP versus all other groups (analysis of variance, Student's t-test). (Adapted from Gangula et al., 1996).

hypertensive disorders during pregnancy may be the result of insufficient CGRP-related compensatory regulatory mechanisms (Fig. 31.4). It is logical, therefore, that hypertensive disorders during pregnancy, including pre-eclampsia, should be amenable to treatment with CGRP or CGRP analogues. In addition, we propose that co-administration of progesterone with CGRP may improve the beneficial vasodilator effects of CGRP in pregnancy associated hypertensive disorders.

31.3. CGRP and Uterine Quiescence

Several neuropeptides, including CGRP are present in primary sensory nerves in the female reproductive tract.[13] CGRP causes relaxation of gastric smooth muscle of rat and guinea pig, inhibits contractions of smooth muscle in both the uterus and fallopian tube, and reduces acetylcholine-stimulated rat uterine contraction.[14,15] However, it is unclear whether CGRP can differentially affect uterine contractile activity in rats during pregnancy, labor, and postpartum. CGRP receptors have been identified in a variety of tissues, including brain, cardiovascular and endothelial tissues, but little is known if CGRP receptors are expressed in rat uterus.[16] We will discuss evidence in rats for 1) the differential effects of CGRP on spontaneous uterine contraction during pregnancy, labor and postpartum. 2) the possible involvement of NO in CGRP-induced uterine relaxation. 3) if uterus expresses CGRP receptors.

Table 31.1. *CGRP$_{8-37}$ induced increases (Δ) in mean arterial pressure (mmHg) in the control and L-NAME-infused pregnant rats (Data adapted from Gangula et al 1997b)*

Day	Control	L-NAME treated
Pregnant		
Day 19	3.8 ± 0.2	7.8 ± 1.1*
Day 20	2.1 ± 0.8	11.4 ± 2.8*
Day 21	4.0 ± 0.6	7.0 ± 0.7*
Postpartum		
Day 1	5.0 ± 1.3	6.3 ± 0.8

Values are mean ± SEM of 3–4 animals per group. Values with an asterisk in L-NAME-treated group differ significantly (p<0.05) from control group on a given day of pregnancy.

31.4.1. CGRP and Spontaneous Uterine Contractility[18]

Addition of CGRP (0.01 - 1.0 μg/ml) to muscle strips caused immediate dose- dependent relaxation of the spontaneous contractile uterine tissue strips day 18 of pregnancy (Fig. 31.5). However, the relaxation of uteri to CGRP was substantially reduced in uterine tissues obtained during spontaneous labor and postpartum. This suggests that the sensitivity of the uterine smooth muscle to CGRP was down regulated in these rats. The relaxation effects of CGRP on uterine strips were profoundly inhibited by 200 μg/ml of CGRP$_{8-37}$, indicating involvement of specific CGRP receptors in uterine muscle.

L-NAME blocked the CGRP-induced relaxation effects, suggesting that the relaxation effects of CGRP appear to be modulated, at least in part, via production of NO. The relaxations of rat uterine tissues by CGRP were also inhibited by LY83583, an inhibitor of soluble guanylate cyclase, confirming that NO-cGMP pathway is involved in CGRP-induced uterine relaxation. CGRP also caused an increase in nitrite produc-

tion in the uterus obtained from rats on day 18 of gestation which was abolished by CGRP$_{8-37}$. In addition, uterine NO generation in response to CGRP was decreased (P<0.05) during spontaneous labor and postpartum indicating reduced sensitivity of uterine tissue to CGRP during labor and postpartum period.

31.4.2. CGRP Receptors in the Rat Uterus[18]

Specific binding of [^{125}I]human CGRP to rat uterus during pregnancy, labor and postpartum and in nonpregnant state indicate a single class of binding sites for CGRP in uterus. The dissociation constant (Kd) values were similar among all groups (1.8-2.4 nM). However, the density of CGRP receptors in rat uterus (Fig. 31.6) on day 18 of gestation was 761.6 ± 83.9 fmol/mg protein which is significantly higher than that in nonpregnant rats (367.9 ± 70.1 fmol/mg protein, p<0.01). CGRP receptor density was significantly decreased at the time of spontaneous labor (132.6 ± 14.2 fmol/mg protein, p<0.01) compared with that of day

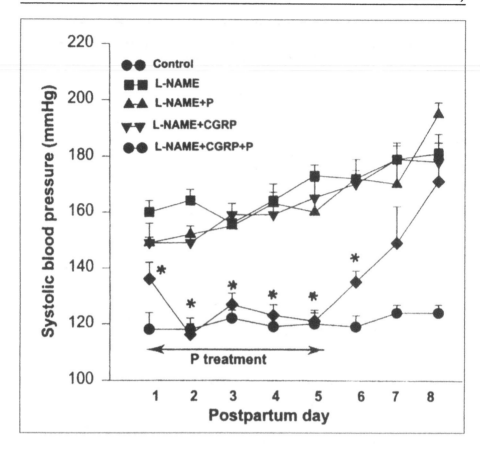

Fig. 31.3. Effects of progesterone (PROG) (4 mg/day per rat) and CGRP (10 µg/day per rat), either alone or in combination, on L-NAME (50 mg/day per rat)-induced elevated systolic blood pressure post partum. L-NAME and CGRP were infused subcutaneously via osmotic pumps from day 17 of pregnancy, and systolic blood pressure was recorded daily up to day 8 postpartum. For clarity, blood pressure values are presented from day 1 to 8 post partum only. Starting on day 1 post partum, five animals from L-NAME- and L-NAME plus CGRP-infused groups received daily subcutaneous injections of progesterone (P treatment) (2 mg per injection, twice a day). Progesterone treatment was stopped on day 6 post partum. Values are expressed as mean ± SEM for five animals in each group. * p < 0.05, L-NAME plus CGRP plus progesterone versus L-NAME, L-NAME plus progesterone, or L-NAME plus CGRP. (Adapted from Gangula et al 1997b).

18 of gestation. CGRP receptors in rat uteri from postpartum animals were also substantially lower (379.9 ± 93.7 fmol/mg protein) than that of day 18 of gestation (p<0.01). These results indicated that CGRP receptors are present in rat uterus, and the concentration of these receptors are

dramatically up-regulated during pregnancy and down-regulated during labor at term.

31.5. Conclusion

CGRP inhibits spontaneous contractions of rat uterus during pregnancy, but not during labor and postpartum, and the inhibitory effects of CGRP appear to be

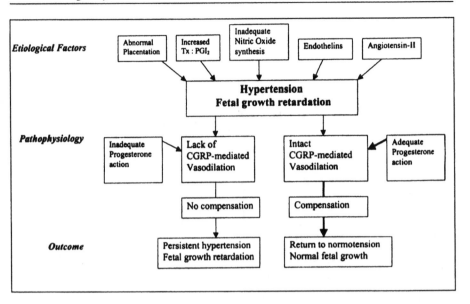

Fig. 31.4 . Schematic diagram to illustrate the possible role of CGRP-related compensatory vascular adaptations that could determine the outcome in hypertensive disorders during pregnancy. There are several etiological factors such as abnormal placentation, increased thromboxane (Tx): prostacyclin (PGI_2) ratio, inadequate NO oxide synthesis, excess endothelin induced vasocontrictions, and increased angiotensin-II mediated pressor responses that could play a role in the initiation of hypertension and/or fetal growth retardation. Once the hypertension is induced, a CGRP-mediated increase in vasodilatory response may occur as a compensatory response to buffer increased blood pressure during pregnancy. If this occurs, the blood pressure returns to normal and fetal growth is restored. A failure in this compensatory CGRP-mediated vasodilation may result in persistence of the hypertensive state and fetal growth retardation. In addition, CGRP-mediated vasodilatory responses appear to be progesterone dependent. Adapted with permission from; Yallampalli C and Wimalawansa SJ, Calcitonin gene related peptide is a mediator of vascular adaptations during hypertension in pregnancy, Trends Endocrinol Metab 1998; 9:113-117.

modulated via activation of NO generation. Furthermore, CGRP receptors are present in rat uterus and the concentration of these receptors increase during pregnancy and decrease during labor, providing evidence that CGRP is likely to be involved in maintaining uterine quiescence during preg5nancy and a decrease in iCGRP and CGRP-µµ5 receptors in uterus at term may contribute to the initiation of labor.

Acknowledgments

The authors' works cited were supported by grants from National Institutes of Health (HL 58144, HD 30273) and Texas Higher Education Coordinating Board Grant 004952–005.

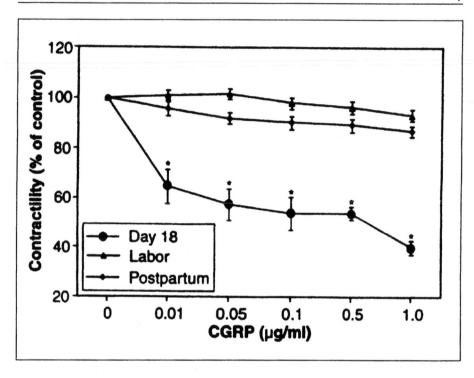

Fig. 31.5. CGRP dose relaxation-response curves for uterine strips from rats on day 18 of gestation, during spontaneous labor and postpartum. Relaxation responses, as percent of control activity at each dose of CGRP, were analyzed by repeated-measures analysis of variance among three groups. Each point represents mean ± SEM for 6 animals. Asterisks indicate that the responses in pregnant rat tissues are statistically significant ($p<0.05$, ANOVA) when compared with the other two groups.

References

1. Stevenson JC, MacDonald DWR, Warren RC et al. Increased concentration of calcitonin gene-related peptide during normal human pregnancy. Br Med J 1986; 293:1329-1330.

2. Gangula PRR, Magness RR, Wimalawansa SJ et al. Pregnancy and sex steroid hormones enhance the calcitonin gene-related peptide concentration in circulation in rats and sheep. Soc Gynecol Invest 1998; Abstract # 379, Atlanta, Georgia.

3. Nelson SH, Steinsland OS, Suresh MS. Possible physiologic role of calcitonin gene-related peptide in the human uterine artery. Am J Obstet Gynecol 1993; 168:605-611.

4. Wimalawansa SJ. Calcitonin gene-related peptide: Molecular genetics, physiology, pathology and therapeutic potentials. Endocr Rev 1996; 17:533-585.

5. Roberts JM, Redman CWG. Pre-eclampsia: More than pregnancy-induced hypertension. Lancet 1993; 341:1447-1451.

6. Gant NF, Daley GL, Chand S et al. A study of angiotensin II pressor response throughout primigravid pregnancy. J Clin Invest 1973; 52:2682-2689.

7. Yallampalli C, Garfield RE. Inhibition of nitric oxide synthesis in rats during pregnancy produces symptoms identical to preeclampsia. Am J Obstet Gynecol 1993; 169:1316-1320.

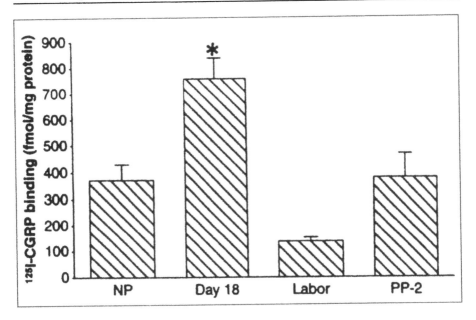

Fig. 31.6. CGRP receptors in uterus from nonpregnant rats (NP) and pregnant rats on day 18 of gestation (Day 18), during labor at term (Labor) and on day 2 postpartum (PP-2). Specific binding sites for CGRP in rat uterus were identified using ^{125}I-human CGRP assay. The results are expressed as ^{125}I-CGRP bound fmol/mg membrane protein (n=3). The data were analyzed with Scatchard's method and an asterisk indicates that CGRP receptors in uterus from the rat on day 18 of pregnancy are statistically different (p<0.01) when compared with the other three groups.

8. Buhimschi I, Yallampalli C, Chwalisz K et al. Preeclampsia-like conditions produced by nitric oxide inhibition: Effects of L-arginine, D-arginine, and steroid hormones. Hum Reprod 1995; 10:2723-2730.

9. Molnar M, Suto T, Toth T et al. Prolonged blockade of nitric oxide synthesis in gravid rats produced sustained hypertension, proteinuria, thrombocytopenia, and intrauterine growth retardation. Am J Obstet Gynecol 1994; 170:1458-1466.

10. Yallampalli C, Dong YL, Wimalawansa SJ. Calcitonin gene-related peptide reverses the hypertension and significantly decreases the fetal mortality in pre-eclampsia rats induced by N^G-nitro-L-arginine methyl ester. Hum Reprod 1996; 11:895-899.

11. Gangula PRR, Wimalawansa SJ, Yallampalli C. Progesterone upregulates vasodilator effects of calcitonin gene-related peptide in N^G-nitro-arginine methyl ester- induced hypertension. Am J Obstet Gynecol 1997a; 76:894-900.

12. Gangula PRR, Supowit SC, Wimalawansa SJ et al. Calcitonin gene-related peptide is a depressor in N^G-nitro-L-arginine methyl ester-induced hypertension during pregnancy. Hypertension 1997b; 29:248-252.

13. Ghatei MA, Gu J, Mulderry PK et al. Calcitonin gene-related peptide in the female rat urogenital tract. Peptides 1985; 6:809-815.

14. Katsoulis S, Conlou JM. Calcitonin gene-related peptides relax guinea pig and rat gastric smooth muscle. Eur J Pharmacol 1989; 161:129-134.

15. Samuelson UE, Dalsgaard CJ, Lundberg JM et al. Calcitonin gene-related peptide inhibits spontaneous contractions in human uterus and fallopian tube. Neurosci Lett 1985; 62:225-230.

16. Shew RL, Papka RE, McNeill DL. Calcitonin gene-related peptide in the rat uterus:Presence in nerves and effects on uterine contraction. Peptides 1990; 11:583-589.

17. Sladek SM, Regenstein AC, Lukins D et al. Nitric oxide synthase activity in pregnant rabbit uterus decreases on the last day of pregnancy. Am J Obstet Gynecol 1993; 169:1205-1291.

18. Dong Y-L, Gangula PRR, Fang L et al. Uterine relaxation responses to calcitonin gene-related peptide and calcitonin gene-related peptide receptors decrease during labor in rats. Am J Obstet Gynecol 1998; 179:497-506.

CHAPTER 32

Characterization of CGRP Receptors in Human Cranial Arteries

Inger Jansen-Olesen, Sergio Gulbenkian and Lars Edvinsson

32.1. Introduction

Calcitonin gene-related peptide (CGRP), one of the most potent vasodilators known today, has been shown to be released from perivascular sensory neurons during a migraine attack. Thus, it is possible that CGRP may be involved in the pathophysiology of migraine (See Goadsby et al, Edvinsson et al, this volume).[1,2] We have in the present study used a range of different techniques in order to show the presence of a CGRP$_1$ receptor in human cranial arteries, and to investigate its role. Specifically, the aims were to characterize the vasodilator effect of human (h) αCGRP, to demonstrate the presence of mRNA encoding a proposed CGRP$_1$ receptor (CRLR; see ref. 3 and Legon, Aiyar et al, this volume) in the human cranial circulation, to show hαCGRP binding sites on cranial arteries and to study the effect of hαCGRP on adenylyl cyclase activity in human cranial arteries.

32.2. Methods

Human cranial arteries were obtained either at neurosurgical operations or at autopsy. For reverse transcript (RT) polymerase chain reaction (PCR) and autoradiography they were immediately frozen in liquid nitrogen. For studies of vasomotor activities in vitro or measurements of cAMP the vessels were immersed into an ice cold buffer solution.

32.3. Results

32.3.1. Reverse Transcriptase Polymerase Chain Reaction

Agarose gel electrophoresis of the PCR products from human cerebral, meningeal and temporal arteries, demonstrated products of the expected size corresponding to mRNA encoding the proposed CGRP$_1$ receptor (Fig. 32.1). DNase was successfully used to eliminate any contaminating DNA, since no bands were detected in negative controls where the reverse transcriptase enzyme was omitted in the first strand cDNA reaction. The experiments showed expression of mRNAs encoding CGRP$_1$ receptors in cerebral, meningeal and temporal arteries.

32.3.2. Receptor Autoradiography

[^{125}I]CGRP binding was performed in the human meningeal and temporal arteries in the absence and presence of unlabelled hαCGRP. An intense binding was observed over the smooth muscle cell layer of both arteries (Fig. 32.2). In control sections the [^{125}I]CGRP labelling was low and homogeneously distributed over the vessel walls.

32.3.3 Vasomotor Responses In Vitro

HαCGRP induced potent relaxations of precontracted human cerebral, middle meningeal and temporal arteries (Table 32.1).

The Calcitonin Gene-Related Peptide Family: CGRP, Amylin, and Adrenomedullin,
edited by David Poyner. ©2000 EUREKAH.COM.

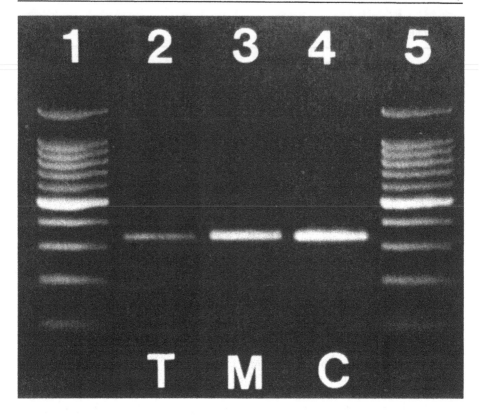

Fig. 32.1. Gel electrophoresis of RT-PCR reaction products of mRNA fragments corresponding to human [1]CGRP$_1$ (339 bp) receptor transcrips. [2]CGRP$_1$-mRNA in human temporal artery (lane 2). CGRP$_1$-mRNA human middle meningeal artery (lane 3). [2]CGRP$_1$-mRNA in human cerebral artery (lane 4). As a negative control, no amplification product occured when reverse transcriptase was omitted in the first-strand cDNA reaction (not shown). Promega's 100 bp DNA Ladder (Promega, SDS, Sweden) was run to confirm molecular size of the amplification product (lane 1 & 5).

Removal of the endothelium did not significantly change the responses to hαCGRP with respect to the amount of maximum relaxation or pIC$_{50}$-values.

A single concentration of hαCGRP$_{8-37}$ (10^{-6} M), given 15 min prior to the administration of agonist, induced neither contraction of vessel segments at the resting level of tension nor relaxation of precontracted vessel segments. HαCGRP$_{8-37}$ (10^{-6} M) induced a shift towards higher concentrations of hαCGRP (cerebral: I$_{max}$: 75 $\pm$ 7% of precon- traction, pIC$_{50}$: 9.13 $\neq$ 0.42; n=9; meningeal: I$_{max}$: 82 $\pm$ 5% of precontraction, pIC$_{50}$: 8.11 $\pm$ 0.21; n=4; temporal: I$_{max}$: 94

$\pm$ 3% of precontraction, pIC$_{50}$: 8.40 $\pm$ 0.27; n=5).

32.3.4. Capsaicin Experiments

Capsaicin (10^{-15}-10^{-4} M) induced a biphasic relaxation of human cerebral arteries (n=4). The first phase of the relaxation amounted to 20 $\pm$ 4% of precontraction and occurred at capsaicin concentrations up to 10^{-11} M. The maximum relaxant response for the second phase of relaxation was found at capsaicin concentrations between 10^{-5} M and 10^{-4} M amounting to 91 $\pm$ 9%. Pretreatment with 10^{-6} M hαCGRP$_{8-37}$ significantly inhibited the first phase (I$_{max}$: 2 $\pm$ 2%) but

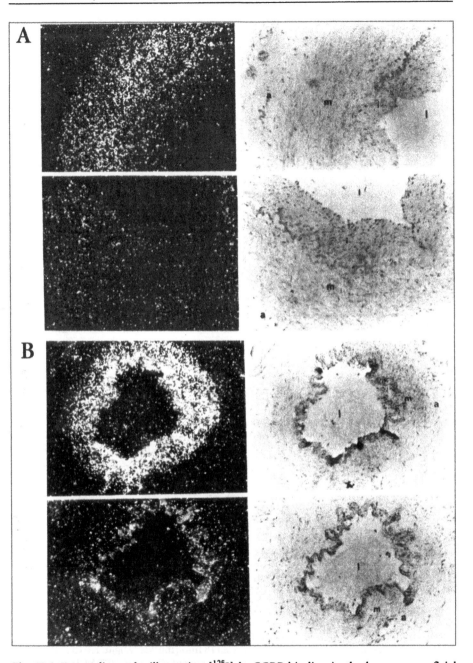

Fig. 32.2. Autoradiographs illustrating [^{125}I]-hαCGRP binding in the human superficial temporal artery (A) and in the middle meningeal artery (B) in the absence (top row; total binding) or in the presence (bottom row; non-specific binding) of unlabelled hαCGRP. Dark field and bright field photographs of the same field are shown in the left and right panels, respectively. Notice that silver grains, visible only in the dark-field microscopy, are preferentially accumulated over the smooth muscle medial (m) layer. Autoradiographs from non-specific binding show a small number of silver grains distributed homogenously over the artery wall. a, adventitia; i, intima; l, lumen.

Table 32.1. Relaxant responses of human (A) cerebral, (B) middle meningeal and (C) temporal arteries to different CGRP agonists.

A. Cerebral arteries	pIC_{50}	I_{max}	n
hαCGRP	9.6 ± 0.2	88 ± 3	30
hβCGRP	9.4 ± 0.3	80 ± 8	17
rαCGRP	9.6 ± 0.2	80 ± 7	5
rβCGRP	10.3 ± 0.3	81 ± 8	5
[CysACM]CGRP	7.4 ± 0.1	80 ± 9	11
B. Middle meningeal arteries	pIC_{50}	I_{max}	n
hαCGRP	8.8 ± 0.4	92 ± 4	8
hβCGRP	8.4 ± 0.2	81 ± 4	21
rαCGRP	8.7 ± 0.2	87 ± 5	6
[CysACM]CGRP	7.3 ± 0.2	70 ± 2	3
C. Temporal arteries	pIC_{50}	I_{max}	n
hαCGRP	9.3 ± 0.3	84 ± 4	18
hβCGRP	8.2 ± 0.2	79 ± 5	20
rαCGRP	9.5 ± 0.3	89 ± 2	8
rβCGRP	8.0 ± 0.7	32 ± 18	2

I_{max} = maximum relaxant response, pIC_{50} = the negative logathm of the agonist concentration eliciting half-maximum relaxation. The data are presented as mean values ± standard error of the mean values (s.e.mean); n = number of vessels examined.

not the second phase (I_{max}: $93 \pm 8\%$) of capsaicin induced relaxation (Fig. 32.3).

32.3.5. Measurement of cAMP

The basal formation of cAMP in human cerebral arteries was 235 ± 17 fmol/mg wet weight (n=4). The adenylyl cyclase activator forskolin given in a concentration of 10^{-6} M induced production of cAMP in the cerebral vessels, amounting to 2967 ± 118 fmol/mg wet weight. HαCGRP given in a concentration of 10^{-9} M increased the cAMP formation to $48 \pm 5\%$ of the production induced by forskolin (Fig. 32.4). When given alone hαCGRP$_{8-37}$ (10^{-6} M) only induced a slight (5%) increase in cAMP production (Fig. 32.4). HαCGRP$_{8-37}$(10^{-6} M) totally inhibited the increase in cAMP

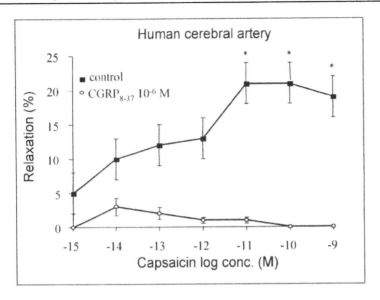

Fig. 32.3. Relaxant response to capsaicin in presence and absence of 10^{-6} M hαCGRP$_{8-37}$ in human cerebral arteries. The arteries were precontracted by 3 x 10^{-6} M prostaglandin F$_{2\alpha}$. Values are given as means ± s.e.mean. Four experiments from two patients.

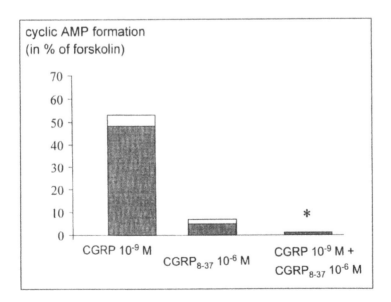

Fig. 32.4. Increase in cyclic AMP levels by 10^{-9} M hαCGRP, and 10^{-9} M hαCGRP in presence and absence of 10^{-6} M hαCGRP$_{8-37}$, given in % relative to the effect induced by 10^{-6} M forskolin in human cerebral arteries. Values are given as means ± s.e.mean. Four to 16 experiments from three to nine patients.

production induced by 10^{-9} M hαCGRP (Fig. 32.4).

32.4. Conclusion

The results clearly indicate the presence of an adenylyl cyclase coupled CGRP$_1$ receptor located in the smooth muscle cell layer of the human cranial arteries.

References

1. Goadsby P, Edvinsson L. Human in vivo evidence for trigeminovascular activation in cluster headache: Neuropeptide changes and effects of acute attack therapies. Brain 1994; 117:427-434.

2. Goadsby P, Edvinsson L, Ekman R. Vasoactive peptide release in the extracerebral circulation of humans during migraine headache. Ann Neurol 1990; 28:183-187.

3. Aiyar N, Rand K, Elshourbagy N, et al. A cDNA encoding the calcitonin gene-related peptide type 1 receptor. J Biol Chem 1996; 19:11325-11329.

CHAPTER 33

CGRP-Mediated Dural Vessel Vasodilation Produces Sensitisation of Trigeminal Nucleus Caudalis Neurones in Rats

D.J. Williamson, M.J. Cumberbatch, S.L. Shepheard and R.J. Hargreaves

33.1. Introduction

CGRP is a potent vasodilator of cranial blood vessels and is thought to be involved in the pathogenesis of migraine headache since plasma levels of CGRP are elevated during migraine attacks.[1] Mechanical distension of the dura-mater, particularly around blood vessels, is intensely painful and it has been proposed that during migraine CGRP released from dural perivascular afferent fibres results in a painful dilation of dural blood vessels. Migraineurs also often report tenderness and sensitivity of the face and scalp during migraine attacks. This may be due to a central convergence of cranial and extracranial sensation, since preclinical anatomical[2] and electrophysiological[3,4] studies show that neurones of the trigeminal nucleus caudalis receive convergent nociceptive inputs from the craniovasculature and extracranial tissues. We have investigated the effects of dural vasodilation on the activity of neurones in the trigeminal nucleus caudalis that receive dural input and how this vasodilation might affect the activity of neurones that receive convergent dural/facial inputs.

33.2. Methods

Male Sprague Dawley rats (250-350 g) were terminally anaesthetised with pentobarbitone (60 mg/kg) and surgically prepared for electrophysiological recording from the trigeminal nucleus caudalis[5] and for simultaneous measurement of middle meningeal artery diameter using intravital microscopy.[6] Adequate anaesthesia (pentobarbitone infusion 18 mg/kg/hr) under neuromuscular blockade (pancuronium bromide, 1 mg/kg/hr) was ensured by continually monitoring blood pressure and the autonomic responses to pinch stimuli. Electrophysiological recordings were made from single neurones within the trigeminal nucleus caudalis which responded to single shock electrical stimulation (10 stimuli at 1 Hz) of the dura mater (1-3 mA, 100-300 µs) and/or to innocuous air-jet stimulation of the vibrissa. Vasodilation of the middle meningeal artery was produced either by intravenous injection of rat-αCGRP (1 µg/kg) or by electrical stimulation of the dura mater (20 stimuli at 1 Hz) which has previously been shown to produce a CGRP-mediated dural vasodilation.[6]

The Calcitonin Gene-Related Peptide Family: CGRP, Amylin, and Adrenomedullin, edited by David Poyner. ©2000 EUREKAH.COM.

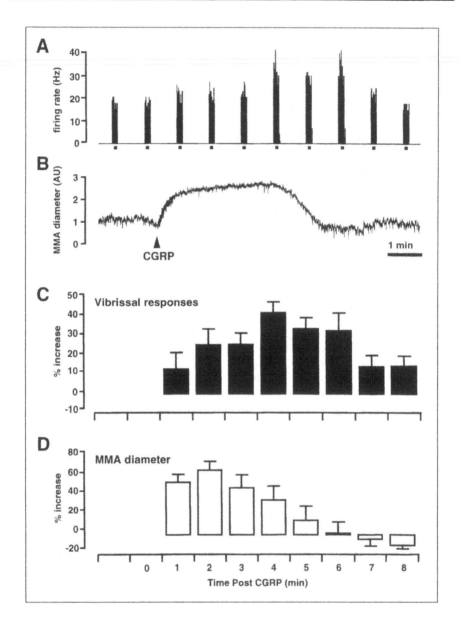

Fig. 33.1. Facilitation of the responses to vibrissal stimulation of neurones in the trigeminal nucleus caudalis which received convergent input from the dura, by vasodilation of the middle meningeal artery (MMA). (Panel A) histograms showing the number of action potentials recorded from a single neurone evoked by air-jet stimulation (10 stimuli at 1 Hz) of the vibrissa at 1 minute intervals and (panel B) trace showing simultaneous MMA diameter and the vasodilation produced by intravenous injection of CGRP (1 µg/kg). (Panels C & D) mean effects of intravenous injection of CGRP (1 µg/kg) on the vibrissal responses and simultaneous MMA diameter of 8 cells. Data is % increase from the mean 3 minute value before injection, bars are mean ± s.e. mean.

The effects of the vasodilation were assessed on the resting activity of the trigeminal neurones or on the activity evoked by vibrissal stimulation.

33.3. Results

Innocuous stimulation of the vibrissa produced a robust and reproducible activation of trigeminal neurones (Fig 33., panel A). Intravenous rat-αCGRP or electrical stimulation of the dura produced a sustained (4-5 minute) vasodilation of the middle meningeal artery, where vessel diameter was increased by approximately 100% (Fig 33.1, panel B). This vasodilation had no effect on the resting activity of the trigeminal neurones however, rat-αCGRP-induced vasodilation facilitated vibrissal responses by 47 ± 8% in 8/14 cells (Fig 33.1 panels C and D) and similarly electrically-induced vasodilation facilitated vibrissal responses by 40 ± 15% in 10/14 cells (data not shown). There was no facilitation of the vibrissal responses in cells that did not receive a convergent dural input (data not shown).

33.4. Conclusion

Dural vasodilation did not cause activation of trigeminal nucleus caudalis neurones per se but it facilitated responses to innocuous vibrissal stimulation and this facilitation only occurred for those neurones which received a convergent input from the dura. This suggests that distension of the meningeal blood vessels can result in a sensitisation of trigeminal nucleus caudalis neurones in rats. If sensitisation of trigeminal neurones receiving convergent dural/facial inputs occurs in man during migraine headache, then this may explain the associated symptoms of extra-cranial pain and facial hypersensitivity. This preparation may provide a novel technique for studying the central mechanisms of migraine headache.

References

1. Goadsby PJ, Edvinsson L. The trigemino-vascular system and migraine: Studies characterising cerebrovascular and neuropeptide changes seen in humans and cats. Ann Neurol 1993; 33:48-56.
2. O'Connor TP, van der Kooy D. Pattern of intracranial and extracranial projections of trigeminal ganglion cells. J Neurosci 1986; 6:2200-2207.
3. Davis KD, Dostovsky JO. Activation of trigeminal brain-stem nociceptive neurons by dural artery stimulation. Pain 1986; 25:395-401.
4. Strassman A, Mason P, Moskowitz MA et al. Response of trigeminal neurones to electrical stimulation of the dura. Brain Res 1986; 379:242-250.
5. Cumberbatch MJ, Hill RG, Hargreaves RJ. Rizatriptan has central antinociceptive effects against durally evoked responses. Eur J Pharmacol 1997;328:37-40.
6. Williamson DJ, Hargreaves RJ, Hill RG et al. Sumatriptan inhibits neurogenic vasodilation of dural blood vessels in the anaesthetised rat—intravital microscope studies. Cephalalgia 1997; 17:525-531.

CHAPTER 34

Mechanism of CGRP Release from Sensory Nerve Terminals in Isolated Trachea of the Rat

G. Oroszi, J. Németh, E. Pintér, T. Kocsy, Zs. Helyes and J. Szolcsányi

34.1. Introduction

It is well established that local release of CGRP and tachykinins from peripheral terminals of afferent nerve fibres occurs in response to capsaicin or electrical nerve stimulation and contributes to the development of neurogenic inflammation in the airways. The aim of the present study was to analyse the mechanism of CGRP release in vitro from capsaicin-sensitive afferent nerve endings in response to electrical or chemical stimulation in the presence or absence of various ion channel blockers.

34.2. Methods

Female Wistar rats (200-250 g) anaesthetised with intraperitoneal sodium pentobarbitone (40 mg/kg) were killed by exsanguination and tracheae were dissected. The tracheae of two animals were superfused (1 ml/min) with oxygenated Krebs solution in an organ bath (1.8 ml) at 37 °C for 60 min. Following equilibration, the solution was collected for three successive eight minute incubations. Electrical field stimulation (40 V, 0.1 ms) with pulses (100, 300, 1200) or chemical stimulation (capsaicin 10 nM-10 µM; resiniferatoxin 0.1 nM-100 nM) were performed at the beginning of the second period to induce CGRP release. Fractions were collected in ice-cold tubes and their CGRP concentrations were determined by radioimmunoassay.[1,2]

34.3. Results

Electrical field stimulation in the range of frequency from 0.1 to 10 Hz liberated CGRP. The release depended on the number of pulses (100 pulses: 0.13 ± 0.02 fmol/mg, 1200 pulses: 0.52 ± 0.04 fmol/mg). Capsaicin and resiniferatoxin evoked significant CGRP release (capsaicin 10 nM: 0.30 ± 0.02 fmol/mg, 10 µM: 4.26 ± 0.17 fmol/mg; resiniferatoxin 0.1 nM: 0.26 ± 0.02 fmol/mg, 100 nM: 3.55 ± 0.15). Tetrodotoxin (1 µM) and lidocaine (5 mM) inhibited the CGRP release evoked by electrical field stimulation (40 V, 0.1 ms, 2 Hz, 50 s) but the effect of capsaicin (10 nM) remained unchanged in the presence of these compounds.

ω-Conotoxin GVIA a selective N-type Ca^{++} channel antagonist (100 and 300 nM) and ω-agatoxin TK a P-type Ca^{++} channel blocker (50 nM) were also ineffective but ω-agatoxin TK in high concentration (250 nM) and Cd^{++} (200 µM) decreased the capsaicin-induced CGRP release.

The Calcitonin Gene-Related Peptide Family: CGRP, Amylin, and Adrenomedullin, edited by David Poyner. ©2000 EUREKAH.COM.

34.4. Conclusion

We conclude that stimulation at frequency as low as 0.5 Hz elicited a maximum peptide release. Resiniferatoxin is 100 times more potent than capsaicin but needs longer time to reach a maximum effect relative to capsaicin. CGRP release in response to electrical field stimulation requires axonal conduction but in the capsaicin-induced peptide release the Na^+, N- or P-type neuronal voltage-operated Ca^{++} channels and axon reflexes are not involved.

Acknowledgments

This work was supported by Hungarian Grants: OTKA T-016945, ETT-372 and the Hungarian Academy of Sciences.

References

1. Helyes Zs, Németh J, Pintér E et al. Inhibition by nociceptin of neurogenic inflammation and the release of SP and CGRP from sensory nerve terminals. Br J Pharmacol 1997; 121;613-615.
2. Németh J, Görcs T, Helyes Zs et al. Development of a new sensitive CGRP radioimmunoassay for neuropharmacological research. Neurobiology 1998; 6:473-475.

CHAPTER 35

Expression and Distribution of CGRP Peptide and Receptor in Human Skin

S. Kapas, K. Pahal and A.T. Cruchley

35.1. Introduction

It is well documented that the role of neuropeptides in the sensory nerves are important for the initiation of neurogenic inflammation and tissue repair. Calcitonin gene-related peptide (CGRP) has been implicated in neurogenic vasodilatation and inflammatory responses in skin. There are morphological contacts between mast cells and sensory nerves containing CGRP.[1] The skin contains immunological cells with inflammatory activity in addition to several bioactive molecules and has a capacity to release neuropeptides as functional molecules. Combined with the possible antimicrobial affect on skin bacteria (see Zihni et al, this volume) CGRP could play an important role in the skin, maintaining a defensive barrier against invading microorganisms. CGRP may be involved in growth and so may have a role in sustaining normal epidermal turnover, wound healing processes, and influencing tumour initiation and proliferation.

This study was designed to investigate and further characterise CGRP and its receptor (CRLR; see Legon, Aiyar et al, this volume) expression in an attempt to better understand its role in skin.

35.2. Materials and Methods

35.2.1. Cell Culture

Keratinocytes were maintained in DMEM supplemented with 10% fetal bovine serum and antibiotics. Thymidine incorporation assays were carried out using confluent, quiescent cultures of cells, in 6-well plates, that were incubated in medium containing 1μCi/ml [^{3}H]thymidine and increasing concentrations of CGRP in the presence of 50μM of isobutyl methyxanthine. After 24hr, DNA synthesis was assessed by measuring the [^{3}H] thymidine incorporated into acid-precipitable material.[2]

35.2.2. Northern Blot Analysis

Cells were serum starved for 24h prior to RNA extraction. Total RNA was extracted using standard protocols. Approximately 20μg of total RNA was electrophoresed in a formaldehyde 1% agarose gel and transferred to Hybond-N nylon membrane. After fixation by ultra violet crosslinking the membrane was hybridised overnight at 42°C with a [α-^{32}P]dCTP labelled CGRP, calcitonin receptor-like receptor (CRLR) or glyceraldehyde phosphate dehydrogenase (GAPDH) probe.

The Calcitonin Gene-Related Peptide Family: CGRP, Amylin, and Adrenomedullin, edited by David Poyner. ©2000 EUREKAH.COM.

35.2.3. Receptor Binding Studies

The binding assay used was a modified form of a previously described method.[3] Cells were placed in 12-well plates and incubated at room temperature for 60min with 0.1nmol/l [^{125}I]CGRP (2000Ci/mmol; Phoenix Pharmaceuticals Inc, USA) and increasing concentrations of unlabelled CGRP in binding buffer. Peptide bound to cells was solubilised in 0.2M sodium hydroxide and determined by γ-spectroscopy.

35.3. Results

35.3.1. Molecular Analysis

The presence of CGRP, CRLR and GAPDH mRNAs in human skin cell lines and tumours were analysed by northern blot analysis (Fig. 35.1). Total RNA (20μg) was electrophoresed in a denaturing gel and transferred to a nylon membrane. The filter was hybridised to a radiolabelled probe corresponding to 450bp fragment of CRLR.

35.3.2. Cellular Analysis

Scatchard analysis of human cell lines revealed all cells had a population of CGRP binding sites with a Kd of 5.6 ± 1.0 nM and a B$_{max}$ range of 10 to 14 ± 2.6 fmol/10^6 cells. CGRP binding was displaced by unlabelled CGRP and ADM but not by βCGRP, CGRP$_{8-37}$ or amylin (Fig. 35.2).

When cells were labelled with [^{3}H]thymidine and incubated in the presence of CGRP there were significant increases in [^{3}H]thymidine uptake. CGRP (from 0.1 nM to 10 nM) caused a doubling in new DNA synthesis: if % increase in [^{3}H]thymidine uptake in control/untreated cells is considered as zero, then 0.1, 1 or 10 nM CGRP caused a 35%, 42% or 40% increase ($P<0.001$).

35.4. Discussion

In this study we have demonstrated immunoreactivity for CGRP and in situ hybridisation signals for CRLR were found in all the epithelial components of the skin (data not shown). It was shown that specific binding to CGRP receptors results in

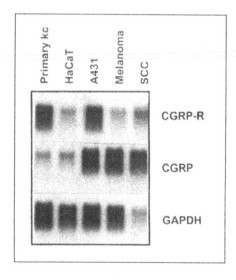

Fig. 35.1. (Above) Northern analysis of CRLR (CGRP-R) in human primary skin keratinocytes, primary kc; 2 keratinocyte cells lines, HaCaT and A431; melanoma and a squamous cell carcinoma.

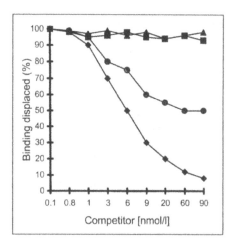

Fig. 35.2. Displacement curves of human [^{125}I]CGRP binding to intact keratinocytes by increasing concentrations of αCGRP (diamonds), ADM (circles), amylin (triangles) or βCGRP (squares).

an increase in new DNA synthesis, a measure of cell proliferation and the cells also produced cAMP in response to CGRP (data not shown). Taken together, these results implicate CGRP as a possible growth regulatory factor of the skin.

As stated above, the role of CGRP in neurogenic inflammation and how the nervous system can have an influence in several skin diseases is documented. However, another function in which CGRP may be implicated in skin physiology is the regulation of growth. Further experiments are required to elucidate the physiological implications, however, it seems clear that CGRP has an important role in cutaneous homeostasis.

References

1. Naukkarinen A, Jarvikallio A, Lakkakorpi J, et al. Quantitative histochemical analysis of mast-cells and sensory nerves in psoriatic skin. J Pathol 1996; 180:200-5.
2. Kapas S, Brown DW, Farthing PM, Hagi-Pavli, E. Adrenomedullin has mitogenic effects on human oral keratinocytes: Involvement of cyclic AMP. FEBS Letts 1997; 418:287-290.
3. Kapas S, Clark AJL Identification of an orphan receptor gene as a type 1 calcitonin gene-related peptide receptor. Biochem Biophys Res Commun 1995; 217:832-838.

The Classification of CGRP, Amylin and Adrenomedullin Receptors: CGRP '98 Consensus View

Ian Marshall

A.1. Introduction

Participants at The Third International CGRP Symposium took the opportunity to discuss some aspects of the receptor nomenclature of this group of peptides. The Symposium did not cover calcitonin and therefore this area is rarely mentioned below although it has important ties with CGRP, amylin and adrenomedullin.

A.2. Current Nomenclature (1998)

The current nomenclature is described in the Trends in Pharmacology (TIPS) 1998 Receptor & Ion Channel Nomenclature Supplement (9th Edition). At present there is no Nomenclature Subcommittee of the International Union of Pharmacology dealing with this area and the scheme in the Supplement represents independent views synthesized to provide a consensus.

Each ligand is most potent at its own receptor i.e. CGRP at CGRP receptors, amylin at amylin receptors and adrenomedullin at adrenomedullin receptors (TIPS classification). Where different CGRPs or amylins have been compared their potencies are usually similar.[1,2] Until it is clear whether these varying forms of the different peptides have essentially equivalent biological effects it will remain important that the particular compound is specified in each research report.

In the TIPS scheme the only selective antagonist mentioned is AC187 (Acetyl-[Asn^{30},Tyr^{32}]-sCT_{8-32}) with an affinity between 8.3 and 9.3 at amylin receptors. No selective antagonists are given for either CGRP or adrenomedullin receptors. However $CGRP_{8-37}$ is mentioned in a footnote as an antagonist against CGRP (range quoted at rat receptor 6.5-8.0) but may also be an antagonist at the amylin receptor (7.0) and adrenomedullin receptor (<6.0). There is also a mention for adrenomedullin$_{22-52}$ as an antagonist of some effects of adrenomedullin.[3] Subtypes of the CGRP receptor have been suggested[4,5] on the basis of the action of the agonist [CysACM] CGRP ($CGRP_2$ selective) and $CGRP_{8-37}$ ($CGRP_1$-selective, pK_i 7.5-8.0).

Currently the canine orphan receptor rdc1 is suggested to be a possible CGRP receptor[6] as is the calcitonin receptor-like receptor (CRLR) isolated from both human (462aa) and rat (463aa).[7,8] The orphan receptor G10d has been proposed as an adrenomedullin receptor.[9] A further site has been identified by radioligand binding although no functional correlate was described. This is the C3 site identified in

The Calcitonin Gene-Related Peptide Family: CGRP, Amylin, and Adrenomedullin, edited by David Poyner. ©2000 EUREKAH.COM.

rat nucleus accumbens and for which salmon calcitonin, amylin and CGRP all have a high affinity[10] and which might be an amylin receptor.

A.3. Discussion of Current Nomenclature

CGRP, amylin and adrenomedullin with their structural similarities cross-react at each other's sites. Therefore it is not possible to say that, e.g., CGRP receptors are those through which the peptide elicits biological effects in a parallel fashion to, for example, adrenaline and adrenoceptors. Thus further criteria are required in addition to the effect of a ligand.

A.3.1. How Can a CGRP Receptor Be Defined?

To what extent can a binding site with functional correlates be distinguished for CGRP? Firstly, a high potency for CGRP would be essential but is not sufficient. Comparison of the relative potency of CGRP with its homologues may also be helpful. However if CGRP and adrenomedullin are both very potent this can suggest that either both receptors are present or that there is a CGRP receptor or an adrenomedullin receptor (or both present) depending on the expectation and interpretation of the experimenter. It is generally accepted that at CGRP receptors amylin is about 100-fold less potent than CGRP. In addition it is likely that at a CGRP receptor $CGRP_{8-37}$, should antagonise the effect of the parent peptide with an affinity usually in the range of 6.0 to 8.0 although values above and below this have been found. Some of the difficulties in interpreting this data have been discussed elsewhere (see Marshall & Wisskirchen, this volume). An alternative view suggests the data falls primarily into two groups in, for example, the rat.[11] However, this scheme does not take into account those situations where CGRP is not antagonised by $CGRP_{8-37}$ at 10^{-6}M or even higher concentration. These might be mediated through a CGRP receptor with an even lower affinity for $CGRP_{8-37}$ than the putative $CGRP_2$ receptor or alternatively through e.g. adrenomedullin receptors. These different

possibilities remain to be resolved and therefore there remains doubt about the interpretation of some of this data.

Some other results with $CGRP_{8-37}$ may be difficult to equate with a $CGRP_1/CGRP_2$ classification. For example in guinea pig basilar artery and ileum differences in the affinity of $haCGRP_{8-37}$ were reported against members of the CGRP family.[1,2] Therefore it remains a possibility that the α- and β-forms of CGRP may act differently in some tissues.

A $CGRP_2$ receptor may have [CysACM] CGRP as a weak agonist[4,5] although relatively few studies have compared this ligand on $CGRP_2$ receptors. As a partial agonist its effect is complicated by differences in receptor reserve between tissues.

There is no differentiation in receptors possible at present in relation to different second messenger or signal transduction pathways. It appears that CGRP, amylin and adrenomedullin can couple via G_s to adenylate cyclase although additional effector mechanisms have been identified in some systems. This is in line with the prevailing view that the receptors all belong to the 7-transmembrane spanning domain group of G-protein coupled receptors.

In summary it may be difficult to classify a set of data unambiguously as consistent with a CGRP receptor, at least in part because a relatively small number of agonists and antagonists are usually employed. It is possible that some complications may arise because of the co-existence of receptors and their subtypes for more than one of these ligands in a given tissue or cell line.

A.3.2. How Can an Amylin Receptor be Defined?

It is clear that amylin can act through both CGRP and calcitonin recptors but unequivocal evidence for its own receptors is harder to find. However, in insulin-treated rat soleus muscle, rat amylin was more potent than either $r\alpha CGRP$ or $r\beta CGRP$ in reducing glycogen synthesis and in competing for $[^{125}I]$-BH-r-amylin specific binding to rat nucleus accumbens membranes.[13] In addition, AC187 was more potent than hαCGRP$_{8-37}$ as an antagonist against both amylin and CGRP, a relative potency reported for some other amylin effects (e.g. hyperglycaemia and hyperlactaemia).[14] Thus this provides evidence for a separate amylin receptor and at one site previously classified as a C3 receptor.[10]

The possibility of amylin receptor subtypes has been raised e.g in the guinea pig vas deferens as AC187 antagonised amylin inhibition with a lower potency than that seen in rat soleus muscle.[2] However, alternative explanations include species differences in receptors and, as the effect was weakly antagonised by hαCGRP$_{8-37}$, the possibility that AC187 was acting at a CGRP$_2$ receptor.

A3.3. How Can an Adrenomedullin Receptor be Defined?

Many effects of this peptide overlap with those of the other members of the group and thus adrenomedullin has been found to act via CGRP receptors e.g. in SK-N-MC cells[15] and via calcitonin receptors.[16] However, there are situations where for example, vasorelaxation to adrenomedullin is not antagonised by hαCGRP$_{8-37}$ even though this did reduce a similar effect evoked by CGRP.[1] Suggested antgonists selective for adrenomedullin responses include ADM$_{22-52}$ and subtypes have been suggested.[3,17]

The cloning and expression of a putative rat adrenomedullin receptor[9] (G10d) led to the human equivalent cDNA being isolated from the lung.[18] However, more detailed studies of the pharmacological characteristics of these receptors suggests that they do not represent authentic adrenomedullin receptors.[19]

A.4. Suggestions for Alterations to Current Nomenclature

Molecular biology is about to transform our understanding of receptors for these peptides. Discussion at CGRP '98 emphasised the crucial question of replication by a group independent of the original investigators. Findings with the RDC1 and G10d clones could not be replicated by others and it was generally agreed that these should no longer be included in any official receptor classification (see Tong et al, this volume). It is hoped that independent replication of the work with RAMPs and RCP (see Foord et al and Rosenblatt et al this volume) will occur very soon.

A.5. Conclusion

The lack of selectivity of ligands has made it difficult to establish an agreed classification of CGRP, amylin and adrenomedullin receptors. The ability to develop transfected cell lines to produce a homogenous receptor population is now near. However, in the meantime, the usefulness of tissue binding or functional assays will be improved if ligands from the different families (rather than from just one) are used in a series of experiments together with several antagonists. Finally the need for the independent replication of important findings remains clear.

References

1. Wisskirchen FM, Burt RP, Marshall I. Pharmacological characterization of CGRP receptors mediating relaxation of the rat pulmonary artery and inhibition of twitch responses of the rat vas deferens. Br J Pharmacol 1998; 123:1673-1683.
2. Tomlinson AE, Poyner DR. Multiple receptors for calcitonin gene-related peptide and amylin on guinea-pig ileum and vas deferens. Br J Pharmacol 1996; 117:1362-1368.
3. Eguchi S, Hirata Y, Iwasaki H et al. Structure-activity relationship of adrenomedullin, a novel vasodilatory peptide, in cultured rat vascular smooth muscle cells. Endocrinology 1994; 135:2454-2458.
4. Dennis T, Fournier A, Cadieux A et al. hCGRP8-37, a calcitonin gene-related peptide antagonist revealing calcitonin gene-related peptide receptor heterogeneity in brain and periphery. J Pharmacol Exp Ther 1990; 254: 123-128.
5. Quirion R, Van Rossum D, Dumont Y et al. Characterization of CGRP$_1$ and CGRP$_2$ receptor subtypes. Ann NY Acad Sci 1992; 657: 88-105.

6. Kapas S, Clark AJL. Identification of an orphan receptor gene as a type 1 calcitonin gene-related peptide receptor. Biochem Biophys Res Commun 1995; 217:832-838.

7. Aiyar NJ, Rand IC, Elshourbagy NA et al. A cDNA encoding the calcitonin gene-related peptide type 1 receptor. J Biol Chem 1996; 271:11325-11329.

8. Han Z-Q, Copock HA, Smith DM et al. The interaction of CGRP and adrenomedullin with a receptor expressed in the rat pulmonary vascular endothelium. J Mol Endocrinol 1997; 18:267-272.

9. Kapas S, Catt KJ, Clark AJL. Cloning and expression of cDNA encoding a rat adrenomedullin receptor. J Biol Chem 1995; 270:25344-25347.

10. Sexton PM, McKenzie JS, Mendelsohn FAO. Evidence for a new subclass of calcitonin/calcitonin gene-related peptide binding site in rat brain. Neurochem Int 1988; 12:323-335.

11. Poyner DR. Molecular pharmacology of receptors for calcitonin-gene-related peptide, amylin and adrenomedullin. Biochem Soc Trans 1997; 25:1032-1036.

12. Jansen I. Characterization of calcitonin gene-related peptide (CGRP) receptors in guinea pig basilar artery. Neuropeptides 1992; 21: 73-79.

13. Beaumont K, Pittner RA, Moore CX et al. Regulation of muscle glycogen metabolism by CGRP and amylin: CGRP receptors not involved. Br J Pharmacol 1995; 115: 713-715.

14. Young AA, Gedulin B, Gaeta LSL et al. Selective amylin antagonist suppresses rise in plasma lactate after intravenous glucose in the rat. FEBS Lett 1994; 343: 237-241.

15. Zimmermann U, Fischer JA, Muff R. Adrenomedullin and calcitonin gene-related peptide receptors interact with the same receptor in cultured human neuroblastoma SK-N-MC cells. Peptides 1995; 16: 421-424.

16. Disa J, Dang K, Tan KB et al. Interaction of adrenomedullin with calcitonin receptor in cultured human breast cancer cells. Peptides 1998; 19: 247-251

17. Zimmermann U, Fischer JA, Frei K et al. Identification of adrenomedullin receptors in cultured rat astrocytes and in neuroblastboma x glioma hybrid cells (NG108-15). Brain Res. 1996; 724:238-45.

18. Hanze J, Dittrich K, Dotsch J et al. Molecular cloning of a novel human receptor gene with homology to the rat adrenomedullin receptor and high expression in heart and immune system. Biochem Biophys Res Commun. 1997; 240:183-8.

19. Kennedy SP, Sun D, Oleynek JJ et al. Expression of the rat adrenomedullin receptor or a putative human adrenomedullin receptor does not correlate with adrenomedullin binding or functional response. Biochem Biophys Res Commun. 1998; 244:832-837.

Index